Maynooth Library

D1321286

Green Separation Processes
Edited by
C. A. M. Afonso,
J. G. Crespo

Further Reading from Wiley-VCH

B. Cornils, W. A. Herrmann (Eds.)

Aqueous-Phase Organometallic Catalysis

2nd Completely Revised and Enlarged Edition

2004

ISBN 3-527-30712-5

J. A. Gladysz, D. P. Curran, I. T. Horváth (Eds.)

Handbook of Fluorous Chemistry

2004

ISBN 3-527-30617-X

W. Ehrfeld, V. Hessel, H. Löwe

Microreactors

2000

ISBN 3-527-29590-9

J. G. Sanchez Marcano, T. T. Tsotsis

Catalytic Membranes and Membrane Reactors

2002

ISBN 3-527-30277-8

660.284
2
CAR

Green Separation Processes

Fundamentals and Applications

Edited by
Carlos A. M. Afonso, J. G. Crespo

WILEY-VCH

WILEY-VCH Verlag GmbH & Co. KGaA

1046482
05.2159
chem

Editors

Professor Dr. Carlos A. M. Afonso
CQFM, Department of Chemical
Engineering
Instituto Superior Técnico
1049-001 Lisbon, Portugal

Professor Dr. João G. Crespo
REQUIMTE/CQFB
Department of Chemistry
FCT/Universidade Nova de Lisboa
Campus da Caparica
2829-516 Caparica
Portugal

■ All books published by Wiley-VCH are carefully produced. Nevertheless, author and publisher do not warrant the information contained in these books, including this book, to be free of errors. Readers are advised to keep in mind that statements, data, illustrations, procedural details or other items may inadvertently be inaccurate.

Library of Congress Card No.: Applied for

British Library Cataloguing-in-Publication Data:
A catalogue record for this book is available from the British Library.

Bibliographic information published by Die Deutsche Bibliothek
Die Deutsche Bibliothek lists this publication in the Deutsche Nationalbibliografie; detailed bibliographic data are available in the Internet at <http://dnb.ddb.de>.

© 2005 WILEY-VCH Verlag GmbH & Co. KGaA, Weinheim

All rights reserved (including those of translation in other languages). No part of this book may be reproduced in any form – by photoprinting, microfilm, or any other means – nor transmitted or translated into machine language without written permission from the publishers. Registered names, trademarks, etc. used in this book, even when not specifically marked as such, are not to be considered unprotected by law.

Printed in the Federal Republic of Germany

Printed on acid-free paper

Typesetting TypoDesign Hecker GmbH, Leimen
Printing Strauss GmbH, Mörlenbach
Bookbinding Litges & Dopf Buchbinderei GmbH, Heppenheim

ISBN-13: 978-3-527-30985-6
ISBN-10: 3-527-30985-3

Foreword

At the heart of Green Chemistry is scientific and technological innovation. This volume contains a collection of important and useful innovations that are of the type that will be essential to enduring that our next generation of products and processes are more benign to human health and the biosphere. What makes these Green Chemistry technologies different from those technologies of the past is that they integrate reduced impact on the environment as a performance criterion of the design. Rather than treating impact of the technology on biological and human systems as an afterthought to be dealt with after introduction and utilization, Green Chemistry technologies as detailed in this book ingrain the goals of sustainability at the outset of the design process.

The impact of this Green Chemistry approach is important on several levels. Certainly, the benefits to protection of the environment are the most evident and can be understood and appreciated in reviewing the many excellent examples in this volume. However, many of the other benefits may be less obvious at first on first analysis. For instance, this collection of technologies taken as a whole demonstrates that it is possible to achieve environmental and economic goals simultaneously. By using the Green Chemistry approaches presented in this book, the benefits of energy efficiency, material minimization, intrinsic hazard reduction, and waste avoidance all can be achieved. Each of these factor have direct linkages to the net profitability of the technology. Too often historically, it has been necessary to achieve these above goals in a decoupled manner that have added costs in the form of material, energy and time. In many ways this historical approach can be viewed as elegant technological "bandages" that sought to repair or make an unsustainable process more legally and socially acceptable. So even in cases where the goals were achieved, the improvements came at significant costs.

The Green Chemistry technologies that have been selected and compiled for this important collection by the editors and that have been commendably portrayed by the authors demonstrate the imperative of using Green Chemistry principles in the design framework. Through this approach and the coupling of environmental and economic goals for societal benefit, environmental protection and sustainability can become autocatalytic in our next generation of products and processes. The editors and authors of this volume have provided important contribution to the advancement of Green Chemistry that will be well utilized and built upon in the future.

Washington, D.C., May 2005 *Paul T. Anastas*

Green Separation Processes. Edited by C. A. M. Afonso and J. G. Crespo
Copyright © 2005 WILEY-VCH Verlag GmbH & Co. KGaA, Weinheim
ISBN 3-527-30985-3

Table of Contents

Green Separation Processes. Edited by C. A. M. Afonso and J. G. Crespo
Copyright © 2005 WILEY-VCH Verlag GmbH & Co. KGaA, Weinheim
ISBN 3-527-30985-3

Preface

Chemistry has been one of the pillars of the wealth and growth of the World economy throughout the twentieth century, based on an increasing understanding of the interactions taking place on a molecular level to enable enhanced production and product quality. Chemistry is, and will certainly continue to be, a primary driver for wellbeing, growth and sustainable development in the economy during this century.

Green(er) Chemistry is the key to sustainable development as it will lead to new solutions to existing problems and will present opportunities for new processes and products by:

- securing access to competitive feedstocks, including the exploration of alternative renewable raw materials to allow a gradual shift from petroleum-based raw materials as required;
- reducing the resource intensity of chemical manufacture and use, including closing materials loops, enhancing reuse and recycling, and reducing waste and emissions;
- developing improved and new functionalities by means of new materials and new formulations based on increasing control of physical properties from the nano to the macro scale;
- increasing control over total production costs through improving materials and energy efficiency and minimizing the impact of chemicals manufacturing on the environment;
- designing engineering solutions to allow for better product quality and fast and flexible responses to market needs.

This book aims to contribute to a better understanding of the new challenges that Chemistry is facing, with a particular emphasis on the need for the development of new processes for product separation and recovery. The contributions to this book are organized into three interlinked sections: "Green Chemistry for Sustainable Development", "New Synthetic Methodologies and the Demand for Adequate Separation Processes" and "New Developments in Separation Processes." The chapters from the first part present the general principles and regulations that support the need for a Green(er) Chemistry for sustainable development, while the second part will introduce novel synthetic methodologies aiming to obtain higher

Green Separation Processes. Edited by C. A. M. Afonso and J. G. Crespo
Copyright © 2005 WILEY-VCH Verlag GmbH & Co. KGaA, Weinheim
ISBN 3-527-30985-3

quality products while respecting those principles. The third part of the book presents a comprehensive discussion of new separation processes, which result from the needs and challenges discussed in the previous sections.

May 2005   *Carlos A. M. Afonso and João G. Crespo*

List of Contributors

Carlos A. M. Afonso
CQFM, Department of Chemical
Engineering
Instituto Superior Técnico
1049-001 Lisbon
Portugal

Bianca Antonioli
TU Dresden
Institute of Inorganic Chemistry
01062 Dresden
Germany

Ching-Ju Monica Chin
Graduate Institute of Environmental
Engineering
National Central University,
Taiwan, R.O.C.

James H. Clark
Centre for Clean Technology
University of York
York YO10 5DD
UK

Anthony A. Clifford
School of Chemistry
University of Leeds
Leeds LS2 9JT
UK

Diana Cook
CookPrior Associates
6 High Street
Sutton in Craven, Keighley
Yorkshire BD20 7NX
UK

João G. Crespo
REQUIMTE/CQFB
Department of Chemistry
FCT/Universidade Nova de Lisboa
Campus da Caparica
2829-516 Caparica
Portugal

Jairton Dupont
Laboratory of Molecular Catalysis
Institute of Chemistry, URFGS
Av. Bento Gonçalves
9150-970 Porto Alegre RS
Brazil

Lieven E.M. Gevers
Center for Surface Chemistry and
Catalysis
Department of Interphase Chemistry
Kasteelpark Arenberg 23
3001 Leuven
Belgium

Green Separation Processes. Edited by C. A. M. Afonso and J. G. Crespo
Copyright © 2005 WILEY-VCH Verlag GmbH & Co. KGaA, Weinheim
ISBN 3-527-30985-3

Karsten Gloe
TU Dresden
Institute of Inorganic Chemistry
01062 Dresden
Germany

Kerstin Gloe
TU Dresden
Institute of Inorganic Chemistry
01062 Dresden
Germany

Jürgen Gmehling
University of Oldenburg
Faculty V
Institute for Pure and Applied
Chemistry
26111 Oldenburg
Germany

Gerhard Jas
Synthacon GmbH
Am Haupttor, Gebäude 4310
06237 Leuna
Germany

Yuhong Ju
Sustainable Technology Division,
National Risk Management Research
Laboratory,
U.S. Environmental Protection
Agency,
26 West Martin Luther King Drive,
MS 443,
Cincinnati, Ohio, 45268
USA

Mazaahir Kidwai
Department of Chemistry
University of Delhi
Delhi-110007
India

Ulrich Kunz
Institut für Chemische Verfahrens-
technik
Technische Universität Clausthal
Leibnizstr. 17
38678 Clausthal-Zellerfeld
Germany

Hiroshi Matsubara
Department of Chemistry
Graduate School of Science
Osaka Prefecture University
Sakai
Osaka 599-8531
Japan

Mirjana Minceva
Laboratory of Separation and Reaction
Engineering
Department of Chemical Engineering
Faculty of Engineering
University of Porto
Rua Dr Robert Frias
4200-465 Porto
Portugal

Richa Mohan
Department of Chemistry
University of Delhi
Delhi-110007
India

Richard D. Noble
University of Colorado
Chemical & Biological Engineering
Dept.
Boulder, CO 80309
USA

Jeremy Noonan
Georgia Institute of Technology
311 Ferst Drive
Atlanta
Georgia 30332-0512
USA

Anna Banet Osuna
Universidade Nova de Lisboa
Instituto de Tecnologia Química e
Biológica
Aptd. 127
2781-901 Oeiras
Portugal

Manuel Nunes da Ponte
New University of Lisbon
Quinta da Torre
2829-516 Caparica
Portugal

Kevin Prior
CookPrior Associates
6 High Street
Sutton in Craven, Keighley
Yorkshire, BD20 7NX
UK

Alírio R. Rodrigues
Laboratory of Separation and Reaction
Engineering
Department of Chemical Engineering
Faculty of Engineering
University of Porto
Rua Dr Robert Frias
4200-465 Porto
Portugal

Ilhyong Ryu
Department of Chemistry
Graduate School of Science
Osaka Prefecture University
Sakai
Osaka 599-8531
Japan

Thomas Schäfer
REQUIMTE/CQFB
Department of Chemistry
FCT/Universidade Nova de Lisboa
Campus da Caparica
2829-516 Caparica
Portugal

Dirk Schmalz
Merck KGaA,
Zentrale Verfahrenstechnik
Ingenieurtechnik
Frankfurter Straße 250
64293 Darmstadt
Germany

Ana Šerbanović
Universidade Nova de Lisboa
Instituto de Tecnologia Química e
Biológica
Aptd. 127
2781-901 Oeiras
Portugal

Sven Steinigeweg
Cognis Deutschland
Process Development
40587 Düsseldorf
Germany

Holger Stephan
Research Center Rossendorf
Insitute of Bioinorganic and Radio-
pharmaceutical Chemistry
01314 Dresden
Germany

Fumio Toda
Department of Chemistry
Okayama University of Science
Faculty of Science
Ridaicho 1-1
Okayama 700-0005
Japan

Costas Tsouris
Oak Ridge National Laboratory
P.O. Box 2008
Oak Ridge
Tennessee 37831-6181
USA

Ivo F.J. Vankelekom
Center for Surface Chemistry and
Catalysis
Department of Interphase Chemistry
Kasteelpark Arenberg 23
3001 Leuven
Belgium

Rajender S. Varma
Sustainable Technology Devision
National Risk Management Research
Laboratory
US Environmental Protection Agency
26 West Martin Luther King Drive,
MS443
Cincinnnati
Ohio, 45268
USA

Sotira Yiacoumi
Georgia Institute of Technology
311 Ferst Drive
Atlanta
Georgia 30332-0512
USA

Tung-yu Ying
Los Alamos Laboratory
Mail stop: J580, ESA-AET
Los Alamos, NM 87545
USA

Part 1
Green Chemistry for Sustainable Development

Green Separation Processes. Edited by C. A. M. Afonso and J. G. Crespo
Copyright © 2005 WILEY-VCH Verlag GmbH & Co. KGaA, Weinheim
ISBN 3-527-30985-3

1.1
Green Chemistry and Environmentally Friendly Technologies

James H. Clark

1.1.1
Introduction

"Green Chemistry" is the universally accepted term to describe the movement towards more environmentally acceptable chemical processes and products [1]. It encompasses education, research, and commercial application across the entire supply chain for chemicals [2]. Green Chemistry can be achieved by applying environmentally friendly technologies – some old and some new [3]. While Green Chemistry is widely accepted as an essential development in the way that we practice chemistry, and is vital to sustainable development, its application is fragmented and represents only a small fraction of actual chemistry. It is also important to realize that Green Chemistry is not something that is only taken seriously in the developed countries. Some of the pioneering research in the area in the 1980s was indeed carried out in developed countries including the UK, France, and Japan, but by the time the United States Environmental Protection Agency (US EPA) coined the term "Green Chemistry" in the 1990s, there were good examples of relevant research and some industrial application in many other countries including India and China [4].

The Americans launched the high profile Presidential Green Chemistry Awards in the mid-1990s and effectively disclosed some excellent case studies covering products and processes [5]. Again, however, it is important to realize that there were many more good examples of Green Chemistry at work long before this – for example, commercial, no-solvent processes were operating in Germany and renewable catalysts were being used in processes in the UK but they did not get the same publicity as those in the United States [2, 4].

The developing countries that are rapidly constructing new chemical manufacturing facilities have an excellent opportunity to apply the catchphrase of Green Chemistry "Benign by Design" from the ground upwards. It is much easier to build a new, environmentally compatible plant from scratch than to have to deconstruct before reconstructing, as is the case in the developed world.

Green Separation Processes. Edited by C. A. M. Afonso and J. G. Crespo
Copyright © 2005 WILEY-VCH Verlag GmbH & Co. KGaA, Weinheim
ISBN 3-527-30985-3

In this chapter I shall start by exploring the drivers behind the movement towards Green and Sustainable Chemistry. These can all be considered to be "costs of waste" that effectively penalize current industries and society as a whole. After a description of Green Chemistry I will look at the techniques available to the chemical manufacturers. This leads naturally into a more detailed discussion about methods of evaluating "greenness" and how we should apply sustainability concepts across the supply chain. It is important that, while reading this, we see Green Chemistry in the bigger picture of sustainable development as we seek to somehow satisfy society's needs without compromising the survival of future generations.

1.1.2
Objectives for Green Chemistry: The Costs of Waste

Hundreds of tonnes of hazardous waste are released to the air, water, and land by industry every hour of every day. The chemical industry is the biggest source of such waste [3]. Ten years ago less than 1% of commercial substances in use were classified as hazardous, but it is now clear that a much higher proportion of chemicals presents a danger to human health or to the environment. The relatively small number of chemicals formally identified as being hazardous was due to very limited testing regulations, which effectively allowed a large number of chemicals to be used in everyday products without much knowledge of their toxicity and environmental impact. New legislation will dramatically change that situation. In Europe, REACH (Registration, Evaluation, Assessment of Chemicals) will come into force in the first decade of the twenty-first century and whilst, at the time of writing, the final form of the legislation has yet to be decided, it is clear that it will be the most important chemicals-related legislation in living memory and that it will have a dramatic effect on chemical manufacturing and use [6]. REACH will considerably extend the number of chemicals covered by regulations, notably those that have been on market since 1981 (previously exempt), will place the responsibility for chemicals testing with industry, and will require testing whether the chemical is manufactured in Europe or imported for use there. Apart from the direct costs to industry of testing, REACH is likely to result in some chemical substances becoming restricted, prohibitively expensive, or unavailable. This will have dramatic effects on the supply chain for many consumer goods that rely on multiple chemical inputs.

Increased knowledge about chemicals, and the classification of an increasing number of chemical substances as being in some way "hazardous", will have health and safety implications, again making the use of those substances more costly and difficult. Furthermore, it will undoubtedly cause local authorities and governments to restrict and increase the costs of disposal of waste containing those substances (or indeed waste simply coming from processes involving such substances). Thus, legislation will increasingly force industry and the users of chemicals to change – both through substitution of hazardous substances in their processes or products and through the reduction in the volume and hazards of their waste.

The costs of waste to a chemical manufacturing company are high and diverse (Fig. 1.1-1) and, for the foreseeable future, they will get worse.

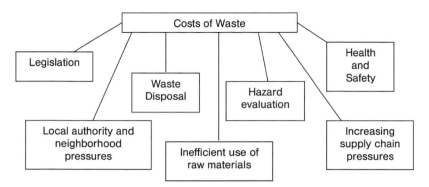

Fig. 1.1-1 The costs of waste.

These costs and other pressures are now evident throughout the supply chain for a chemical product – from the increasing costs of raw materials, as petroleum becomes more scarce and carbon taxes penalize their use, to a growing awareness amongst end-users of the risks that chemicals are often associated with, and the need to disassociate themselves from any chemical in their supply chain that is recognized as being hazardous (e.g. phthalates, endocrine disrupters, polybrominated compounds, heavy metals, etc.; Fig. 1.1-2)

1.1.3
Green Chemistry

The term Green Chemistry, coined by staff at the US EPA in the 1990s, helped to bring focus to an increasing interest in developing more environmentally friendly chemical processes and products. There were good examples of Green Chemistry research in Europe in the 1980s, notably in the design of new catalytic systems to replace hazardous and wasteful processes of long standing for generally important synthetic transformations, including Friedel–Crafts reactions, oxidations, and various base-catalyzed carbon–carbon bond-forming reactions. Some of this research had led to new commercial processes as early as the beginning of the 1990s [4].

In recent years Green Chemistry has become widely accepted as a concept meant to influence education, research, and industrial practice. It is important to realize that it is not a subject area in the way that organic chemistry is. Rather, Green Chemistry is meant to influence the way that we practice chemistry – be it in teaching children, researching a route to an interesting molecule, carrying out an analytical procedure, manufacturing a chemical or chemical formulation, or designing

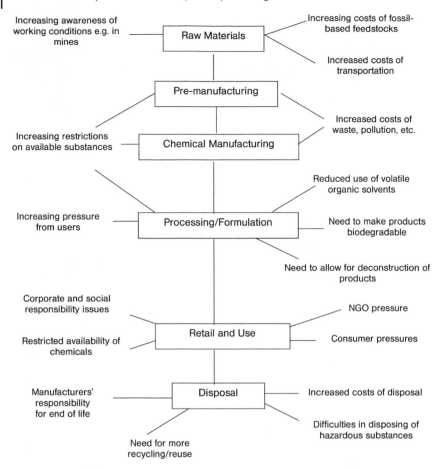

Fig. 1.1-2 Supply chain pressures.

a product [7]. Green Chemistry has been promoted worldwide by an increasing but still small number of dedicated individuals and through the activities of some key organizations. These include the Green Chemistry Network (GCN; established in the UK in 1998 and now with about one thousand members worldwide) [8] and the Green Chemistry Institute (established in the USA in the mid 1990s, now part of the American Chemical Society and with "chapters" in several countries around the world) [9]. Other Green Chemistry Networks or other focal points for national or regional activities exist in other countries including Italy, Japan, Greece and Portugal and new ones appear every year. The GCN was established to help promote and encourage the application of Green Chemistry in all areas where chemistry plays a significant role. (Fig. 1.1-3)

At about the same time as the establishment of the GCN, the Royal Society of Chemistry (RSC) launched the journal "Green Chemistry". The intention for this journal was always to keep its readers aware of major events, initiatives, and edu-

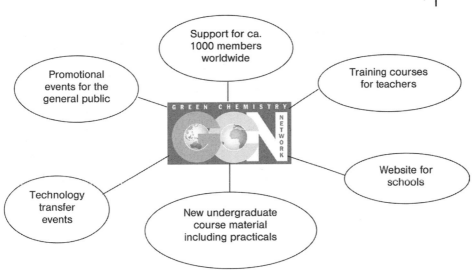

Fig. 1.1-3   The roles of the Green Chemistry Network.

cational and industrial activities, as well as leading research from around the world. The journal has gone from strength to strength and has a growing submission rate and subscription numbers, as well as having achieved one of the highest impact factors among the RSC journals (Fig. 1.1-4).

Green Chemistry can be considered as a series of reductions (Fig. 1.1-5). These reductions lead to the goal of triple bottom-line benefits of economic, environmental, and social improvements [11]. Costs are saved by reducing waste (which is becoming increasingly expensive to dispose of, especially when hazardous) and energy use (likely to represent a larger proportion of process costs in the future) as

Fig. 1.1-4   The first issue of *Green Chemistry*.

well as making processes more efficient by reducing materials consumption. These reductions also lead to environmental benefit in terms of both feedstock consumption and end-of-life disposal. Furthermore, an increasing use of renewable resources will render the manufacturing industry more sustainable[12]. The reduction in hazardous incidents and the handling of dangerous substances provides additional social benefit – not only to plant operators but also to local communities and through to the users of chemical-related products.

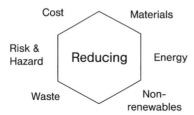

Fig. 1.1-5 "Reducing": The heart of Green Chemistry.

It is particularly important to seek to apply Green Chemistry throughout the lifecycle of a chemical product (Fig. 1.1-6) [13, 14].

Scientists and technologists need to routinely consider lifecycles when planning new synthetic routes, when changing feedstocks or process components, and, fundamentally, when designing new products. Many of the chemical products in common use today were not constructed for end-of-life nor were full supply-chain issues of resource and energy consumption and waste production necessarily considered. The Green Chemistry approach of "benign by design" should, when applied at the design stage, help assure the sustainability of new products across their full lifecycle and minimize the number of mistakes we make.

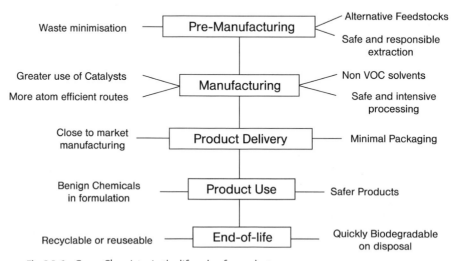

Fig. 1.1-6 Green Chemistry in the lifecycle of a product.

Much of the research effort relevant to Green Chemistry has focused on chemical manufacturing processes. Here we can think of Green Chemistry as directing us towards the "ideal synthesis" (Fig. 1.1-7) [3, 15].

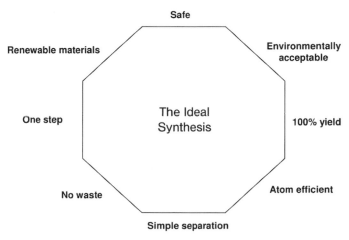

Fig. 1.1-7 Features of the "ideal synthesis".

Yield is the universally accepted metric in chemistry research for measuring the efficiency of a chemical synthesis. It provides a simple and understandable way of measuring the success of a synthetic route and of comparing it to others. Green Chemistry teaches us that yield is not enough. It fails to allow for reagents that have been consumed, solvents and catalysts that will not be fully recovered, and, most importantly, the often laborious and invariably resource- and energy-consuming separation stages such as water quenches, solvent separations, distillations, and recrystallizations. Green Chemistry metrics [16] are now available and commonly are based on "atom efficiency" whereby we seek to maximize the number of atoms introduced into a process into the final product. These are discussed in more detail later in this chapter. As indicated, simple separation with minimal input and additional outputs is an important target. An ideal reaction from a separation standpoint would be one where the substrates are soluble in the reaction solvent but the product is insoluble. The process would, of course, be further improved if no solvent was involved at all! Some of the worst examples of atom inefficiency and relative quantities of waste are to be found in the pharmaceutical industry. The so-called *E* factor (total waste/product by weight) is a simple but quite comprehensive measure of process efficiency and commonly shows values of 100+ in drug manufacture [17]. This can be largely attributed to the complex, multistep nature of these processes. Typically, each step in the process is carried out separately with work-up, isolation, and purification all adding to the inputs and amount of waste produced. Simplicity in chemical processes is vital to good Green Chemistry. Steps can be "telescoped" together for example, reducing the number of discrete stages in the process [18].

To achieve greener chemical processes we will need to make increasing use of technologies, some old and some new, which are becoming proven as clean technologies.

1.1.4
Environmentally Friendly Technologies [3]

There is a pool of technologies that are becoming the most widely studied or used in seeking to achieve the goals of Green Chemistry. The major "clean technologies" are summarized in Fig. 1.1-8. They range from well-established and proven technologies through to new and largely unproven technologies.

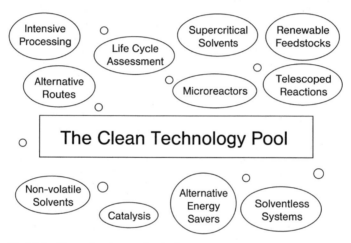

Fig. 1.1-8   The major clean technologies.

Catalysis is truly a well-established technology, well proven at the largest volume end of the chemicals industry. In petroleum refineries, catalysts are absolutely fundamental to the success of many processes and have been repeatedly improved over more than 50 years. Acid catalysts, for example, have been used in alkylations, isomerizations and other reactions for many years and have progressively improved from traditional soluble or liquid systems, through solid acids such as clay, to structurally precise zeolite materials, which not only give excellent selectivity in reactions but are also highly robust, with modern catalysts having lifetimes of up to 2 years! In contrast, the lower volume but higher value end of chemical manufacturing – specialties and pharmaceutical intermediates – still relies on hazardous and difficult routes to separate soluble acid catalysts such as H_2SO_4 and $AlCl_3$ and is only now beginning to apply modern solid acids. Cross-sector technology transfer can greatly accelerate the greening of many highly wasteful chemical processes [19]. A good, if sadly rare, example of this is the use of a zeolite to catalyze the Friedel–Crafts reaction of anisole with acetic anhydride (Scheme 1.1-1).

Scheme 1.1-1

In comparison to the traditional route using $AlCl_3$, the zeolite-based method is more selective. However, anisole is highly activated and the method is not applicable to most substrates – zeolites tend to be considerably less reactive than conventional catalysts such as $AlCl_3$.

Many specialty chemical processes continue to operate using traditional and problematic stoichiometric reagents (e.g. in oxidations), which we should aim to replace with catalytic systems. Even when catalysts are used, they often have low turnover numbers due to rapid poisoning or decomposition, or cannot be easily recovered at the end of the reaction. Here we need to develop new longer-lifetime catalysts and make better use of heterogenized catalysts, as well as considering alternative catalyst technologies (e.g. catalytic membranes), and to continue to improve catalyst design so as to make reactions entirely selective to one product [20].

Another good example of greener chemistry through the use of heterogeneous catalysis is the use of TS1, a titanium silicate catalyst for selective oxidation reactions [21] such as the 4-hydroxylation of phenol to the commercially important hydroquinone (Scheme 1.1-2).

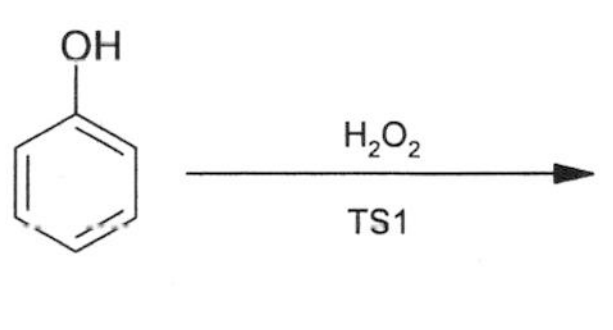

Scheme 1.1-2

TS1 has also been used in commercial epoxidations of small alkenes. A major limitation with this catalyst is its small pore size, typical of many zeolite materials. This makes it unsuitable for larger substrates and products. Again like many zeolites, it is also less active than some homogeneous metal catalysts and this prevents it from being used in what would be a highly desirable example of a green chemistry process – the direct hydroxylation of benzene to phenol. At the time of writing, commercial routes to this continue to be based on atom-inefficient and wasteful processes such as decomposition of cumene hydroperoxide, or via sulfonation (Scheme 1.1-3).

Of course, the direct reaction of oxygen with benzene to give phenol would be 100% atom efficient and based on the most sustainable oxidant – truly an ideal synthesis if we can only devise a good enough catalyst to make it viable!

The increased use of catalysis in the manufacture of low volume, high value chemicals will surely extend to biotechnology and, in particular, the use of enzymes

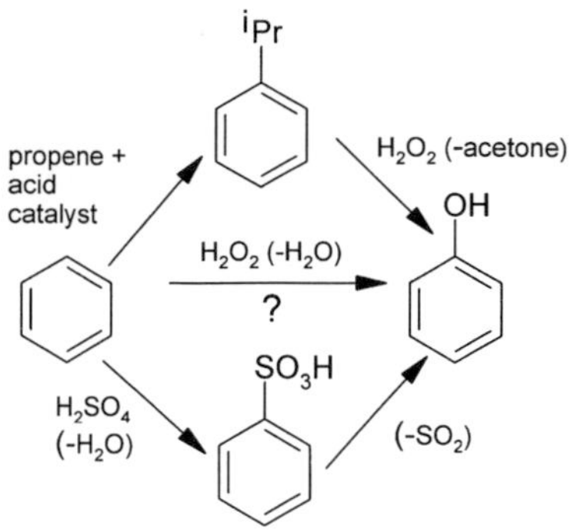

Scheme 1.1-3

[3, 22]. Enzymes provide highly selective routes to chemical products, often under mild conditions and usually in environmentally benign aqueous media. Drawbacks to their more widespread introduction include slow reactions, low space–time yields and, perhaps most importantly, a lack of familiarity with and even suspicions of the technology from many chemical compounds.

The replacement of hazardous volatile organic compounds (VOCs) as solvents is one of the most important targets for countless process companies including those operating in chemical manufacturing, cleaning, and formulation [23]. Some VOCs such as carbon tetrachloride and benzene have been widely prohibited and re-placed but other problematic solvents, notably dichloromethane (DCM), continue in widespread use. While in many cases other, less harmful, VOCs are used to re-move the immediate problems (e.g. ozone depletion) due to such compounds as DCM, more fundamental technology changes have included the use of non-orga-nic compounds such as supercritical carbon dioxide or water, the use of non-vola-tile solvents such as ionic liquids (molten salts), and the total avoidance of solvent (e.g. through using a surface-wetting catalyst in a reaction, or simply relying on interfacial reaction occurring between solids). All of these alternative technologies have been demonstrated in numerous organic reactions such as those examples shown in Scheme 1.1-4.

Carbon dioxide has also been successfully introduced into some dry-cleaning processes and various consumer formulations now no longer contain a VOC sol-vent.

Green Chemistry needs to be combined with more environmentally friendly technologies if step-change improvements are to be made in chemical manufac-turing processes. Synthetic chemists have traditionally not been adventurous in their choice of reactors – the familiar round-bottomed flask with a magnetic stirrer remains the automatic choice for most, even when the chemistry they plan to use is innovative e.g. the use of a non-volatile ionic liquid solvent or a heterogeneous catalyst as an alternative to a soluble reagent. However, an increasing number of re-search articles describing green chemical reactions are based on alternative reac-tors including [3, 24].

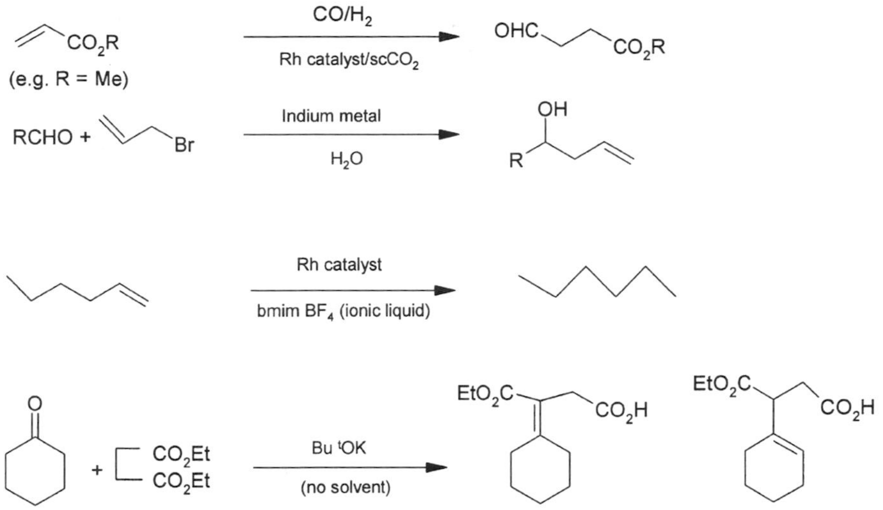

CO$_2$R
(e.g. R = Me)

CO/H$_2$
——————→
Rh catalyst/scCO$_2$

OHC CO$_2$R

RCHO + Br

Indium metal
——————→
H$_2$O

OH
R

Rh catalyst
——————→
bmim BF$_4$ (ionic liquid)

O
+ CO$_2$Et / CO$_2$Et

Bu ^{t}OK
——————→
(no solvent)

EtO$_2$C CO$_2$H

EtO$_2$C CO$_2$H

Scheme 1.1-4

- continuous flow reactors (a technology that dominates the petrochemical industry but is little utilized in specialty chemical manufacturing)
- microchannel reactors whereby reaction volumes are kept small and scale is highly flexible thus reducing hazards and risk
- intensive processing systems such as spinning disc reactors which combine the benefits of low reaction volumes with excellent heat transfer and mixing characteristics
- membrane reactors that can maintain separation of aqueous and non-aqueous phases, hence simplifying the normally waste-intensive separation stages of a process

These alternative reactor technologies can be combined with Green Chemistry methods including, for example, catalytic membrane reactions and continuous flow supercritical fluid reactions.

Energy has often been somewhat neglected in the calculations of resource utilization for a chemical process. Batch processes based on scaled-up reaction pots can run for many hours or even days to maximize yield and often suffer from poor mixing and heat transfer characteristics. As the cost of energy increases and greater efforts are made to control emissions associated with generating energy, energy use will become an increasingly important part of Green Chemistry metrics calculations. This will open the door not only to better designed reactors such as those described earlier but also to the use of alternative energy sources. Of these, two of the more interesting are:

- ultrasonic reactors
- microwave reactors[25].

Both are based on the use of intensive directed radiation that can lead to very short reaction times or increased product yields and also to more selective reactions [3]. Examples of the use of these reactors are shown in Scheme 1.1-5.

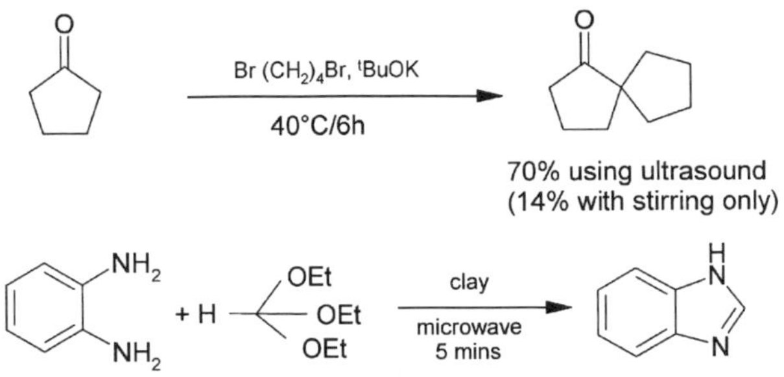

70% using ultrasound
(14% with stirring only)

Scheme 1.1-5

A lifecycle approach to the environmental performance and sustainability of chemical products demands a proper consideration of pre-manufacturing and specifically the choice of feedstocks. Today's chemical industry is largely based on petroleum-derived starting materials, a consequence of the rapid growth in the new petroleum-based energy industry in the early twentieth century. This industry was based on an apparently inexhaustible supply of cheap oil, which we could afford to use on a once-only basis for burning to produce energy. Petrochemicals was a relatively small (around 10%) part of the business, generating a disproportionately high income and helping to keep energy costs down, which in turn maintained ultra-high demand for the raw material even when extraction became more difficult and transportation more controversial. The parallel and mutually supportive growth in petro-energy and petrochemicals from the petro-refineries of the Middle East, Americas, Africa, and elsewhere is surely past its peak. It now seems likely that as we try to tackle the inevitable decline of oil as an energy source, so shall we attempt to seek alternatives for the manufacture of at least some of the many chemicals we use today. While forecasts seem to change every day and political parties can selectively use bits of the overwhelming amount of conflicting data to suit their own agenda, no one will argue that these changes must occur in the twenty-first century – "one hundred years of petroleum" is beginning to look about right.

The use of sustainable, plant-based chemicals for future manufacturing can involve several approaches (Fig. 1.1-9)· [3, 7 14].

Many of the earlier plans in this area were based on the bulk conversion of large quantities of biomass into the type of starting materials that the chemical industry has grown up on (CO, H_2, C_2H_4, C_6H_6, etc.). On one hand the logic behind this approach is clear – the manufacturing industries are equipped to work with such simple small molecules. On the other hand, it is perverse to consume resources and generate waste in removing functionality from albeit a soup of molecules, just so that we can then apply our chemical technology toolkit to consume more resources and generate more waste in converting the intermediate simpler molecules into ones we can use in the many industries that use chemicals. The scale of operation,

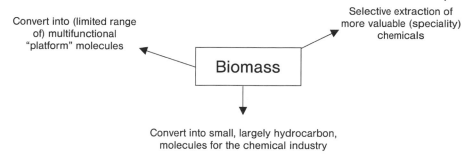

Fig. 1.1-9 Approaches to the use of plant-based chemicals.

and the added costs of the extra steps, will always make this technology expensive and of limited appeal except in those situations where a large volume of waste bio-mass is in close proximity to suitable industrial plant.

Nature manufacturers an enormous array of chemicals to perform the many functions that its creatures need to survive, grow, and propagate. A tree contains some 30 000 different molecules ranging from simple hydrocarbons to polyfunc-tional organics and high molecular weight polymers. Many of these molecules ha-ve immediate and sometimes very high value, for example as pharmaceutical inter-mediates. The selective extraction of compounds from such complex mixtures is, however, often impractical and uneconomic and may lead to a very high environ-mental impact product as a result of enormous inputs of energy and outputs of waste. The extraction of families of compounds with high value themselves or through Green Chemistry modification is a more likely approach to take advanta-ge of some of nature's gifts of sustainable and interesting molecular entities.

The third approach of using a large proportion of biomass to produce so-called "platform molecules" is worth close consideration. Here, we need to learn how to make best use of a number of medium-sized, usually multifunctional, organic mo-lecules that can be obtained relatively easily by controlled enzymatic fermentation or chemical hydrolysis. The simplest of these is (bio) ethanol; others include levu-linic acid, vanillin, and lactic acid. These are chemically interesting molecules in the sense that they can be used themselves or can quite easily be converted into ot-her useful molecules – building on rather than removing functionality – as can be seen, for example, with lactic acid (Scheme 1.1-6).

One of these products, polylactic acid, has become the basis of one of the best re-cent commercial illustrations of the potential value of this approach. Cargill-Dow now manufacture polylactic acid polymer materials using a starch feedstock. The materials are finding widespread use as versatile, sustainable, and (importantly) biodegradable alternatives to petro-plastics. [14, 26].

Making more direct use of the chemicals in biomass and the functionality they contain, rather than reducing them to simpler, smaller starting materials for synthesis, makes sense from a lifecycle point of view as well as economically (Fig. 1.1-10).

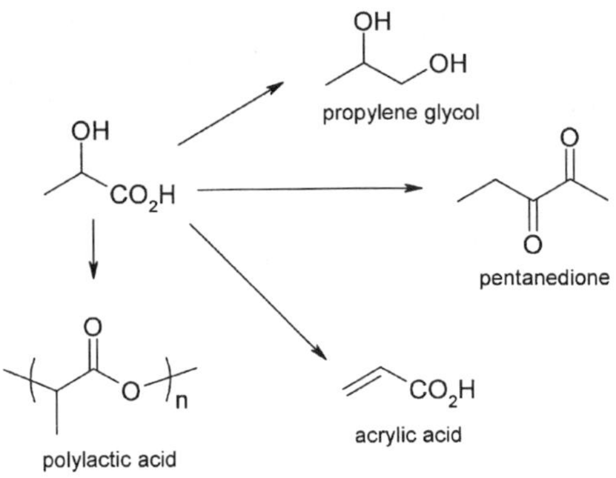

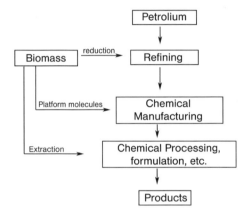

propylene glycol

pentanedione

polylactic acid

acrylic acid

Scheme 1.1-6

Petrolium

Biomass —reduction→ Refining

Platform molecules → Chemical Manufacturing

Extraction → Chemical Processing, formulation, etc.

Products

Fig. 1.1-10 The use of biomass chemicals in traditional chemical industry processes.

1.1.5
Green Chemistry Metrics

In its short history, Green Chemistry has been heavily focused on developing new, cleaner, chemical processes using the technologies described earlier in this chapter. Increasing legislation will force an increasing emphasis on products but it is important that these in turn are manufactured by green chemical methods. Industry is becoming more aware of these issues and some companies can see the business edge and competitive advantage that Green Chemistry can bring. However, the rate of uptake of Green Chemistry into commercial application remains very small. While the reasons for this are understandably complex, and also dependent on the economic vitality of the industry, it is important that the advantages offered by Green Chemistry can be quantified. Legislation or supply-chain pressures may persuade a company that the use of a chlorinated organic solvent is undesirable, but how can they select a genuinely "greener" alternative? How can a company add environmental data to simple cost and production factors when comparing routes to a particular compound? Can the environmental advantages of using a renewable feedstock compared to a petro-chemical be quantified? In order to make Green Chemistry happen, we need to see the concept mature from an almost philosophical belief that it is the "right thing to do" to one that can give hard, reliable data to prove its merits.

These needs and "reality checks" have led to the emergence of Green Chemistry-related metrics, although they are very new and by no means widely applied or tested. The ultimate metric can be considered to be lifecycle assessment (LCA), but full LCA studies for any particular chemical product are difficult and time consuming.[13, 14]. Nonetheless we should always "think LCA", if only qualitatively, whenever we are comparing routes or considering a significant change in any product supply chain. Green Chemistry metrics [16, 25] are most widely considered in comparing chemical process routes, including limited, if easy-to-understand, metrics such as atom efficiency and attempts to measure overall process efficiency such as *E* factors, mass intensities, and mass efficiency [27]. As with LCA, these metrics have to be applied with definite system boundaries, and it is interesting to note that for process metrics these boundaries generally do not include feedstock sources or product fate. Energy costs and water consumption are also normally not included, although given the increasing concerns over both of these it is difficult to believe that they can be ignored for much longer. At the product end of the lifecycle we are used to testing for human toxicity and this will become much more prevalent through REACH [6]. We will also need to pay more attention to environmental impact, and here measures of biodegradability, environmental persistence, ozone depletion, and global-warming potential are all important metrics. Last, but not least, we are moving towards applying Green Chemistry metrics to feedstock issues. As we seek "sustainable solutions" to our healthcare, housing, food, clothes, and lifestyle needs, so we must be sensitive to the long-term availability of the inputs that go into the supply chain for a product [14]. With increasing pressures from the feedstock and product ends, and increasing restrictions and controls on the intermediate processing steps, chemistry must get greener!

References

1 P.T. Anastas and J.C. Warner, *Green Chemistry, Theory and Practice*, Oxford University Press, Oxford, **1998**.

2 P.T. Anastas and R.L. Lankey, *Green Chem.*, **2000**, *2*, 289.

3 J.H. Clark and D.J. Macquarrie, *Handbook of Green Chemistry & Technology*, Blackwell, Oxford, **2002**.

4 J.H. Clark, *The Chemistry of Waste Minimisation*, Blackie Academic, London, **1995**.

5 www.epa-gov/greenchemistry

6 A.M. Warhurst, *Green Chem.*, **2002**, 4, G20; and see also www.europa.eu.int/comm./enterprise/chemicals/chempol/reach/explanatory-note.pdf and www.pond.org/downloads/eurge/wwfreebreachnewopforindustry.pdf

7 M. Lancaster, *Green Chemistry, an Introductory Text*, Royal Society of Chemistry, Cambridge, **2002**.

8 www.chemsoc.org/gcn

9 www.gci.org

10 www.rsc.org/greenchem

11 J. Elkington, *Australia CPA*, **1999**, 69, 18.

12 C.V. Stevens and R.G. Vertie, eds., *Renewable Resources*, J. Wiley & Sons, Chichester, **2004**.

13 T.E. Graedel, *Streamlined Life-Cycle Assessment*, Prentice Hall, New Jersey, **1998**.

14 A. Azapagic, S. Perdan and R. Clift, *Sustainable Development in Practice*, J. Wiley & Sons, Chichester, **2004**.

15 J.H. Clark, *Green Chem.*, **1998**, 1, 1.

16 D.J.C. Constable, A.D. Curzons and V.L. Cunningham, *Green Chem.*, **2002**, 4, 521.

17 R.A. Sheldon, *Chemistry and Industry*, **1997**, 1, 12.

18 See, for example, G.D. McAllister, C.D. Wilfred and R.J.K. Taylor, *Syn. Lett*, **2002**, 2, 1291 and C.W.G. Fishwick, R.E. Grigg, V. Sridharan and J. Virica, *Tetrahedron*, **2003**, 4451.

19 R.A. Sheldon and H. van Bekkum, eds., *Fine Chemicals through Heterogeneous Catalysis*, Wiley-VCH, Weinheim, **2001**.

20 P.M. Price, J.H. Clark and D.J. Macquarrie, *J.Chem. Soc., Dalton Trans.*, **2000**, 101.

21 US Patent, 4,410,501 (1983) and F. Maspero and U. Romano, *J. Catal.*, **1994**, 146, 476.

22 www.europabio.org/upload/documents/150104/becas_report_en.pdf

23 D.J. Adams, P.J. Dyson and S.J. Tavener, *Chemistry in Alternative Reaction Media*, J. Wiley & Sons, Chichester, **2004**.

24 S.J. Haswell, *Green Chem.*, **2003**, *5*, 240.

25 M. Nüchter, B. Ondruschka, W. Bonrathard and A. Gurn, *Green Chem.*, **2004**, *6*, 128.

26 E.S. Stevens, *Green Plastics*, Princeton University Press, Princeton, NJ, **2002**.

27 M. Eissen, K. Hungerbühler, S. Dirks and J. Metzger, *Green Chem.*, **2004**, *6*, G25.

1.2
Sustainable Development and Regulation

Diana Cook and Kevin Prior

1.2.1
Introduction

Sustainable development has become the accepted orthodoxy for global economic development and environmental protection since the end of the twentieth century.

Sustainable development means many things to many people and the range of actions and their implications is as varied. This work uses the most often quoted and accepted definition from the report *Our Common Future* [1] (also know as the Brundtland Report): "Sustainable Development is development that meets the needs of the present without compromising the ability of future generations to meet their own needs."

This definition leads to the consideration of three major aspects of sustainable development: environment, economy, and community. The UK's headline indicators (see Appendix to this chapter) adopted as part of the country's Sustainable Development Strategy in 1999 [2] give a good indication of the breadth of the scope of sustainable development.

This sets global society a number of challenges as to how to balance the competing needs of feeding, clothing, and housing a growing global population, whilst managing limited resources efficiently and generating sufficient wealth to meet the reasonable economic aspirations of the societies at large. This must all be done without damaging the earth's eco-system. There is overall consensus on the objectives, with passionate debate [3] about the actual steps that are needed to achieve them.

The achievement of sustainable development will require action by the international community, national governments, commercial and non-commercial organisations, plus individual action by citizens. The international community has passed a number of milestones in moving forward the sustainable development agenda. The most notable are:

1987: The World Commission on Environment and Development (The Brundtland Commission chaired by Gro Harlem Brundtland) produced the report *Our*

Green Separation Processes. Edited by C. A. M. Afonso and J. G. Crespo
Copyright © 2005 WILEY-VCH Verlag GmbH & Co. KGaA, Weinheim
ISBN 3-527-30985-3

Common Future. That report produced and popularized the current sustainable development definition that is quoted above.

1992: The Earth Summit, also known as the UN Conference on Environment and Development (UNCED), was held in Rio de Janeiro to reconcile worldwide economic development with protecting the environment. The Summit brought together 117 heads of states and representatives of 178 nations, who agreed to work towards the sustainable development of the planet.

2002: A further meeting, The World Summit on Sustainable Development, was held in Johannesburg (South Africa) to review progress in the ten years since UNCED.

These meetings produced statements of intent from the world's political leaders and other stakeholders. The basic questions still remain:

– How to create wealth without permanently damaging the environment?
– How to enable a fair distribution of wealth, with equality of access to health and education?

The outcome of each society's response to these questions is its laws, regulations and other government interventions.

This chapter examines whether regulation is aiding or hindering the goal of sustainable development in relation to its environmental aspects. It focuses primarily on the European Union (EU), whilst also drawing on examples from the other areas of the world.

1.2.1.1
Sustainable Development and the European Union

The prime function of the EU was to create a common market for goods and services. This role has grown as the necessary rules to enable a common market to operate have developed. The EU has well-developed economic, social, and environmental policies, which require that environmental protection must be integrated into other EU policies. This is with the overt objective of promoting sustainable development. The economics and corresponding policies of the EU are described elsewhere [4].

The inclusion of environmental protection is particularly relevant when considering a common market or identifying potential conflict between environmental protection and economic growth. In fact one of the objectives of the EU's Sixth Environmental Action Programme is the decoupling of economic growth from resource usage.

1.2.1.2
Why Regulation is Required to Achieve Sustainable Development

In order to appreciate the need for regulations in relation to achieving sustainable development, it is necessary to explore briefly the link between the environment and some of the basic features of the operation of a market economy. This section highlights the main features that underpin the need for regulation, and explores some of the instruments that governments use to achieve the goals of sustainable development. Other authors discuss this topic more fully [5]. Later sections comment on how different approaches can aid or hinder sustainable development.

The underlying principle behind the operation of a market economy is that the market provides all the incentives to the participants to operate as efficiently and effectively as possible. However, this is not always the case; a market can be "imperfect" and this is particularly so for the environment. Externalities (where the full costs of an operation are not included in the price of the goods or services) are often viewed as the most widespread cause of market failures in relation to the environment. In such situations market prices often do not reflect the full cost of the environmental resources. Such externalities can be either negative or positive.

A typical example of a negative externality is the situation where the discharge of industrial wastewater from a manufacturing operation into a watercourse results in a downstream user incurring extra costs for a situation that is beyond his control. Unless there is government intervention, the upstream manufacturer does not have to pay the full costs of his operation.

This is the basis of the "polluter pays principle", which was incorporated into the Maastricht Treaty and states that those who are responsible for environmental pollution, resource depletion, and social cost should pay the full cost of their activities. If these costs are ultimately passed on to the consumer, then there is an incentive to reduce the levels of environmental damage and resource depletion.

It thus becomes essential to internalize the full environmental cost into the price of goods and services in order to achieve sustainable development. Government intervention, in the form of regulations, and a variety of economic instruments aim to enable this process by ensuring that the social costs and benefits are included in the prices that are charged.

The environment also has many of the features of a "common" or "public good" in economic terms. Overall, society would be better off with clean air and water and abundant natural resources, but in a market economy this does not occur without some form of intervention. In most situations in a market economy, people who have not paid for goods do not receive them; it is not possible to exclude people from the benefits of clean air because they have not paid for it.

Governments may also intervene by supplying information on the better use of environmental resources in situations where producers lack the knowledge to make the best business decisions in relation to the environment. The UK government's "Envirowise" scheme is a typical example.

1.2.1.3
Environmental Policy and Innovation

One of the most serious implications of the fact that goods and services frequently do not bear the full environmental costs of their production is the effect on innovation in environmental technologies that could contribute to sustainable development. There is less incentive for firms to undertake research and development into technologies, process improvements, or systems that could contribute to sustainable development if the rate of return on the investment does not reflect fully the costs involved i.e. the costs are still externalized. This can result in a vicious circle of under-investment that market forces alone cannot resolve. A related argument is that dischargers will do the minimum to meet the enforced standards rather than look further to see what else could be done to reduce their environmental impact.

Conversely, there are also situations where tight environmental regulation has forced the development of new technology to enable the standards to be met; for instance fuel additives and water-treatment processes.

There is also a view that regulation stifles innovation in general. However the authors of a report [6] to the UK Royal Commission on Environmental Pollution recently explored the claim that EU legislation put the European chemical industry at a disadvantage compared with the US and Japanese industries. The report concluded that in most cases regulation plays a "modulating role" and can both inhibit and stimulate innovation. It is thus the *rate* of innovation rather than the *quantity* of innovation that appears to be affected.

As well as the investment in the innovation activity itself, there is also the issue that even when new environmental technologies are launched, their uptake is slow [7]. The European Commission feels that economic barriers are a particular problem unless true environmental costs are taken into account. They also cite poor access to finance, long investment cycles, and poor dissemination of new technologies as issues. The Commission has now adopted an action plan [8] to overcome the perceived barriers, including utilizing financial instruments to share the risks of investing in environmental technologies.

1.2.2
Environmental Policy Instruments

This section briefly examines the range of policy instruments that might be used to achieve a particular objective or counterbalance a perceived market failure.

1.2.2.1
"Command and Control" Regulation

Environmental policy instruments in the Organization for Economic Co-operation and Development (OECD) countries have, historically, been typified by "command and control" regulation, such as the European Union's Integrated Pollution Prevention and Control Directive (IPPC) regulations. Under such regulations, potential polluters are "commanded" to comply with particular standards and then "controlled" by tight monitoring and enforcement activity. Such approaches have been successful in reducing air and water pollution: for instance in greenhouse gas and sulfur dioxide emissions and in the improvement in the level of treatment and the proportion of the population connected to wastewater treatment plants [9]. However, concerns have been raised about their effectiveness in other areas.

As well as the concerns discussed above in relation to the effect on innovative activity, the OECD [10] comments that such regulations are costly to implement and enforce. Typically, they also do not take into account the broader environmental impacts that result from general patterns of production and consumption and how these change.

They are also relatively ineffective in controlling diffuse sources of pollution (such as the effect of agricultural fertilizers on surface water quality and consequently drinking water quality). Some regulations are perceived to work well and in a cost effective manner. The Montreal Protocol, aimed at phasing out all the main ozone-depleting chemicals, has achieved nearly a 90% reduction in the production of ozone-depleting substances in Western Europe and also reductions in their production and use in Central and Eastern Europe. These achievements are reflected in a gradual fall in the concentration of chlorine-containing ozone-depleting substances in the troposphere [9].

IPPC addresses the effects of pollution once a manufacturing process has been carried out. However, sustainable development also needs to address the control of substances in the overall supply chain. These are both the substances that are used as raw materials for the manufacture of other substances and the resulting products themselves. Controlling the input and use of chemicals in the environment is a central theme of a holistic view of sustainable development and has key implications for the social aspects of sustainable development.

The EU is, at the time of writing, reviewing its regulatory control of chemicals with its REACH (Registration, Evaluation, Authorization and Restriction of Chemical) proposals [11]. Although the current EU system is often seen as fragmented and cumbersome, even the amended proposals published in October 2003 have prompted vigorous debate. Arguments on the one hand center again on the perception that the proposals will have a negative impact on innovation, will be costly to implement, and damage the competitiveness of the European chemical industry. Environmental organizations, however, counter that REACH will result in significant environmental and health benefits and encourage innovation. UK retailers [12] also welcome the REACH proposals, despite the imperfections in the regulations, because they see that they could provide a single robust system that both suppliers and customers can use and that will share the burden of the costs involved.

1.2.2.2
Government Subsidies

Government intervention in the form of subsidies to particular industries and the impact of such subsidies on sustainable development is contentious. The OECD [10] highlights subsidy policies such as those to agriculture, fisheries, and peat and coal production as having a particularly harmful environmental impact. Similarly, if subsidies are used instead of taxes to try to reduce a particular pollution, then these can distort the polluting industry by attracting an inefficient high level of new entrants.

However, in some situations the source of pollution is so diffuse and comes from such a wide variety of sources that the "polluter pays" principle is not practical and therefore society at large has to pay. On a related aspect, although subsidies are not allowed for pollution abatement technologies in general in the EU, there are exceptions. A positive use of subsidies permits assistance to encourage Small and Medium Size Enterprises (SMEs) in particular to go beyond mandatory standards, use renewable energy and/or install energy-saving equipment.

1.2.2.3
Alternative Approaches

As well as regulations, governments are now using and developing more innovative and effective means of intervention to achieve environmental policy requirements. Economic instruments such as taxes and trading schemes are designed to change behavior by giving industrialists a financial incentive to operate differently and more flexibility in how they achieve the required environmental objectives. Table 1.2-1 summarizes some examples of the tools available to governments and their relationship to the market failures discussed above.

Taxes and trading schemes   These are designed to internalize environmental costs and send the message that if sustainable development is to be achieved then the polluter and ultimately the user of the products and services concerned must pay the full cost of the goods or services. These include those associated with the previously hidden environmental costs. It would be unreasonable to implement the full environmental costs through such systems immediately, principally because of

Table 1.2-1   Examples of policies to address market failures.

Market failure	Tax	Trading schemes	Tax credits/ Public spending	Voluntary agreements	Regulation
Negative Externalities	Climate-change levy Landfill tax Fuel duty	Emissions trading scheme Landfill permits	Reduced rate of VAT on grant-funded installation of central heating and heating appliances	Pesticides EU CO_2 from cars agreement	Integrated pollution prevention and control Water quality legislation
Positive Externalities or Public Good			Tax relief for cleaning up contaminated land		Habitats and species protection legislation
Information Failures	Differential rates of fuel duty				Environmental impact assessment directive

costs sunk into existing capital equipment. Hence the timescale for adjustment is likely to be lengthy. The phasing in of the Climate Change levy and Landfill Tax in the UK are examples of this approach.

Although environmental taxes and trading schemes are similar in many respects, there are some key differences:

- Taxes fix the value assigned to pollution, but do not specify the amount.
- Trading schemes operate the other way round by fixing the amount that can be discharged, but enable the market to put a value on the pollution load.

Both approaches should allow the market to find the cheapest way of meeting the environmental objectives. The EU will be ready to implement trading schemes in 2005.

The perception of environmental taxes and trading schemes amongst industrialists is often that they are simply more measures imposed by governments to raise revenue rather than to achieve particular environmental policy objectives or replace inefficient regulations with more effective alternatives. They thus need to be accompanied by appropriate targeted information to ensure industrialists understand the reasons behind the schemes.

The OECD [10] feel that there is little evidence that such taxes reduce international competitiveness, although they qualify their view with the comment that the sectors most exposed to competition often receive exemptions and reductions.

Tax credits and public spending   Schemes utilizing tax credits typically offer tax incentives to industrialists to invest in sustainable and resource-efficient technologies. Public spending is often used to encourage directly specific environmental innovation and to promote relevant research and development. Another facet of this is the deliberate utilization of green procurement policies by local and national governments to drive the market towards sustainable development.

Voluntary agreements   These are agreements between industry and public authorities to meet environmental objectives. They often contain an element of coercion in the form of a threat of tax or regulation, and the OECD [10] regards voluntary agreements as ineffective on their own and recommends reassessing such practices. The UK government has a voluntary agreement with the pesticides sector to reduce the impact of pesticides on the environment, and feels that good progress is being made against the targets. However, the UK government has continued to investigate the use of economic instruments that could be used if the voluntary agreement fails or takes too long.

Another aspect of voluntary agreements is the voluntary action undertaken by individuals, typically in the recycling of household waste. Voluntary actions by individuals are also more successful in solving significant environmental problems when they are backed up by a cost implication. The recent introduction of a pilot program in the Republic of Ireland that involves householders paying for the disposal of household waste by weight has seen a 50% reduction in the weight of household refuse to be disposed of.

There is thus a balance to be achieved between intervention and non-intervention and the type of intervention to be used. There is no single solution and the optimum approach aims to mix and balance the tools available to match the economies and circumstances of the country in question.

This section has set the scene principally in the European arena on the relationship between the desire for sustainable development and the broad regulatory means by which governments pursue their objectives for sustainable development in a market economy. It has also highlighted the strengths and weaknesses of different tools and how these can help or hinder sustainable development. The next section examines what these authors judge to be the likely challenges and trends in the future for the environmental aspects of sustainable development and regulation.

1.2.3
Future Trends and Challenges

This section summarizes some key areas in the future relationship between sustainable development and regulation and their implications for producers, researchers, and society as a whole.

There will be a continuing balance between the needs of the environment and the social and economic needs of society.

Although the OECD [10] stated that environmental performance in OECD countries had improved in several respects since 1990, it argued that this improvement had come at the expense of the economic aspects of sustainable development. The OECD also argued that the cost of achieving these improvements could have been less (or the scale of improvements greater) if more cost-effective means of government intervention had been used. Industrialists will increasingly be expected to offer cost-effective proactive solutions to the environmental consequences of their operations rather than simply minimizing the negative effects. Balanced against this is the view that it is neither possible nor economically desirable to achieve zero environmental impact.

Government intervention is likely to make use increasingly of a variety of economic instruments as well as "command and control" regulations.

The latter will probably principally be confined to controlling localized point sources of pollution where it is important that emissions are kept below particular levels. "Command and control" regulations, by relying on fixed emission standards, do not take account of the fact that the costs of reducing pollution are different for different firms. They are thus inefficient in achieving the overall aim of reducing pollution whilst retaining economic and social development. Similarly, governments will need to resolve conflicts in some areas of environmental policy, typically for subsidies, and review other less-effective policy instruments such as voluntary agreements.

There are price implications for both producers and consumers as environmental costs continue to be included in the price of goods and services.

As government economic instruments and regulations force producers throughout the supply chain to consider the environmental costs of their operations, producers have to contend with several challenges. How do they cost these aspects into goods and services using a traditional accounting system that is not designed to cope with such costs? This basic problem is frequently the reason why projects with an environmental benefit do not get taken up. Producers also have to contend with the sunk costs of existing capital equipment, which means that investing in new equipment needs to take place at the appropriate point in the investment cycle.

There is then the challenging issue of how much of these costs can be passed on as a price rise in the supply chain, ultimately to the domestic consumer. The domestic consumer may not appreciate the environmental implications or accept the price rise as their contribution in overall responsibility for improving the environment. The implications here are for education throughout the supply chain.

Ultimately price elasticity comes into play. The fall in revenue of around 8% since 1999 (OECD [10]) from environmental taxes is an interesting illustration of how environmental taxes can result in a change in consumer behavior. The OECD attribute about a third of this decrease to a decrease in the sale of petrol because of higher prices.

The general public and NGOs tend to distrust the chemical industry and the use of chemicals and this is likely to increase unless there is action by industries that use chemical themselves.

In many parts of the world, the public perceive the chemical industry as polluting and are concerned about the impact of chemical products on health, safety, and the environment [7]. This poor perception has become more significant since the mid-1980s. This is despite the fact that measurable improvements have been achieved in the reduction of emissions. There is thus a link between the public's attitude and their behavior as consumers. A further contributor here is the increase in freedom of environmental information: for instance easier access to public registers of pollutants.

There is a complex set of issues here in relation to sustainable development. Stakeholder pressure could force the removal of some substances or products such that the outcome is perceived overall as a positive contribution towards sustainable development. Conversely, stakeholder behavior could force the continued use of substances or practices that are not sustainable in the long term.

Similarly, pressures from retailers reacting to consumers could have an impact on REACH and its implementation. This was summed up [12] by one retailer who, although recognizing the economic success of the chemical industry, felt "it has mismanaged the whole concept of trust in the last twenty years".

There is a continuing need for the development of green metrics that demonstrate to all where progress is being made and where further action is required.

Table 1.2-2 (in the Appendix to this chapter) describes the headline indicators that the UK government uses to measure the country's progress towards sustainable development. Whatever the merits or otherwise of these chosen indicators, they start to track and highlight the complexities of sustainable development. Although the European Environment Agency measures numerous parameters to track environmental performance, the European Commission is still developing a set of metrics comparable to those used by the UK to track sustainable development. Measures that are developed need to be based on outcomes, i.e. environmental quality objectives. In the short and medium term, it may be necessary to control specifically some inputs: for example emissions of greenhouse gases.

Operators of industrial concerns in the European Union will shortly find themselves specifically liable under the "polluter pays principle" when the European Environmental Liability Directive is implemented.

The EU Environmental Liability Directive [13] came into force in 2004 and member states have three years to implement it into national law. This Directive holds operators whose activities have caused environmental damage financially liable for remedying the damage. It also has a preventative benefit as well in that operators whose activities are deemed to have caused an imminent threat of environmental damage are also liable to take preventative action.

Although existing policy instruments and some national civil liability laws already cover some aspects of this Directive, this is the first comprehensive approach and one that in particular will cover damage to biodiversity that has not previously had protection. It is wide ranging in its implications and is likely to result in greater involvement by citizens and NGOs, who will now be able to require the competent authorities to act.

Again, the underlying principle behind this Directive is to move the financial expenditure associated with environmental protection and sustainable development to the operator responsible rather than the costs being met by society in general. In terms of a market economy, the costs have been internalized.

European Union industrialists will come under increasing pressure to consider the environmental lifecycle of their products "from cradle to grave".

The EU is committed to introducing an Integrated Product Policy (IPP [14]) which aims to promote sustainable development by reducing the negative environmental impacts of products throughout their lifecycle "from cradle to grave".

Whereas until now much of the effort related to environmental policies has concentrated on large point sources of pollution either at the start of the product's life (manufacturing) or its end (waste disposal), IPP aims to integrate policies across the whole lifecycle. It is market orientated and aims to use incentives to move the market towards more sustainable options.

One of the first areas that the EU is working on is the provision of lifecycle information and a handbook (due 2005) on best practice in lifecycle analysis (LCA) along with a gradual expansion of environmental labeling. At the moment there is no commitment to create regulations for IPP. However, regulation is likely to remain a potential tool, in addition to the range of economic instruments discussed above along with various guidelines.

The timescale for the implementation of IPP is long and the impact of the current proposals reduced from the original intentions. However, the successes achieved in dealing with major point sources of pollution have enabled the scope of environmental policy to be broadened to cover the whole lifecycle of a product. This introduces new challenges and opportunities to achieve sustainable development.

Producers will see more end-of-life Directives such as The Waste Electrical and Electronic Equipment Directive (2002/96/EC) plus the End of Life Vehicles Directive (2000/53/EC). There will be mounting pressure for LCAs to be carried out on all new products and substances. Not only will the manufacturing process have to be green but also the use and disposal of the product or substance will have to be benign. As a result of this, the growing science of green product design is likely to spread from overtly consumer goods to all aspects of the supply chain.

Table 1.2-2 UK headline indicators for sustainable development.

Indicator	Description
Economic output	GDP per head (UK) measured against an index based on 1970 figures
Investment	Total (measured against an index based on 1970 figures) and social (railways, hospitals schools, etc.) investment (UK)
Employment	Percentage of people of working age in work measured against an index based on 1970 figures (UK)
Poverty and social exclusion	Selected indicators of poverty and social exclusion
Education	NVQ Level 2 qualifications measured as percentage of population at age 19 (UK)
Health	Life expectancy (years) and expectancy of good or fairly good health (GB)
Housing conditions	Percentage of households in non-decent housing both social and private sector (England)
Crime – robbery, vehicle and burglary	Number of recorded crimes (England and Wales)
Climate change	Emission of greenhouse gases measured in million tonnes C both as a basket of six gases and carbon dioxide (UK)
Air quality	Days when pollution is moderate or higher (UK), measured for both rural and urban sites
Road traffic – total traffic volumes	– Measured in billion vehicle kilometers
Traffic per unit of GDP	– Vehicle kilometers per unit of GDP
River-water quality	Rivers of good or fair chemical quality
Wildlife – farmland birds – woodland birds	Population of wild birds (UK) measured against an index based on 1970 figures
Land use	Percentage of homes built on previously developed land (England)
Waste – Household waste – All arisings and management	Measured in kg per person

1.2.4
The Implications for Green Separation Processes

The European Commission takes a broad view of what constitutes "environmental technologies". It includes integrated technologies that prevent pollutants being generated in the production process as well as new materials, energy- and resource-efficient processes, environmental know-how, and new ways of working. The preceding sections in this chapter have identified that environmental policy instruments in OECD countries and Europe are moving in the future towards a greater use of economic instruments and less reliance on "command and control" regulation that specify the use of a particular technique. Thus green separation processes that enable a greater degree of environmental control to take place as part of the manufacturing process are likely to be more attractive.

As producers start to realize the financial implications of green taxes such as those aimed at reducing the amount of waste that goes to landfill, separation processes that enable producers to reduce, recycle, or not produce the waste in the first place, become more viable.

Separation processes that enable valuable materials to be recovered from waste streams, especially as fossil fuels and primary ores are exhausted, are also likely to be in increasing demand. A similar logic applies to technologies that maximize yields and recoveries of target materials for minimum input of fossil fuels and other utilities or can lead to the use of renewable or alternative feedstocks.

In the short and medium term, industry will still require end-of-pipe processes, but IPPC and IPP will encourage the use of in-pipe, in-process technologies and techniques along with appropriate management systems.

1.2.5
Conclusion

The way that regulation contributes to sustainable development is changing significantly. Europe and the OECD countries are moving towards a regulatory system that combines the historic "command and control" with a system that makes greater use of economic instruments. The debate is more about what economic instruments will be used for what pollution load and how quickly rather than whether it happens.

These regulatory strategies have the potential to help sustainable development by encouraging industrialists to take proactive steps to prevent pollution before it happens rather than by policing the pollution that has already occurred. This is likely to be balanced with a greater use of consumer-based environmental taxes to encourage individual citizens to appreciate the real costs of sustainable development.

Whether or not one signs up to sustainable development as a politically desirable policy, there is a belief among the G7 countries that the global demand for oil will increase by 50% between 2004 and 2025. Those same commentators believe that the perceived gap between oil supplies and demand will be met by technology, some of which has not yet been invented!

These technologies might fall within the definition of "Green Chemical Technologies". They may use alternative feedstocks to synthesize existing desirable molecules, make more efficient use of fossil-fuel (oil) derived materials, and/or, finally, create substances which give the same effect (utility) than those used currently but with less environmental or public health impact.

Whatever the consequences of regulation, stakeholder pressure driven by rising oil prices, will create a greater demand for green product design. Manufacturers will seek greater resource efficiency, reuse/recyclability and alternative feedstocks. This is likely to happen long before any regulation on IPP is enshrined in legislation.

References

1 World Commission on the Environment and Development (WCED), *Our Common Future* Oxford, Oxford University Press, **1987**, p. 43.

2 *Quality of Life Counts: Indicators for a Strategy for Sustainable Development in the UK 2004 Update,* Department for Environment, Food and Rural Affairs, London, **March 2004**.

3 Bjørn Lomborg, *The Skeptical Environmentalist, Measuring the Real State of the World,* Cambridge University Press, Cambridge, **2001**.

4 Ali M. El-Agraa, *The European Union: Economics and Policies,* 7th edition, Prentice Hall Financial Times, **2004**.

5 Richard G. Lipsey, K. Alec Crystal, *Economics,* 10th edition, Oxford University Press, Oxford, **2004**.

6 Surya Mahdi, Paul Nightingale, and Frans Berkhout, *A Review of the Impact of Regulation on the Chemical Industry,* Executive Summary of the final report to the Royal Commission on Environmental Pollution, SPRU-Science and Technology Policy Research, University of Sussex, **November 2002**.

7 Diana Cook and Kevin Prior, eds., *Facilitating the Uptake of Green Chemical Technologies,* Crystal Faraday Partnership Ltd, Rugby, **2003**.

8 *Stimulating Technologies for Sustainable Development: An Environmental Technologies Action Plan for the European Union,* Com(2004) 38 final, Brussels, **28 January 2004**.

9 European Environment Agency, *Europe's Environment: the third assessment, summary,* Office for Official Publications of the European Communities, Luxembourg, **2003**.

10 OECD, *Implementing Sustainable Development: Key Results 2001–2004,* OECD, Paris, **2004**.

11 European Commission White Paper, *Strategy for a Future Chemicals Policy,* **February 2001** as amended **October 2003**; See also ENDS, *REACH caught up in EU's competitiveness agenda,* **November 2003**, 346, 51.

12 ENDS, *Retailers voice support for REACH chemicals reform,* **January 2004**, 348, 31.

13 European Parliament and of the Council of 21st April 2004, Directive 2004/35/EC on *"Environmental Liability with regard to the Prevention and Remedying of Environmental Damage",* published in the Official Journal L 143, **30 April 2004**.

14 Communication from the Commission to the Council and the European Parliament, *Integrated Product Policy: Building on Environmental Life-Cycle Thinking,* Commission of the European Communities, **June 2003**; ENDS, *The wheels of integrated product policy grind slow,* **July 2003**, 342, 28.

Part 2
New Synthetic Methodologies and the Demand for Adequate Separation Processes

Green Separation Processes. Edited by C. A. M. Afonso and J. G. Crespo
Copyright © 2005 WILEY-VCH Verlag GmbH & Co. KGaA, Weinheim
ISBN 3-527-30985-3

2.1
Microreactor Technology for Organic Synthesis

Gerhard Jas, Ulrich Kunz, and Dirk Schmalz

2.1.1
Introduction

During the early 1990s microreaction technology (MRT) was introduced as a new tool for chemical research and production [1]. The first international conference (IMRET 1) dedicated to this emerging area was held in February 1997 in Frankfurt, Germany, setting the starting point for an accelerated development of the promising young technology, which in recent years has become widely accepted in various fields of industrial research [2]. From the start, MRT has been considered an innovative and revolutionary tool in chemical synthesis [3]. The hope developed that the use of microreactors in industrial research environments, namely in the pharmaceutical and fine chemical industries, would enhance the transfer from research level to scale-up or even production level, thus speeding up chemical research, dramatically shortening the time lines in chemical and process development and significantly reducing production costs. Meanwhile, MRT has been the subject of an impressive but realistic benchmarking against conventional approaches [4].

The aim of this contribution is to present a short summary of this exciting technology. We first describe some characteristic key features of microreactors. Secondly we outline a shortcut regarding the fabrication of various types of microreactors and we then present some selected examples from the growing number of exciting applications in organic synthesis. We subsequently discuss the capabilities offered by microstructured devices for the work-up of reactions mixtures today. Finally, we complete the survey with a discussion of how microreactors will contribute to a greener chemistry and a reflection on the bottlenecks limiting the widespread use of microreaction technology.

Green Separation Processes. Edited by C. A. M. Afonso and J. G. Crespo
Copyright © 2005 WILEY-VCH Verlag GmbH & Co. KGaA, Weinheim
ISBN 3-527-30985-3

2.1.2
Key Features of Microreactors

Basically, microreactors are defined as miniaturized reactors to be used in continuous production processes. From an industrial point of view, the main advantages of production with the aid of microreactors compared to conventional production are as follows:

- isothermal reaction conditions because of excellent heat-transfer capabilities;
- efficient mixing properties because of small reaction volumes;
- exact process control because of small hold-up volumes;
- enhanced safety because of precise control and very small material transport;
- safe handling of unstable intermediates due to precise reaction control;
- reliable reproducibility because of precisely controlled reaction conditions;
- rapid synthesis of kilogram amounts of development compounds because of the continuous synthesis approach;
- accelerated process development;
- reduced scale-up risk because numbering-up replaces scale-up;
- high flexibility towards product quantities;
- improved yields and selectivities because of reduced byproducts;
- better product quality;
- easy installation and handling of microreactors, and
- overall lowered production and development costs.

Based upon specific demands towards strong miniaturization and continuous synthesis, microreactors exhibit a characteristic but individual 3D architecture, with one or more reaction channels, ranging from a few micrometers up to several 100 micrometers in diameter, and reaction volumes in the nanoliter-to-microliter range. The overall length of the reaction channels may differ between the various types but is usually some centimeters. The materials microreactors are made from include metals, ceramics, polymers, glass, silicon and combinations of these materials. Production methods vary from micromachining with mechanical tools to photolithography by synchrotron radiation or electrochemical methods, depending on the material, the size of the device and the quantities to be produced. In most cases the first examples of microstructured devices were just a proof of concept, namely that it is possible to prepare a microdevice using already existing or new production methods. With few exceptions the classical production principles are very similar. First, a plate of the appropriate material is taken, and then mechanical forces, laser beams or etching techniques, form channels. Thereafter, the channels are closed to form tunnels. Closing can be done by simply glueing, soldering or welding a cover plate or by chemical methods such as chemical vapor deposition. Using a microstructured plate as a cover during this operation can lead to the formation of stacks of such plates [5]. Meanwhile, a huge number of different reactor types are commercially available ranging from single mixers and heat exchangers to fully integrated systems with variable residence time units. Photos, technical illustrations and exploded views of different reactors can be found elsewhere in the literature [1–5, 9].

The first applications of MRT were heat exchangers with higher heat-transfer capabilities than previously achieved [6]. Values up to 10 kW m^{-2} K^{-1} allow a more or less instantaneous cooling and/or heating of reaction mixtures. As a direct consequence, reactions can be carried out under isothermal conditions and with well-defined residence times. At the beginning, many applications followed this strategy, indicating safe handling of highly exothermic reactions as one of the main benefits of the technology. Not surprisingly, some reactions could be conducted that cannot be performed by conventional methods, e.g. the direct fluorination of organic compounds by elemental fluorine as one of the most impressive examples [12]. Impressed by these findings, workers soon developed other interesting fields of application (see below).

One of the outstanding characteristic features of a microreactor is its ratio of wall height to channel width, called *aspect*. A high aspect value means a high surface area, resulting in high heat- and mass-transfer capabilities. Furthermore, the flow inside the channels usually has laminar characteristics with Reynolds numbers in the range of 1 to 1000. These two characteristics are the main differences compared to conventional chemical reactors. Generally, conventional chemical reactors have a low volume-to-surface ratio and flow inside is in the turbulent regime. Figure 2.1-1 describes the specific surface area of different chemical reactors dependent on the diameter of the reaction channels.

Conventional alternatives to a microreactor, such as non-catalytic tubular reactors, exhibit about 100–150 m^2 m^{-3} surface area to reactor volume, a value very similar to conventional heat exchangers. Using these reactors with porous catalysts filling the tubes can increase the surface area dramatically up to 10^8 m^2 m^{-3}. Typically, the surface area of microreactors is in the range of 10^4 to 10^5 m^2 m^{-3}. This is the surface area only of the microreactor walls, which in general are non-porous.

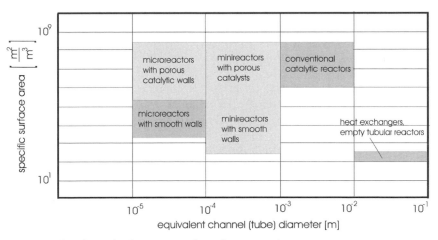

Fig. 2.1-1 The relationship between specific surface area and channel diameter.

This imposes some restriction on potential applications of microreactors, particularly with regard to their use in heterogeneous catalysis (see below). Another interesting aspect of MRT is enhanced mass transfer due to dramatically lowered mixing times (<<1 s compared to some 10 s for conventional reactors) and short diffusion times to such an extent that the influence of mass transfer on the reaction kinetics can be almost neglected.

The common practice with conventional chemical reactors is to adjust the operating conditions to values that allow safe running. For example, to lower the heat output during operation a solvent to create heat capacity is used or highly active catalysts are diluted with inert material. This is done in addition to methods to increase the surface to volume ratio, for example by using a tubular reactor with many tubes in parallel. Unfortunately, this approach quickly reaches limits in construction principles. For that reason large numbers of tubes with a diameter smaller than 1 cm are not in general use. Since conventional reactor design principles do not allow the achievement of some operating regimes, chemists (and economists!) were driven by the hope of filling this gap by the rapid application of microreactors in industrial processes.

Given the characteristic features of microreactors described above, high production rates in continuous processes (particularly in fast exothermic and endothermic reactions) can be obtained. Further benefits are improved selectivity and yields, combined with higher product quality arising from suppressed side reactions and byproducts. Since reaction conditions can be controlled very exactly, MRT permits the use of reactions involving unstable intermediates (e.g. highly reactive organometallic compounds) and the safe use of hazardous reagents. Last but not least, MRT offers the chance to construct smaller production plants compared to batch production plants. Owing to the small reaction volumes of microreactors, their use in production can improve the safety of the production process, especially if highly poisonous chemicals are involved. The small size that is obtainable with a microreactor-based chemical production plant led to the idea of building microplants at the places where the chemicals are needed. Thus, the transport of toxic materials over long distances can be avoided [10].

2.1.3
Applications of Microreactors

The high heat-transfer rate of micro heat exchangers raised the possibility of preparing microreactors for highly exothermic chemical reactions. Examples are oxidation reactions with molecular oxygen [11] or direct fluorination in micro falling-film reactors [12]. These applications impressively demonstrated that microreactors are tools for the performance of chemical reactions in reaction regimes not accessible even in state-of-the-art conventional reactors.

Another field of application was the design of micromixers. Initially, this was limited to fluid-phase reactions because even small particles would plug a microchannel and micromixers were used purely as mixing devices, but integration with heat exchangers and residence time units soon became recognized as being of great advantage for applications in organic synthesis using standardized microreactors [13]. Meanwhile, MRT has proved to be of benefit in almost all kind of chemistry including liquid reactions, gas–liquid reactions, photochemical and electrochemical reactions and gas-phase reactions. The state of the art of synthesis applications in microreactors has been summarized in a very comprehensive review [14].

2.1.3.1
Microreactors in Organic Synthesis

From a general point of view, chemical reactors can be divided between three principles of operation: batch and semi-batch operations and continuous operation. This general differentiation is equally valid for microreactors, as illustrated in Fig. 2.1-2.

Fig. 2.1-2 Available microreaction technology and current fields of application.

From a technical point of view, microplates with large numbers of holes/chambers may be also considered as microreactors operating as batch or semi-batch reactors. Because of the small reactor volume of only some microliters these reactors are not designed for chemical production. The purpose is the preparation of small quantities of new substances, for example as candidates for drug discovery. These microreactors are mainly applied in combinatorial chemistry programs and have been reviewed in detail [15]. Even small sieves to allow separation of solids can be integrated into the microchambers [16]. Thus, microreactors are available for homogeneous liquid-phase reactions as well as for heterogeneous reactions, the later ones mostly applied batchwise or semi-batchwise. Nevertheless, when talking about microreactors most people mean continuously operated flow-through devices.

In principle, MRT can be applied to industrial synthesis in very different fields: the synthesis of basic chemicals, commodities, specialties, fine chemicals and functional chemicals. The most interesting approach lies in the synthesis of high-value-added compounds, such as pharmaceuticals, fine chemicals and functional chemicals [17].

One of the most striking and convincing examples for the application of MRT is the nitration of a Viagra™ intermediate [18] on a manufacturing scale (Fig. 2.1-3). Several years of chemical development were needed to solve the problem in a conventional batch approach. A serious problem was the decarboxylation of the reaction product owing to the lack of temperature control under highly exothermic conditions. The problem could be efficiently and quickly solved using a standardized microreactor system that allowed very precise temperature control. As a consequence, the reaction could be conducted at a maximum temperature (90 °C instead of 50 °C) to give the product in 73% reproducible yield without decarboxylation. It is important to note that the reaction could be optimized within hours on a laboratory scale and immediately transferred to a production run with constant product quality, thus demonstrating impressively how the time lines in process development and production can be significantly shortened [16].

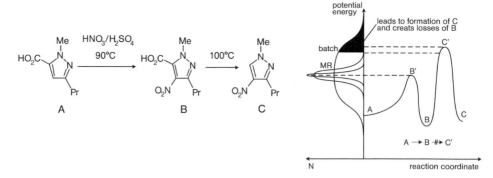

Fig. 2.1-3   The nitration of a Viagra™ intermediate.

Further convincing examples can be given from metallorganic reactions, e.g. lithiations. Usually, in the conventional batch approach, these reactions have to be carried out under cryogenic reactions, often with restricted selectivity and strong limitations on scale-up. In microreactors, lithiation reactions and subsequent reaction with electrophiles can often be conducted at temperatures ranging from –20 to 0 °C. In most cases the yield and selectivity can be increased, altogether leading to reduced investment and operating cost. A typical example is the lithiation reaction of bromoanisole, which leads to an unstable organolithium intermediate. The conventional batch approach led to significant decomposition and low yields when the reaction was scaled-up. A two-stage microreactor system offered a convenient solution to the problem since the lithiated intermediate could be generated and reacted under precisely controlled conditions (Fig. 2.1-4) [24]. The process did not need strict temperature supervision owing to the thermo-controlled system, thus saving costs for laboratory staff and minimizing the risk to loose material in case of process failure. Access to kilogram amounts was achieved in less than 40% of the time required for the batch synthesis. A consistent, scale-independent, reproducible yield of 88% could be obtained, avoiding an evaluation at intermediate levels (e.g. 100 g and 500 g).

Today, the use of microreactors for the synthesis of organic fine chemicals is just beginning to be investigated. A keynote lecture was presented with the focus on or-

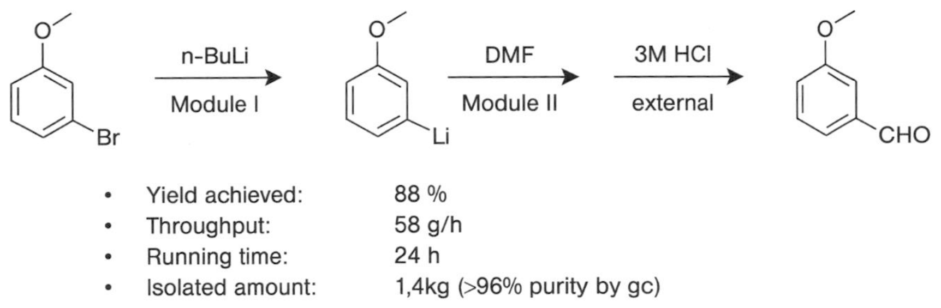

- Yield achieved: 88 %
- Throughput: 58 g/h
- Running time: 24 h
- Isolated amount: 1,4kg (>96% purity by gc)

Fig. 2.1-4 Handling of an unstable organometallic intermediate in a two-stage system.

ganic chemistry using microreactors [19]. In this presentation it was demonstrated that micromixers could raise selectivity in Friedel–Crafts alkylation reactions of aromatic compounds, shifting the yield to the desired monoalkylated product. The selectivity of cycloaddition of an N-acyliminium ion to styrene could be shifted to the desired cycloadduct, avoiding polymeric byproducts. The reason for this improved selectivity is the exact 1:1 ratio of the reactants on a microscale, generated by micromixers.

The continuous operation of microreactors applied to a broad range of heterogeneous reactions with immobilized or polymer-supported reagents or catalysts is rather limited because of problems in manufacturing such microreactors. An interesting and promising approach is the development of a microreactor with an irregular channel geometry. This irregular channel structure is not very common in MRT, but promises to be much cheaper in production, having a comparable beneficial effect on mass transfer. Inside this type of microreactor a monolithic polymer phase – obtained by precipitation polymerization in a porous carrier material – is used to anchor reactants or catalysts [20]. In this microreactor, ion exchange resins act as a support for reactants and simultaneously as an absorbent for byproducts formed during reaction. By this combination organic synthesis is coupled with a purification step, contributing to a greener chemistry. An exciting application is the hydrogenation reaction of double and triple bonds as well as the reduction of aromatic nitro compounds to the corresponding anilines by means of immobilized Pd(0) particles [21].

2.1.3.2
Applications of MRT in Process Development

With MRT the synthesis of increased amounts of a compound is reduced to a simple question of the rate of throughput. Once a synthesis has been optimized in a microreactor, small-scale production will be only a matter of numbering up the microreactor systems or a problem of scaling out the microreactor [22]. Not surprisingly, chemists in process development departments, particularly in the pharmaceutical industry, recognized MRT as a versatile tool for their daily work as it enabled fast and low-cost kilogram-scale synthesis, e.g. for early clinical studies, in an early stage of the discovery and development process. The synthesis of drug candidates in these early phases is usually characterized by high attrition rates and usually there is no time to establish fully developed commercial processes. In that sense, MRT leads to bypassing the time-consuming efforts towards process optimization from laboratory to pilot scale, as conventional process development is mainly guided by looking for safer process alternatives and avoiding hazardous conditions. Despite these fundamental advantages, the use of MRT in process development is still in its early stages. Nevertheless, a number of tracking examples, including highly exothermic reactions, reactions at elevated temperatures, reactions involving unstable intermediates and reactions involving hazardous reagents, e.g. the use of ethyl diazoacetate in a ring-enlargement reaction (see Fig. 2.1-5), have been reported by Johnson & Johnson chemists and compared to the corresponding conventional batch processes [23]. In all cases MRT proved to be superior compared to the conventional approaches, since it offered an efficient and safe scale-up and eventually, reduced process research times. Although there is still a lack of examples showing the overall capabilities of MRT in the whole development process, it can be concluded from the reported examples that MRT is suitable for shortening the development times significantly.

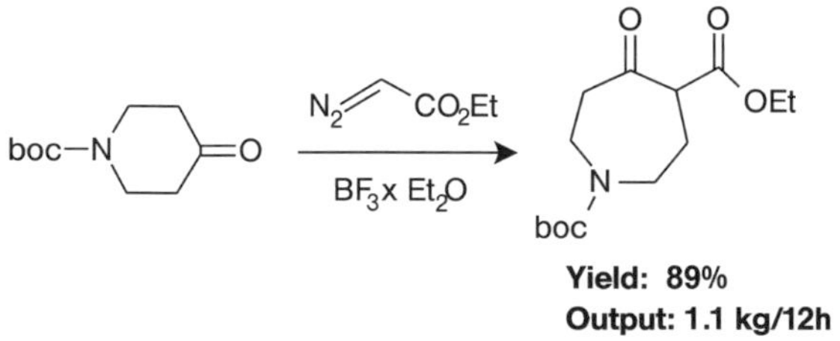

Yield: 89%
Output: 1.1 kg/12h

Fig. 2.1-5 Small-scale production with MRT in process development using a hazardous reagent.

In addition to aspects of early small-scale production, the development process itself can take advantage of MRT. It is interesting to note that already standardized microreaction systems are commercially available that allow sequential operations and, moreover, coupling to in-line analytical facilities [24]. Obviously, these devices are high-throughput tools for the combinatorial optimization of continuous reactions, e.g. the fully automated optimization of a reaction by the aid of DoE (Design of Experiments) can be realized within a few hours.

2.1.3.3
MRT in Industrial Production

Whereas it has been shown in recent years that a considerable number of industrially relevant production processes can be advantageously transferred to MRT in a laboratory scale, to date only a few industrial production processes relying on MRT have been disclosed, mainly because of the common confidentiality reasons.

Nevertheless, it is interesting to note that a number of companies, namely some key players in the fine and specialty chemical industry, have reported great efforts towards the implementation of existing MRTs or the development of their own approaches. Generally, MRT is considered to be one of the key technologies towards economic success [25]. There is a broad consensus that a great proportion of existing production processes could benefit from switching to MRT with little or no modification [26].

One of the first companies to take advantage from MRT in production was Merck, Germany. It was shown in a laboratory that using interdigital micromixers in the addition of a reactive organometallic compound to a carbonyl compound gave a 25% higher yield than the corresponding batch process. Based on these results, Merck launched a production plant in 1998 replacing a stirred batch reactor with five parallel micromixers that had been modified for the production process [27].

An interesting industrial application is the synthesis of pigments by means of a diazo coupling reaction as was successfully realized at the Clariant company (Frankfurt, Germany) based on standardized microreaction technology [28]. The pigments produced in the microreactor exhibited higher color strength, brightness and transparency, and the particle size distribution could be lowered by a factor six. This example is an impressive demonstration that MRT can be advantageously used for unorthodox problem solving. Usually, MRT has strong limitations when the reaction is not homogeneous because of insufficient solubility of the starting materials or precipitation (e.g. inorganic salts) in the course of the reaction. Nevertheless, by careful control of the reaction conditions these obstacles can be overcome. It is interesting to note that MRT can even be applied to polymerization processes.

One of the reasons why MRT has not been integrated to a greater extent in chemical production is that process control – as well as aspects to be discussed in the following sections – can be a serious problem. Despite all the advantages shown, numbering-up the technology needs a lot of effort and may be crucial, particularly with regard to flow control and flow distribution in the whole system. Furthermore, although many microreactors and micromixers are available, each production plant has to be designed individually towards a specific process. This might cause economic problems because most of the existing batch production plants have been built as multipurpose plants. Thus, a first step towards a broadened application of MRT in production would be the development of multipurpose plants based upon standardized microreactors as reported by CPC engineers [29]. The pilot plant – based on external parallelization of standardized microreactors – is designed for a broad range of organic reactions and will have production capacities in the range of 20 tons per year.

2.1.4
Microstructured Unit Operations for Workup

While many suitable applications of microreactors for organic synthesis exist, there is still a great demand for micro-scaled workup methods such as distillation, adsorption or extraction. Currently, only a few preliminary examples have been realized. An example of a stripping column has been presented [30]. In a stripping process a volatile compound is removed from a liquid by contact with a gas stream. To enhance the efficiency of this process, packed towers are used to increase the surface area through a higher level of convective mass transport. A microstructured stripping column can reduce the film thickness, leading to an increase in mass-transfer coefficient by a factor of ten. The device was fabricated by etching channels in silicon. Two channels – one for the gas phase and one for the liquid phase – were linked with small holes between the channels to ensure good contact between the two phases. As a test example the removal of toluene from water with nitrogen was chosen, demonstrating that the prepared micro stripping column can raise productivity compared to conventional packed towers.

Extraction on a micro scale was reported for the transfer of iron ions between aqueous hydrochloric acid and tributyl phosphate/xylene solutions [31]. In this device the two immiscible liquids were conducted through two channels, which were in partial contact with each other. Surface tension ensures that the two liquids stay in the corresponding channels and do not mix. Mass transport in liquids with diffusion coefficients in the range of 10^{-9} m^2 s^{-1} require path lengths of 30 to 100 µm to achieve transfer in about 1 s. It was shown that the described device, which was fabricated from silicon and glass, could perform the extraction effectively.

Another example of microextraction more related to organic chemistry was the transfer of acetone in hexane/water mixtures [32]. In this study, mixing by a T-shaped glass micromixer was first used to create small droplets (about 20 µm) to increase the exchange surface area. Additionally, a settler was used to allow phase separation. The chosen example resulted in a drop-free organic phase, with only some drops found in the water phase. Coalescence of the phases readily occurred during the settling period. The authors mention that this spontaneous droplet combination would not occur in many systems of industrial interest.

The problem of spontaneous separation of micromixer-produced droplets was investigated using dodecane and water as the model system [33]. To enhance spontaneous separation the emulsion from the micromixer was fed into a rectangular channel fabricated from aluminum foil as a spacer between two flat plates of glass and/or PTFE. In the case of PTFE coalescence occurred readily whereas glass did not promote the separation. This study demonstrated that the wall material and its surface properties (wettability) can have a big influence on coalescence behavior.

The ability of micromixers to create small droplets to increase the exchange surface is well established, while the field of coalescence in microdevices needs much more research. Micromixing will make mass transfer very rapid, but after this stage the coalescence of the droplets has to occur. The small droplets produced by micromixers often give rise to stable emulsions but, for an effective extraction, separation of the two liquids into two well-defined pure phases is necessary. Only micromixing together with microcoalescence can speed up extraction processes.

An interesting device for application in distillation was fabricated by milling on a silicon substrate [34]. The chamber was closed by a glass plate using anodic bonding. Methanol/water mixtures were used as a model system. In this device wall effects achieve separation of gas and liquid. The liquid is collected near the wall whereas the gas is withdrawn at a central cavity. Liquid moves to the wall by surface forces and gravity. The rectangular separation chamber was equipped with a liquid inlet, and a liquid outlet was located in the lower part of the device whereas the gas phase leaves at the top.

The given examples indicate that much more effort will be needed to develop micro unit-operation devices and to couple microreactors with such devices to get micro production plants.

2.1.5
Industrial Needs Relating to MRT

Beyond the facts described in Section 2.1.2, with continuous and automated working micro systems it is possible to speed up process development with less material consumption [35]. Many companies working in the field of organic fine and specialty chemicals are using MRT today. But now as ever, state-of-the-art processes for the optimization of organic syntheses in fine and special chemistry are batchwise operations. These operations allow synthesis under a wide range of parameters and allow the performance of different unit operations, such as extraction processes, distillations and crystallization, in the same reactor. Additionally, there is the possibility for liquid/solid separation processes and drying processes in external apparatus. To show the importance of these unit operations in this special field of organic chemistry more than 50 different syntheses from Merck KGaA were analyzed regarding the various process steps. As shown in Fig. 2.1-6, only 22% of the overall time was spent in reaction processes, but 78% in downstream processes. The necessary downstream times include between 10 and 14% of extraction, distillation, crystallization and filtration steps. The drying times were determined at 22%.

Continuous technologies such as microreactors have the disadvantage that for every different unit operation a new reactor is needed. For reactions there are many different examples and papers, e.g. in 2003 at IMRET 7, 90% of the papers related to reactions, but only 1% to distillation, 3% to crystallization and 6 % to extraction processes. These limited numbers of articles reflect the poor commercial availability of workup operations.

In most cases it is only possible to give a definite value to product quality when the product has been crystallized, separated and dried because specifications are only available for isolated products. When MRT is used today, the technology has to be changed so that the workup is carried out using batch technology. To obtain the complete benefit of MRT these missing unit operations must be developed.

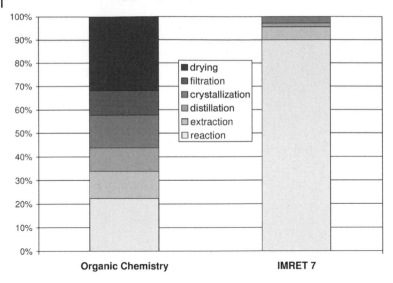

Fig. 2.1-6 Share of unit operations in fine and specialty chemistry (more than 50 different syntheses) compared with contributions at IMRET 7, 2003.

Microstructured channels could be utilized, but it would be equally possible to use milliscaled plants to perform extractions or distillations. Even these plants improve the mass transfer compared with a traditional batch reactor. The scientific reasons are explained in Section 2.1.5.

As long as these unit operations are not available, process development departments will have to develop their own unit operations to close the gap. Figure 2.1-7 shows an example of an extraction unit developed and used by the central process development department of Merck KGaA for extractions and hydrolysis on a laboratory scale. In this special case an organic solution was hydrolyzed with a weak acid. Compared with batch reactors the mass transfer is very good and there is no problem with settling the phases.

A standardized connection technology is important from the practical point of view. Before 2002 it was quite impossible to use equipment from different companies together in one microreaction plant; there was no connection technology that was suitable for all individual components. However, there are some developments in this field. The platform for modular microreaction technology of DECHEMA is developing a standardized connection system [36]. Another organization is NeSSI (New Sampling/Sensor Initiative). This is a vendor-neutral and non-affiliated effort that operates out of the Center for Process Analytical Chemistry (CPAC) at the University of Washington in Seattle. The system is based on commercial components and includes sensors and interfaces for human/machine and a standardized management system for communication between sensors and actuators [37].

Because of the continuous concept of MRT it is very important to have a measuring method for temperature on-line in the channels and not indirectly at an outlet

Fig. 2.1-7 Extraction unit from Merck KGaA, for example to perform hydrolysis.

or at the surface of the outer wall. Additional pressure sensors and flow meters must be available to detect blocking and set the right stoichiometry. Today there is no measurement technology commercially available for measuring temperature, pressure and flow in a microscaled channel.

Microreaction technology can improve the process development. But to use the complete potential of this technology more unit operations must be developed. Additional suitable sensors and actuators must be made available.

2.1.6
How can Microreactors Contribute to a Greener Chemistry?

When microreactors are used to perform chemical reactions in reaction regimes not attainable by conventional methods there is potential for greener chemistry. For example the amount of solvent can be reduced if efficient cooling can be performed by micro heat-exchangers. More concentrated solutions of chemicals can be produced, leading to savings of energy during evaporation to recover the pure substances. Microreaction technology can even open new ways to chemicals, such as direct fluorination of hydrocarbons [12]. This is not possible with regularly available "macro-technology" because the high heat of reaction would cause an explosion.

Using microreactors in situ where a hazardous chemical is needed can reduce the risk of transportation, so this can also be considered to contribute to greener chemistry.

If microscaled workup methods were available the general use of microreactors would speed up, but micro unit operations alone will not contribute to greener chemistry, because in purification processes such as adsorption the equilibrium concentrations govern the amounts of solvents that have to be used. This is independent of the scale of the system. The same holds for distillation. Microreactors only increase the surface area for phase contact. This can speed up the process, but equilibrium values cannot be shifted by microreactors. To evaporate a given amount of liquid requires the same amount of heat, independently of the size of the system. So for the most common unit operations, extraction and distillation, MRT will not bring much benefit. In separation processes such as settling/demixing in microscaled settlers the use of wall effects (surface tension) can shorten the time of separation and can perhaps contribute to a sharper phase separation line, which can improve separation of the two phases. This could contribute to a saving of solvent.

The possibility of reaching higher selectivity in micromixed reaction mixtures also contributes to greener chemistry because this saves reactants, which are not converted to undesired by-products that have to be removed from the product. In addition waste removal or deposition can be reduced.

Another interesting field for the application of microreactors in organic chemistry is the use of electrochemical synthesis. In electrochemical synthesis the yield is governed by several factors, one of which is the conductivity of the reaction mixture because the current passing through the reaction mixture determines the rate in an electrochemical process. In macroscale reactors a salt often has to be added to raise electron conductivity because the separation of the electrodes is greater then in microreactors. Using microstructured electrodes the gap can be reduced, giving the potential to improve electrochemical synthesis. The addition of conductive additives can be reduced or avoided, simplifying workup and/or decreasing side reactions of the additives.

2.1.7
Conclusions and Outlook

Obviously, MRT has moved on from the proof-of-concept stage as is demonstrated by the ever-increasing number of publications dealing with modern chemistry applications. In general, fabrication methods for microstructured devices for chemical applications are well established and, consequently, in recent years a number of companies have begun to commercialize MRT. Not only are single reactors or mixers available but also toolboxes with several microstructured devices to build custom-made micro-plants as well as a variety of system solutions [38–41]. Even microstructured modules of sizes up to several tens of centimeters are now available. Nowadays, the time has come to move the conventional batch methodologies towards continuous synthesis, a goal that is partly driven by the strong efforts of drug discovery departments in the pharmaceutical industry to set up – in conjunction with modern combinatorial chemistry approaches – real-time synthesis with online analyses, purification and bioassays. Consequently, research with MRT will be more and more chemistry driven and – in our opinion – as standard reactors become available MRT will be a versatile tool at the laboratory level. This is particularly caused by the fact that early scale-up to kilogram quantities of compounds is accessible without any effort, a real need in the pharmaceutical industry. Therefore, we must expect that modern synthesis methods such as enantio- and diastereoselective reactions, as well as catalytic reactions, will be adapted to MRT very soon and even complex molecules will become available by means of this technology.

Evaluation studies in industry have shown that microreactors can significantly contribute to improved chemical production. In principle, a variety of chemical reactions in the liquid phase or the gas phase can be carried out readily for the production of fine and specialty chemicals and active pharmaceutical ingredients. Nevertheless, it has to be said that so far no widespread use of microreactors has occurred. The main reasons for this is the lack of microstructured unit operations that can be connected easily with the reactor modules. In only a few examples has purification been implemented directly in the microreactor, as mentioned earlier [20,21]. This lack arises because in the past the development of most microreactors was carried out by workers with backgrounds not in chemistry but in physics, mechanics or related disciplines. Indeed, the world of MRT was not born from the desire of chemists to close gaps in chemical production but emerged from origins in which the focus was the production process of small channels, with methods more related to semiconductor fabrication. To focus simply on synthesis does not really demonstrate all the advantages of MRT, since a chemical production plant consists of several unit operations, of which so far only heat exchange, mixing and chemical reaction can be delivered satisfactorily on the micro scale. Modules for distillation/rectification, extraction, absorption, adsorption, phase separation of two liquids, handling of liquids with suspended particles or multiphase reactors are at an early state of development and have to be developed further. Only if all parts of a chemical production plant can be offered as a sturdy package, robust in practical application, will industry consider choosing a plant based on microstructured devices as a production tool. The benefits of the high production rates of microreactors will only bring added value to chemical production if completely integrated systems can be offered.

It is only a question of time before MRT will be seen to have a bright future, since the intrinsic advantages of microreactors are well recognized by industry [42]. As an intermediate solution some companies have already established programs using minireactors with miniaturized workup modules for the production of fine chemicals. This is a slightly easier because the integration of these miniplants into existing equipment is facilitated by the possibility of using standard, readily available components, such as capillary tubes. Scale-down of existing technology to the mini format will lead to industrial applications of future truly microreactor-based continuously operated production plants with integrated unit operations, analysis and process control.

References

1 W. Ehrfeld, ed., *Proceedings of the International Conferences on Microreaction Technology*, Volumes 1, 3, 5, Springer, Berlin, **1997, 1999, 2001**.

2 V. Hessel, S. Hardt, H. Löwe, *Chemical Micro Process Engineering*, Wiley-VCH, Weinheim, **2004**.

3 P. D. I. Fletcher, S. J. Haswell, E. Pombo-Villar, B. H. Warrington, P. Watts, S. Y. F. Wong, X. Zhang, *Tetrahedron* **2002**, *58*, 4735–4757.

4 H. Pennemann, P. Watts, S.J. Haswell, V. Hessel, H. Löwe, *Org. Proc. Res. Dev.* **2004**, *8*, 422–439.

5 (a) W. Ehrfeld, V. Hessel, V. Haverkamp, in *Ullmann's Encyclopedia of Industrial Chemistry*, Wiley-VCH, Weinheim, **2002**, Vol. 22 pp. 1–32; (b) W. Ehrfeld, V. Hessel, H. Löwe, *Microreactor: New Technology for Modern Chemistry;* Wiley-VCH, Weinheim, **2000**.

6 (a) K. Schubert, J. Brandner, M. Fichtner, G. Lindner, U. Schygulla, A. Wenka, *Microscale Thermophys. Eng.* **2001**, *5*, 17; (b) K. Schubert, W. Bier, J. Brandner, M. Fichtner, C. Franz, G. Lindner, in W. Ehrfeld, I. H. Rinard, R. S. Wegeng, eds., *Proceedings of the Second International Conference on Microreaction Technology*, March 9–12, New Orleans, **1998**, 88–95.

7 G. Wießmeier, K. Schubert, D. Hönnicke, in W. Ehrfeld, ed., *Microreaction Technology*, Springer, Berlin, **1997**, 20–26.

8 A. Kursawe, E. Dietzsch, S. Kah, D. Hönnicke, M. Fichtner, K. Schubert, G. Wießmeier, in W. Ehrfeld, ed., *Microreaction Technology: Industrial Prospects*, Springer, Berlin, **1999**, 213–223.

9 T. Schwalbe, V. Autze, G. Wille, *Chimia* **2002**, *56*, 636–646.

10 V. Hessel, W. Ehrfeld, K. Golbig, C. Hofmann, S. Jungwirth, H. Löwe, T. Richter, M. Storz, A. Wolf, O. Wörz, J. Breyss, in W. Ehrfeld, ed., *Microreaction Technology: Industrial Prospects*, Springer, Berlin, **1999**, 151–164.

11 K. F. Jensen, I. M. Hsing, R. Srinivasan, M. A. Schmid, M. P. Haraold, J. J. Lerou, J. F. Ryley, in W. Ehrfeld, ed., *Microreaction Technology*, Springer, Berlin, **1997**, 2–9.

12 V. Hessel, W. Ehrfeld, K. Golbig, V. Haverkamp, H. Löwe, M. Storz, C. Wille, A. E. Gruber, K. Jähnisch, M. Baerns, in W. Ehrfeld, ed., *Microreaction Technology: Industrial Prospects*, Springer, Berlin, **1999**, 526–540.

13 T. Schwalbe, G. Wille: *Cytos continuous chemistry – a coherent chemical synthesis technology from research through to production*, 7th International Conference on Microreaction Technology, Book of abstracts, 1–3, Sept 7–10, Lausanne, Switzerland, **2003**.

14 K. Jähnisch, V. Hessel, H. Löwe, M. Baerns, *Angew. Chem.* **2004**, *116*, 410–451; *Angew. Chem. Int. Ed.* **2004**, *43*, 406–446.

15 N. K. Terrett, *Combinatorial Chemistry*, Oxford University Press, London, **1998**.

16 G. Mayer, J. Tuchscheerer, T. Kaiser, K. Wohlfart, E. Ermantraut, J. M. Köhler, in W. Ehrfeld, ed., *Microreaction Technology*, Springer, Berlin, **1997**,112–119.

17 T. Schwalbe, V. Autze, G. Wille, *Chimia* **2002**, *56*, 636–646 and references cited therein.

18 G. Panke, Th. Schwalbe, W. Stirner, S. Taghavi-Moghadam, G. Wille, *Synthesis* **2003**, 2827–2830.

19 J. Yoshida, A. Nagaki, S. Suga: *Highly selective reactions using microstructured reactors*, 7th International Conference on Microreaction Technology, Book of abstracts, 1–3, Sept 7–10, Lausanne, Switzerland, **2003**.

20 A. Kirschning, C. Altwicker, G. Dräger, J. Harders, N. Hoffmann, U. Hoffmann, H. Schönfeld, W. Solodenko, U. Kunz, *Angew. Chem.* **2001**, *113*, 4118–4120.

21 W. Solodenko, H. Wen, S. Leue, F. Stuhlmann, G. Sourkouni-Argirusi, G. Jas, H. Schoenfeld, U. Kunz, A. Kirschning, *Eur. J. Org. Chem.* **2004**, 3601–3610.

22 (a.) S. Taghavi-Moghadam, A. Kleemann, K. Golbig, *Org. Proc. Res. Dev.* **2001**, *5*, 652–658; (b) T. Laird, *Org. Proc. Res. Dev.* **2001**, *5*, 612.

23 X. Zhang, S. Stefanick, F.J. Villani, *Org. Proc. Res. Dev.* **2004**, *8*, 465–470.

24 T. Schwalbe, V. Autze, M. Hohmann, W. Stirner, *Org. Proc. Res. Dev.* **2004**,*8*, 440–454.

25 U.-H. Felcht, *Chem. Eng. Technol.* **2002**, *25*, 345.

26 *Chem. Eng. News*, **2004**, *82* (27), 18–19.

27 H. Krummradt, U. Kopp, J. Stoldt, in W. Ehrfeld (ed.), *Microreaction Technology – IMRET 3: Proceedings of the 3rd International Conference on Microreaction Technology*, Springer, Berlin, 181, **2000**.

28 C. Wille, V. Autze, H. Kim, U. Nickel, S. Oberbeck, Th. Schwalbe, L. Unverdorben, *IMRET 6 – 6th International Conference on Microreaction Technology*, New Orleans, Louisiana, March 10–14, **2002**.

29 A. Franke, M. Leitgeb, G. Jas, *From Vision to Realization – Continuous Processing in the Pharmaceutical and Fine Chemicals Industry*, Crystal Faraday Symposium, London, **2004**.

30 S. H. Cypes, J. R. Engstrom, *Construction, analysis and evaluation of a microfabricated stripping column*, 7th International Conference on Microreaction Technology, Book of abstracts, 1–3, Sept 7–10, Lausanne, Switzerland, **2003**.

31 I. Robins, J. Shaw, B. Miller, M. Harper, *Solute transfer by liquid/liquid exchange without mixing in micro-contactor devices*, 1st International Conference on Microreaction Technology, Book of abstracts, Feb. 23–25, Frankfurt, Germany, **1997**.

32 D. Kirschneck, A. Wojik, R. Marr, *Microextraction characteristics of the n-hexane/water/acetone-system*, 7th International Conference on Microreaction Technology, Book of abstracts, Sept 7–10, Lausanne, Switzerland, 38–40, **2003**.

33 Y. Okubo, M. Toma, H. Ueda, T. Maki, K. Mae, *Microchannel devices for the coalescence of dispersed droplets for use in rapid extraction processes*, 7th International Conference on Microreaction Technology, Book of abstracts, Sept 7–10, Lausanne, Switzerland, **2003**, 41–43.

34 K. I. Sotowa, K. Kusakabe, *Design of microchannels for use in distillation devices*, 7th International Conference on Microreaction Technology, Book of abstracts, Sept 7–10, Lausanne, Switzerland, **2003**, 158–159.

35 K. Bofinger, M. Grund, Merck KgaA, *Faster to robust processes with less materials*, Achema, Frankfurt, Germany, **2002**.

36 www.microchemtech.de: Home page of the "platform for modular microreaction technology", DECHEMA.

37 M. Weiss, *NESSI Progresses from Vision to Reality*, Control Magazine, February **2003**.

38 Ehrfeld Mikrotechnik BTS, Mikroforum Ring 1, D-55234 Wedelsheim, Germany (www.ehrfeld.com).

39 Institut für Mikroverfahrenstechnik, Karlsruhe, Germany (www.fzk.de).

40 Institut für Mikrotechnik Mainz GmbH, Carl-Zeiss-Straße 18–20, D-55129 Mainz, Germany (www.imm-mainz.de).

41 CPC-Cellular Process Chemistry Systems GmbH, Heiligkreuzweg 90, D-55130 Mainz, Germany (www.cpc-net.com).

42 A. M. Rouhi, *Chem. Eng. News* **2004**, *82*, 18–19.

2.2
Solventless Reactions (SLR)

Rajender S. Varma and Yuhong Ju

2.2.1
Introduction

Chemistry in the new millennium is widely adopting the concept of "green chemistry" to meet the fundamental scientific challenges of protecting human health and the environment while simultaneously achieving commercial viability. Green chemistry and engineering are new concepts for pollution prevention that have appeared in the past decade driven by the U.S. Pollution Prevention Act of 1990. Organic synthesis is one of the most rapidly developing segments of chemistry, with numerous new reactions and methodologies appearing in chemical literature every day. To design and conduct chemical reactions with "green" experimental protocols is an enormous challenge that chemists have to confront to improve the quality of the environment for present and future generations. The emerging area of green chemistry envisages minimum hazard as the performance criterion for the design of new chemical processes. Target areas for achieving this goal are the exploration of alternative reaction conditions and reaction media to accomplish the desired chemical transformations with minimized by-products or waste, and elimination of the use of conventional organic solvents, wherever possible.

Traditional chemical syntheses or transformations generally require volatile and often hazardous organic solvents as reaction media to facilitate mass and heat transfer, and to isolate and purify desired products from reaction mixtures. An ideal chemical reaction may strive to have the following features to accommodate the principles of green chemistry [1]: proceeds at ambient temperature and pressure; requires less or no organic solvent or utilizes safer solvents; generates desired product with high atom efficiency; and produces no waste. To encompass these basic principles, the number of publications related to green chemistry and engineering has grown rapidly in the past decade, featuring such aspects as solventless (dry media) [2], solid-supported [2, 3] and solid/solid reactions, the use of room temperature ionic liquids [4], supercritical carbon dioxide [5], and water [6] as alternative reaction media combined with microwave irradiation [7], fluorous solvents, and catalysis [8].

Green Separation Processes. Edited by C. A. M. Afonso and J. G. Crespo
Copyright © 2005 WILEY-VCH Verlag GmbH & Co. KGaA, Weinheim
ISBN 3-527-30985-3

Ionic liquids are a new and remarkable class of "designer solvents" [9] with negligible vapor pressure for chemical syntheses because their properties such as solubility, density, refractive index, and viscosity can be tuned to suit specific requirements [10, 11]. Water is a relatively more environmentally friendly alternative for organic synthesis, but its application has been limited so far [6]. Supercritical carbon dioxide is yet another clean solvent but it generally requires a high-pressure reactor [12].

An ultimate goal in green chemistry is to eliminate or minimize the use of volatile organic solvents in modern organic syntheses. Development of new synthetic methodology under solventless condition is an important area of research with growing popularity [13] because solvent-free reactions reduce or eliminate solvent usage, simplify synthesis and separation procedures, prevent waste, and avoid the hazards and toxicity associated with the use of solvents. Generally, these reactions can be broadly classified in three types [14]:

- reactions between neat reagents (gases/solids, solids/liquids, liquids/liquids, and solids/solids [14c]
- reactions between reactants supported on solid mineral supports such as silica, alumina and clays [3]
- reactions carried out under phase-transfer catalysis (PTC) conditions free of solvent [14b].

PTC reactions, especially under microwave irradiation, have been discussed at length recently [14b] and are not included in this chapter. Reactions between neat reactants, with and without solid inorganic supports and with microwave assistance are described herein with the salient advantages of the processes as they relate to environment, energy consumption, and recycling of materials.

2.2.2
Solventless (Neat) Reactions (by Mixing or Grinding)

Solventless organic reactions, neat reagents (solid–solid or solid–liquid) react together in the absence of a solvent and have been well reviewed as a fast developing technology [14a, 15]; the postulated model by Scott et al. for such solventless reactions was proposed for better understanding the mechanism [16]. Some selected representative reactions under solvent-free conditions are discussed below with emphasis on product separation.

2.2.2.1
Solvent-free Robinson Annulation

Robinson annulation is a reaction sequence consisting of a Michael addition of a cycloalkanone α-carbanion to an α,β-unsaturated ketone derivative, followed by aldol condensation to form the final product, a 2-cyclohexenone ring fused to the starting alkanone. This methodology has found extensive application in organic synthesis, especially in the areas of steroid, terpene, and alkaloid chemistry of natural products for the construction of six-membered ring systems. One elegant solvent-free route to the enantioselective synthesis of optically active enedione compounds in high yield is shown in Scheme 2.2-1 [17]. Kalluraya and Rai have also reported the synthesis of sydnone-containing 3'-arylcyclohexenone derivatives by simply grinding chalcones and ethyl acetoacetate in the presence of potassium carbonate under solvent-free conditions via Robinson annulation [18]

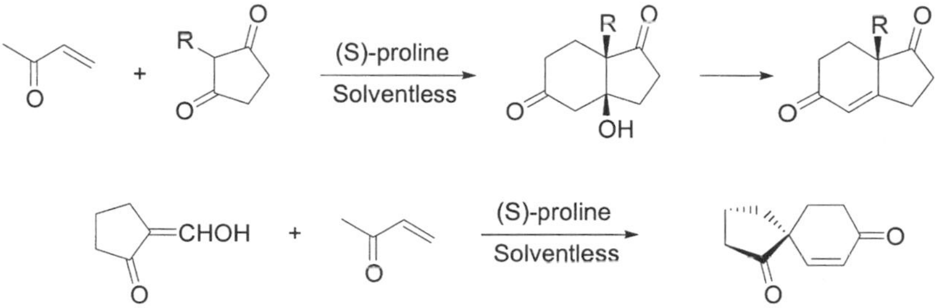

Scheme 2.2-1 Enantioselective Robinson annulation.

A solvent-free self-assembly of fullerene C$_{60}$ using high-speed vibration milling has been described [19]. This simple, efficient, and green pathway to the synthesis of a supramolecular complex in the solid state opened a new avenue for the application of solventless reactions.

2.2.2.2
Chemoselective, Solvent-free aldol Condensation Reactions

The aldol reaction is widely used in synthetic organic chemistry to build carbon–carbon bonds and often requires stoichiometric amount of noxious reagents such as lithium-containing strong bases and polar organic solvent [20]. Raston and Scott [21a] accomplished the reaction by simply grinding together the solid reactants in the presence of sodium hydroxide at room temperature (Scheme 2.2-2). No organic solvent is used in the reaction and only a small amount of acidic aqueous waste is produced. The separation of crude product can be achieved by quenching the reaction mixture with aqueous HCl followed by simple filtration. An attractive characteristic of this solventless reaction is that the intermediate formed by grinding the reactants with base catalyst NaOH is highly stable and can be stored for months with no loss of product quality, thus minimizing waste and attaining a great degree of chemoselectivity for generating only one major α,β-unsaturated carbonyl compound.

The same research group accomplished a benign synthesis of Kröhnke pyridines via sequential solventless aldol condensation and Michael addition (Scheme 2.2-3); [21b] these are prominent building blocks in supramolecular chemistry characterized by their π-stacking ability, H-bonding, and coordination. The reaction has been

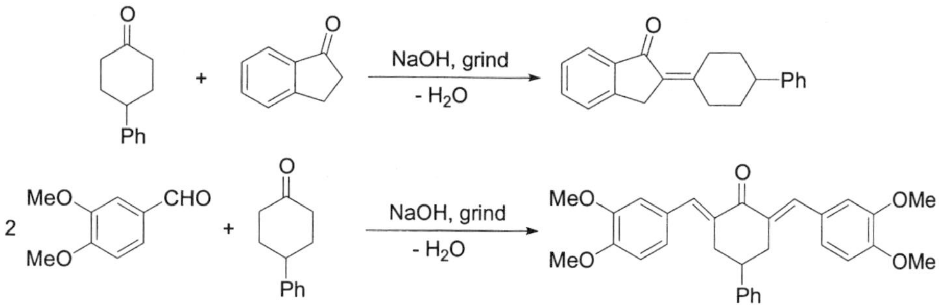

Scheme 2.2-2 Chemoselective aldol condensation.

accomplished by grinding with a pestle and mortar at room temperature for only 15 min, followed by treatment with ammonium acetate and acetic acid to afford the Kröhnke pyridine in high yield while the conventional method usually gave moderate to low yields.

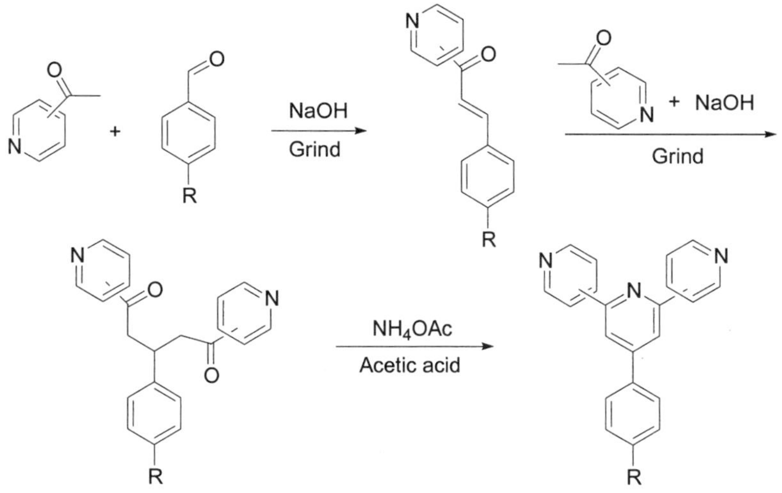

Scheme 2.2-3 Solvent-free synthesis of pyridines.

2.2.2.3
Knoevenagel Condensation Free of Solvent and Catalyst

Knoevenagel condensation [22] is a commonly used synthetic method for forming carbon–carbon bonds. Amines, ammonium salts, or various Lewis acids usually catalyze the reaction, which is strongly solvent-dependent. An improved Knoevenagel condensation reaction of aromatic aldehydes and malononitrile was reported by Ren et al [23]. The experimental protocol entails grinding the aldehydes with malononitrile using a glass mortar and pestle at room temperature for 1 h and allowing the mixture to stand overnight. The real advantage of this method is that it is a truly solvent-free process, as the pure products were isolated by simply stirring the resulted solid mixture in water, filtering, and drying at room temperature with excellent yields. This overall procedure is simple, efficient, economical, and environmentally benign in view of its absolute "green" feature: the complete absence of solvents both during the course of the reaction and subsequent product separation. The analogous reaction carried out under microwave irradiation will be discussed in the next section, dealing with cases for which organic solvents are required to extract the products.

A successful example of the preparation of 3-carboxycoumarins in quantitative yields is depicted below (Scheme 2.2-4) [24]. This approach provides a convenient, clean, and efficient synthetic method of synthesizing an important class of biologically active coumarins by avoiding the use of polar aprotic solvent such as dimethylformamide (DMF), minimizing waste generation, and involving no heating. Further, purification is possible using only an aqueous wash and simple filtration to generate high-purity coumarins instead of tedious chromatographic separation. The recovered water-soluble catalyst ammonium acetate (NH$_4$OAc) in the filtrate can be recycled for further use.

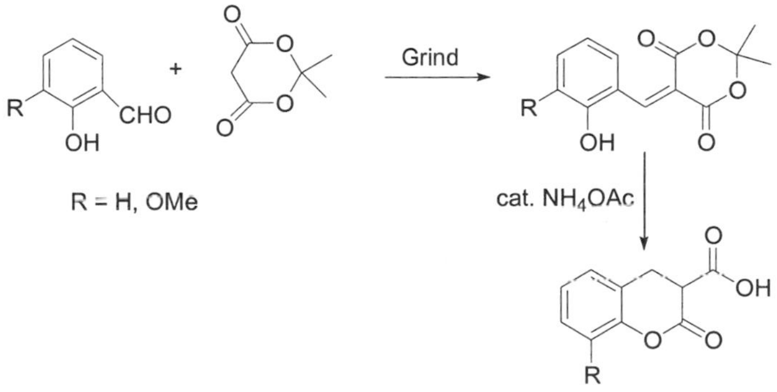

Scheme 2.2-4 Synthesis of 3-carboxycoumarins.

2.2.2.4
Solventless Oxidation Using the Urea–Hydrogen Peroxide Complex (UHP)

The utility of metal-based reagents such as potassium permanganate (KMnO$_4$), manganese dioxide (MnO$_2$), chromium trioxide (CrO$_3$), potassium dichromate (K$_2$Cr$_2$O$_7$), potassium chromate (K$_2$CrO$_4$), peracids, and peroxides in oxidation processes is compromised for several reasons, including the potential danger of handling metal complexes, inherent toxicity, cumbersome product isolation, and waste-disposal problems. A solvent-free oxidation of a variety of organic groups using an inexpensive, safe, and easily handled reagent, urea–hydrogen peroxide (UHP), has been discovered by Varma and Naicker (Scheme 2.2-5) [25]. This general solid-state oxidative protocol is applicable in oxidizing hydroxylated aldehydes and ketones (to phenols), sulfides (to sulfoxides and sulfones), nitriles (to amides), and *N*-heterocycles (to *N*-oxides),

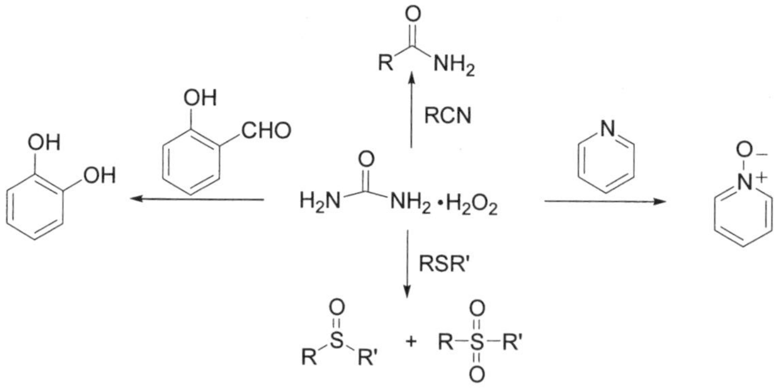

Scheme 2.2-5 Solventless oxidations using urea–hydrogen peroxide (UHP).

2.2.2.5
Expeditious Synthesis of 1-Aryl-4-methyl-1,2,4-triazolo[4,3-*a*]quinoxalines

The 1,2,4-triazole nucleus is an integral part of therapeutically active compounds that exhibit antibacterial, antifungal, sedative, antitumor, and CNS stimulative activities [26]. Traditional syntheses of this heterocycles require toxic reagents, harsh reaction conditions, and prolonged reaction times and involve multi-step reactions. Varma et al., as part of a broader program to develop environmentally friendlier synthetic methods, have reported an oxidative transformation of arenecarbaldehyde 3-methylquinoxalin-2-yl-hydrazones to 1-aryl-4-methyl-1,2,4-triazolo[4,3-*a*]quinoxalines using the relatively more friendly non-metallic iodobenzene diacetate, PhI(OAc)$_2$, as an oxidant [27]. The environmentally benign nature of this green oxidative protocol is apparent as it involves grinding the two solid substrates using a pestle and mortar; a mildly exothermic reaction results in the formation of a yellowish eutectic melt from hydrozones and iodobenzene diacetate and the reaction is completed in a few minutes. The resulting solid residues can be washed with a small amount of hexane and further purified by recrystallization or filtration through a small pad of silica gel to yield analytically pure products. This rapid, solventless transformation is general for the synthesis of 1-aryl-4-methyl-1,2,4-triazolo[4,3-a]quinoxalines from arenecarbaldehyde 3-methylquinoxalin-2-yl-hydrazones (Scheme 2.2-6).

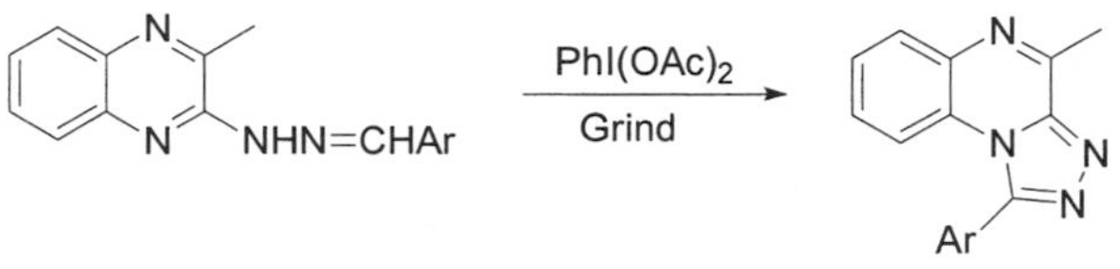

Ar = C₆H₅, p-CH₃C₆H₄, p-ClC₆H₄, p-OCH₃C₆H₄,
m-OCH₃C₆H₄, p-N(CH₃)₂C₆H₄,

Scheme 2.2-6 Synthesis of 1-aryl-4-methyl-1,2,4-triazolo[4,3-a]quinoxalines.

2.2.2.6
Solventless Wittig Olefination

Wittig olefination is a well-known reaction for constructing carbon–carbon bonds in organic synthesis [28]. A few solvent-free Wittig reactions [29, 30] have been developed which are explored later in this chapter in conjunction with microwave irradiation [31, 32] or using a special stirring device such as a high-speed steel mill [33]. Thiemann et al. investigated exothermic solvent-free Wittig olefination with stabilized alkyl (triphenylphosphoranylidene)acetates (Scheme 2.2-7) [34], which furnishes high yields. The structures of the stabilized phosphoranes are shown in Scheme 2.2-8. Flash column chromatography is employed to purify the products in order to separate *E*- and *Z*-isomers from the reaction mixture.

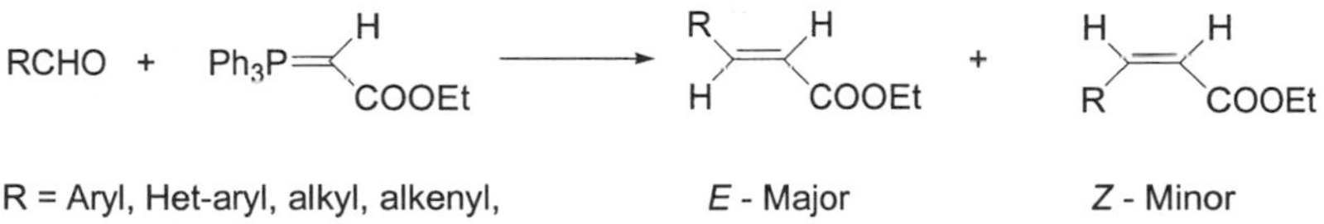

R = Aryl, Het-aryl, alkyl, alkenyl,   *E* - Major   *Z* - Minor

Scheme 2.2-7 Wittig olefination using neat reactants.

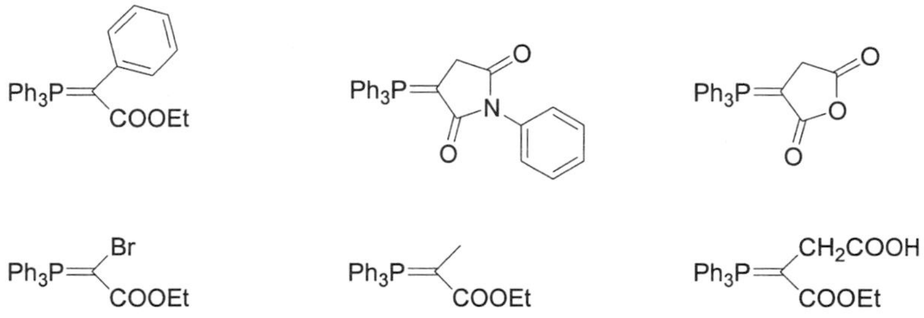

Scheme 2.2-8 Stabilized phosphorans.

2.2.3
Solventless Microwave-assisted Reactions

Microwave (MW) irradiation has attracted considerable attention since the 1980s [2, 35, 36] for rapid and efficient syntheses of a variety of organic compounds because of the selective absorption of microwave energy by polar molecules [37]. Microwaves (0.3–300 GHz), with relatively large wavelengths (between 1 mm and 1 m), lie in the electromagnetic radiation region between radio (Rf) and infrared (IR) frequencies. Microwave energy, long used for heating food materials [38], is now finding new and potentially useful applications in chemical technology.

Microwaves are a non-ionizing form of radiation energy that cannot break chemical bonds but can transfer energy selectively to various substances. Some materials (such as hydrocarbons, glass, and ceramics) are nearly transparent to microwaves and therefore behave as good insulators in a microwave oven since they are heated only to a very limited extent; metals reflect MW energy; molecules with dipole moment (many types of organic compounds) and salts absorb MW energy directly. Microwaves couple directly with molecules in the reaction mixture producing a rapid rise in temperature, dipole rotation and ionic conduction being the two most important fundamental mechanisms for transfer of energy to the molecules. Essentially, polar molecules try to align themselves with the rapidly changing electric field of the microwave and the coupling ability, among other features, is determined by the polarity of the molecules. Consequently, there are some significant differences between conventional chemical reactions in the liquid phase and the same reactions conducted under microwave irradiation.

In view of the increasing emphasis on pollution prevention to protect the environment and to address the diverse nature of the chemical entities involved, the use of MW irradiation to accelerate chemical synthesis under solvent-free conditions has gained popularity. Initially, commercial MW systems for chemical reactions were not readily available and inexpensive household MW ovens could be used for synthetic purposes in the absence of solvents. The main advantage of solvent-free MW reactions is the reduction or elimination of the use of volatile and toxic organic solvents (used 50- or 100-fold in reactions) thus enabling the neat reactants to undergo facile transformations at atmospheric pressure. This facilitates easy isolation of products and promotes safety associated with handling the reaction in open vessels thus culminating in chemical processes with unique attributes such as faster reaction kinetics, higher product yields, and safer manipulation. The application of microwave enhancement in chemical synthesis has been extended to various disciplines of chemistry such as polymerization [39], drug discovery and multi-step synthesis [40]

2.2.3.1
Microwave-assisted Solventless Synthesis of Heterocycles

The first example of a reaction between two solids, under solvent-free and catalyst-free conditions, was demonstrated [41] when the reaction of neat 5- or 8-oxobenzopyran-2(1*H*)-ones, with a variety of aromatic and heteroaromatic hydrazines, provided rapid access to several synthetically useful heterocyclic hydrazones (Scheme 2.2-9).

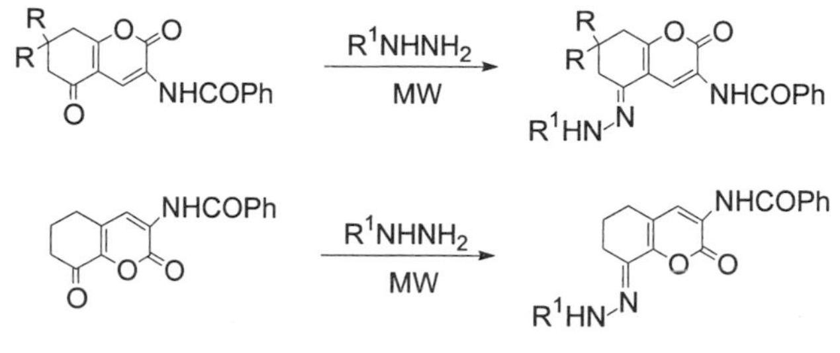

R = H, Me; R^1 = Ph, 3-CF$_3$-C$_6$H$_4$, 4-Nitrophenyl, 2,5-Difluorophenyl,
Pyridin-2-yl, 6-Chloropyridazin-3-yl

Scheme 2.2-9 Solvent-free preparation of hydrazones using microwaves.

Hydrazones are important synthons in organic synthesis; general syntheses of hydrazones from carbonyl compounds and hydrazine or hydrazine hydrate are well documented [42]. The above solventless route using an unmodified household microwave oven expedited the formation of hydrazones in minutes without any catalyst. The reaction proceeds smoothly even when both the starting materials are solids and the reaction occurs below the melting points of the starting materials. This can be explained by the lowering of the melting point with the formation of the eutectic melt [43]. The noteworthy green feature is that the produced hydrazones can be simply isolated in high yields (65–98%) and purity by washing with methanol and filtration.

Pyrazolo[3,4-*b*]quinolines and pyrazolo[3,4,-*c*]pyrazoles have been synthesized by microwave irradiation of β-chlorovinylaldehydes and hydrazines in the presence of *p*-toluenesulfonic acid (*p*-TsOH) (Scheme 2.2-10) [44]

The classic Biginelli three-component condensation [45,46] of aldehyde, urea or thiourea, and a dicarbonyl compound generating 3,4-dihydropyrimidin 2(1*H*)-one has been accomplished under solventless conditions to synthesize spiro-fused heterocycles in higher yields with shortened reaction time of a few minutes (Scheme 2.2-11) [47]. Purification was achieved by adding crushed ice to the reaction mixture, separating the solid via filtration and washing with cold water, followed by

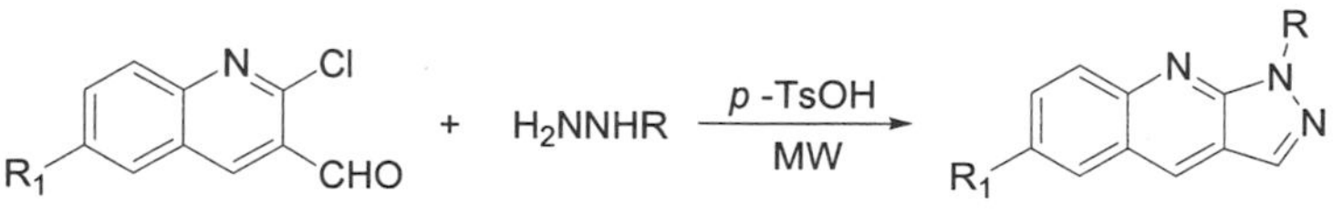

R1 = H, Me, OMe; R = H, Ph

Scheme 2.2-10 MW-assisted synthesis of pyrazolo[3,4-*b*]-quinolines.

recrystallization from *n*-hexane/ethyl acetate (3/1, v/v). The overall usage of solvent is greatly reduced.

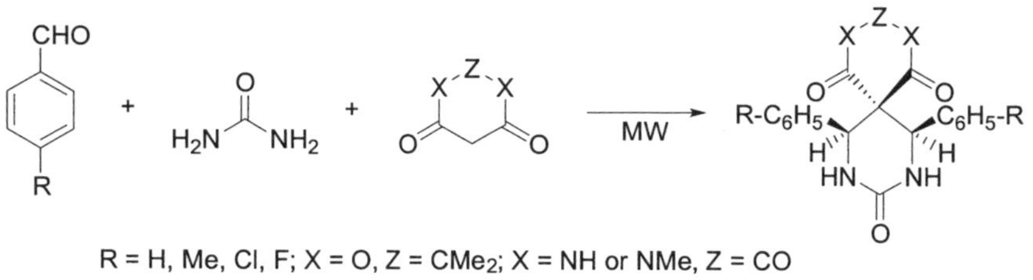

R = H, Me, Cl, F; X = O, Z = CMe$_2$; X = NH or NMe, Z = CO

Scheme 2.2-11 Solvent free synthesis of heterobicycles.

A one-pot MW-enhanced synthesis of selective glycine receptor antagonists, 3-aryl-4-hydroxyquinolin-2(1*H*)-ones, has been developed via the amidation of malonic ester derivatives with anilines and subsequent cyclization of the intermediate malondianilides under solvent-free conditions (Scheme 2.2-12) [48]

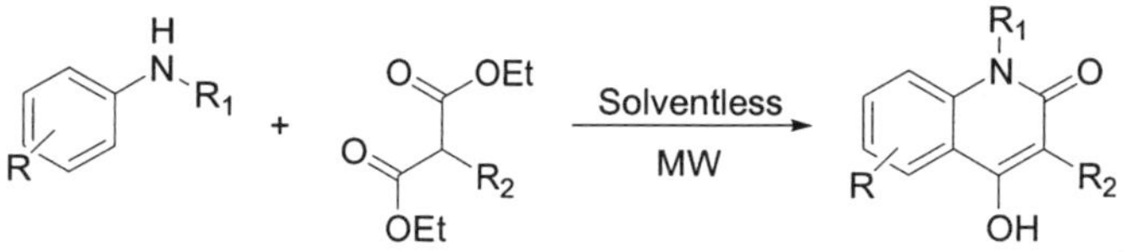

Scheme 2.2-12 One-pot MW-assisted synthesis of 3-aryl-4-hydroxyquinolin-2(1*H*)-ones.

2.2.3.2
Microwave-assisted Solventless Condensations

Knoevenagel condensation reactions of aldehydes with 2-imino-1-methyl-imidazolidine-4-one occurs rapidly under solvent-free reaction conditions at 160–170°C using a focused MW oven (Scheme 2.2-13) [49].

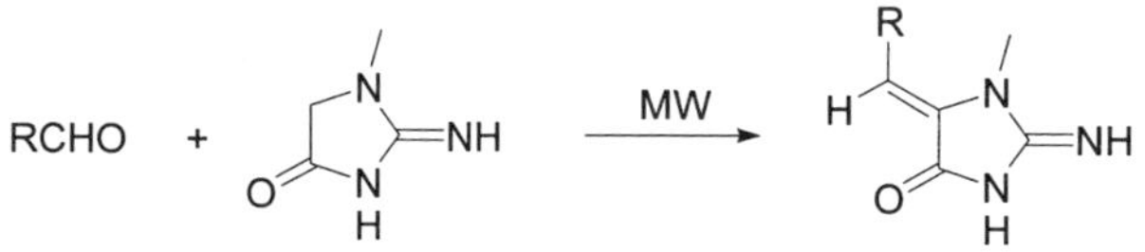

Scheme 2.2-13 Knoevenagel condensation reaction.

The classical Pechmann approach for the synthesis of coumarins via microwave promotion has now been extended to a solvent-free system wherein salicylic aldehydes undergo Knoevenagel condensation with a variety of ethyl acetate derivatives in the presence of piperidine (Scheme 2.2-14) [50].

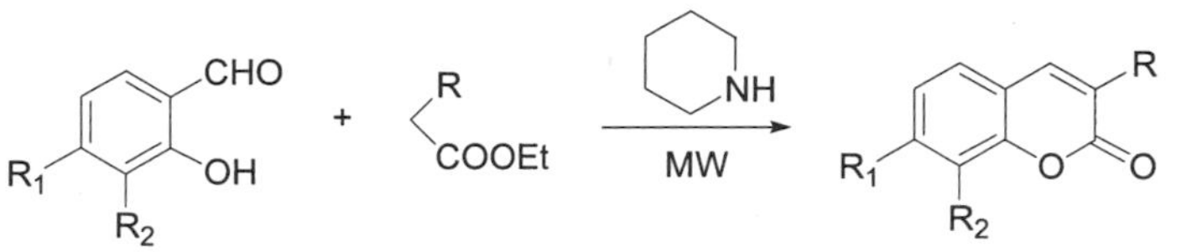

Scheme 2.2-14 Synthesis of coumarins facilitated by MW irradiation.

The Wittig olefination, a widely used transformation for the construction of carbon–carbon double bonds, has been improved. Some difficult reactions of stable phosphorus ylides with ketones in the absence of solvent are accelerated by MW irradiation using a domestic MW oven and the procedure affords improved yields compared with reactions carried out by conventional heating methods (Scheme 2.2-15) [32]. Although the separation employs an organic solvent to obtain pure olefin, it is still a useful protocol in the total synthesis. Similarly, several phosphonium salts, stabilized as well as non-stabilized, have been prepared in a domestic MW oven [51]

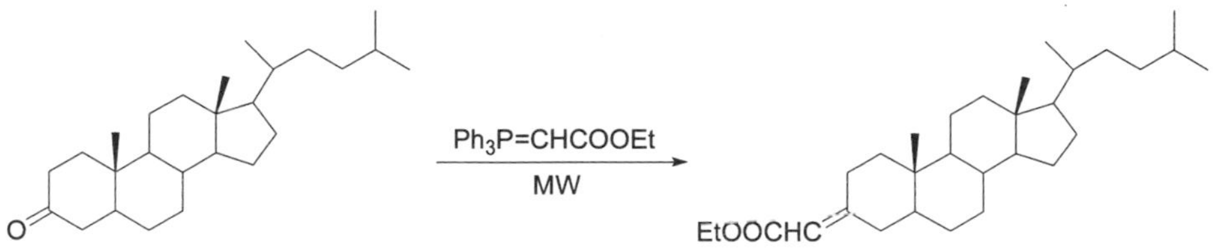

Scheme 2.2-15 MW-expedited Wittig olefination reaction.

Imines, enamines, nitroalkenes, and N-sulfonylimines are conventionally prepared by condensation reactions in which liberated water is removed azeotropically to drive the reaction to completion. Most of these reactions are catalyzed by p-toluenesulfonic acid, titanium(IV) chloride, or montmorillonite K10 clay and require the use of a Dean–Stark apparatus and a large excess of volatile aromatic hydrocarbons such as benzene or toluene for azeotropic water removal. Varma et al. have successfully demonstrated the significance of a MW approach to generate nitroalkenes (Scheme 2.2-16), valuable synthetic precursors and building blocks in the

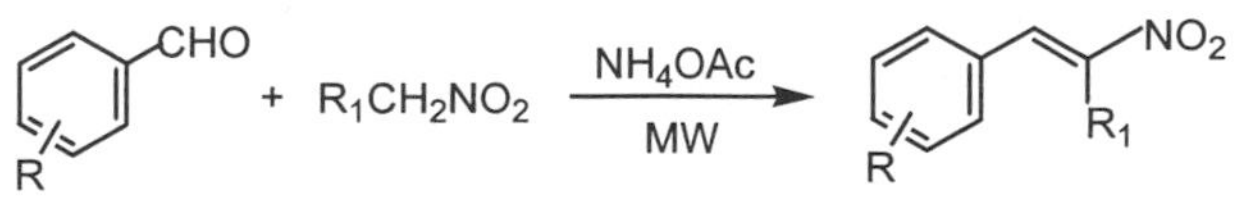

R_1 = H, R = H, p-OH, m,p-(OCH₃)₂, m-OCH₃-p-OH, 1-naphthyl, 2-naphthyl
R_1 = CH₃, R = H, p-OH, p-OCH₃, m,p-(OCH₃)₂, m-OCH₃-p-OH

Scheme 2.2-16 MW-expedited synthesis of nitroalkenes.

preparation of nitroalkanes, *N*-substituted hydroxyamines, amines, ketones, oximies and α-substituted oximines and ketones, from nitroalkanes and carbonyl compounds [52]

2.2.3.3
Microwave-assisted Solventless Oxidation

Binaphthol has played an important role in asymmetric syntheses as a chiral auxiliary. A rapid oxidative coupling of β-naphthols occurs in the presence of iron(III) chloride hexhydrate ($FeCl_3 \cdot 6H_2O$) using a focused MW-oven under solvent-free conditions; the procedure is superior to conventional heating in which case naphthols bearing electron-withdrawing groups are very slow to convert to binaphthols (Scheme 2.2-17) [53]. Solventless MW-assisted oxidations have been used for the oxidation of benzylic bromides to the corresponding aldehydes using pyridine *N*-oxides [54].

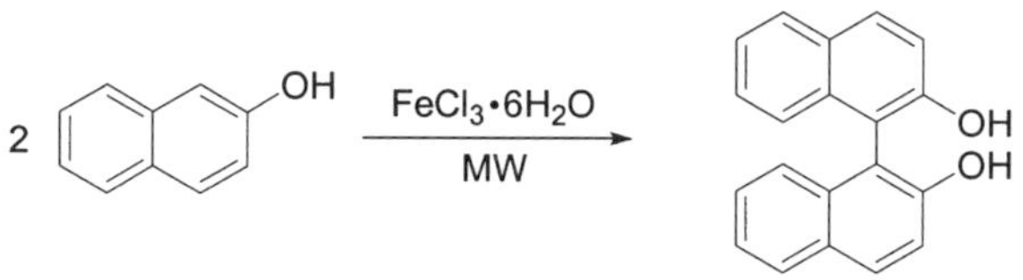

Scheme 2.2-17 Oxidative coupling of β-naphthol with iron(III) chloride.

The oxidation of benzylic alcohols to carbonyl compounds is an important transformation in organic synthesis [55, 56]. An efficient method for the oxidation of benzylic alcohols with [hydroxyl(tosyloxy)iodo]benzene (HTIB), Koser's reagent, under solvent-free MW-irradiation conditions has been described by Lee et al. (Scheme 2.2-18) [57]. The salient environmentally more friendly features of the solvent-free reaction are rapid reaction kinetics, experimental simplicity, and higher product yields, although the purification process still requires the traditional flash column chromatography. Several biologically useful α-keto esters have been synthesized in high yields from the corresponding α-hydroxy esters.

2.2.3.4
Amination of Aryl Halides without a Transition Metal Catalyst

A rapid and direct MW-assisted amination has been published that assembles the *N*-aryl moiety under solvent-free conditions (Scheme 2.2-19) [58]. Amination of halo-(pyridine or pyrimidine) with piperidine and pyrrolidine derivatives to form a carbon–nitrogen bond occurs without using any transition metal catalyst such as copper reagents or a palladium complex.

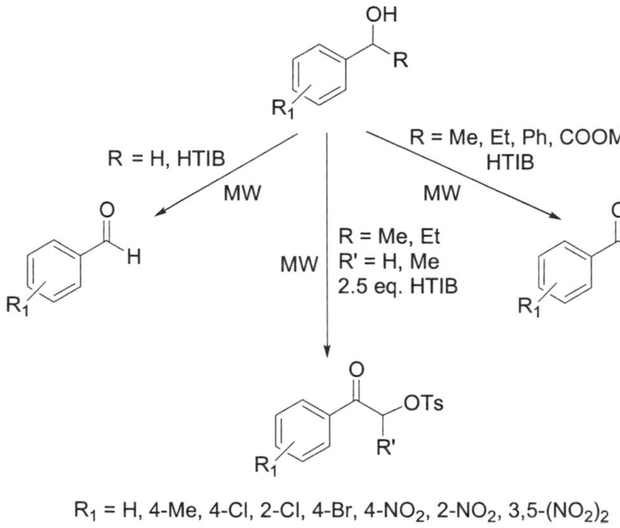

R$_1$ = H, 4-Me, 4-Cl, 2-Cl, 4-Br, 4-NO$_2$, 2-NO$_2$, 3,5-(NO$_2$)$_2$

Scheme 2.2-18   Efficient oxidation of benzylic alcohol with HTIB.

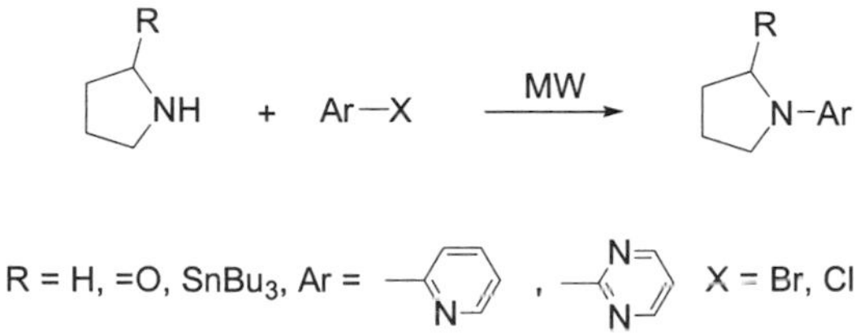

R = H, =O, SnBu$_3$, Ar = ,   X = Br, Cl

Scheme 2.2-19   Direct amination of aryl halides with amines.

2.2.3.5
Microwave-accelerated Transformation of Carbonyl Functions to their Thio Analogues

Among expeditious chemical transformations that can be accomplished under solventless conditions, the conversion of carbonyl compounds to the corresponding thio analogues is especially useful. The conventional synthesis of thioketones involves the reaction of substrates with phosphorus pentasulfide under basic conditions, hydrogen sulfide in the presence of acid or Lawesson's reagent. Using the MW approach, no acidic or basic medium is used and the carbonyl compounds are simply admixed with neat Lawesson's reagent and irradiated under solvent-free conditions. This benign approach is general and is applicable to the high yield conversion of ketones, flavones, isoflavones, lactones, amides, and esters to the corresponding thio analogues: thioketones, thioflavonoids, thiolactones, thioamides and thionoesters (Scheme 2.2-20). This eco-friendly, solvent-free protocol uses comparatively much smaller amount of Lawesson's reagent and avoids the use of the large excesses of dry hydrocarbon solvents such as benzene, xylene, triethylamine, or pyridine that are conventionally used [59]

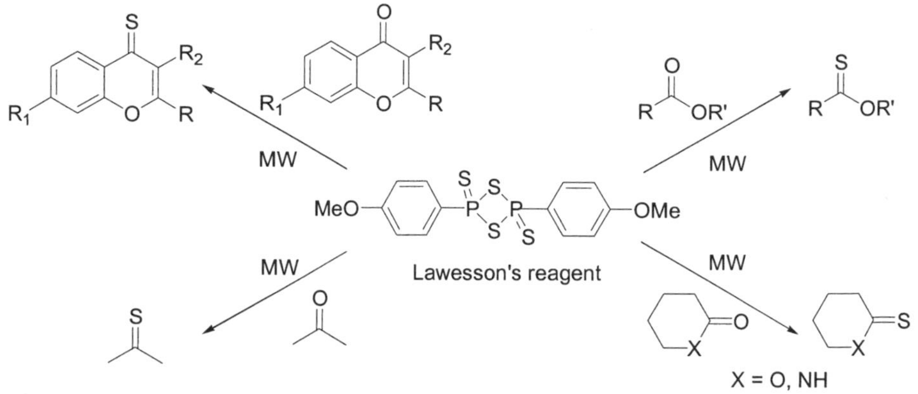

Scheme 2.2-20 Solventless synthesis of thioketones, thiolactones, thioamides, and thionoesters using Lawesson's reagent.

2.2.4
Microwave-assisted Solventless Reactions on Solid Supports

Heterogeneous reactions facilitated by supported reagents on inorganic mineral oxide surfaces have received widespread attention as attested by several books, reviews, and account articles [60–68]. The use of MW-irradiation techniques for the acceleration of organic reactions had a profound impact on these heterogeneous reactions since the appearance of the initial report on the application of microwaves for chemical synthesis in polar solvents [35]. The approach has blossomed into a useful technique for a variety of applications in organic synthesis and functional group transformations, as is testified by the large number of publications and review articles on this theme [69–78]. Microwave-accelerated organic reactions under solvent-free conditions on mineral supports have dominated the scene when commercial MW systems for chemical reactions were not readily available and inexpensive household MW ovens could be used for synthetic purposes in the absence of solvents. The combination of MW irradiation with solvent-free conditions provides cleaner chemical processes by eliminating the use of volatile, toxic organic solvents, allowing the solid reactants to undergo facile transformation at atmospheric pressure, facilitating the easy isolation of products and handling the reaction safely with open vessels, thus providing chemical processes with unique attributes such as faster reaction kinetics, higher product yields, greater selectivity and safer manipulations. More importantly, the limitations of microwave-assisted reactions in solvents, namely the development of high pressure and the need for specialized equipment such as sealed vessels and reflux condensers, were overcome by this solid-state strategy, which enables organic reactions to proceed rapidly at atmospheric pressure even in an open vessel.

The application of MW irradiation with the use of catalysts or mineral-supported reagents, under solvent-free conditions, gained increasing popularity due to its broad applicability with open reaction flasks at atmospheric pressure. These reactions conducted with the help of reagents immobilized on porous solid supports have salient advantages such as good dispersion of active reagent sites, selectivity, and ease of reaction work-up. Other major factors responsible for their popularity are the ready availability of inexpensive household microwave ovens that can be safely used for solventless reactions and the opportunity to work with open vessels, thus avoiding the risk of high-pressure development. The reactions appear to occur at relatively low bulk temperatures although higher localized temperatures may be reached during MW irradiation. Unfortunately, accurate recording of temperature has not been made in the majority of such studies. This solventless MW strategy on solid supports has been the most widely practiced approach in laboratories around the globe in spite of the relatively poor understanding of the reasons for such dramatic rate acceleration. The recyclability of some of these solid supports, in some selected processes, renders them environmentally friendlier "green" protocols, especially in cases where reactants/reagents are simply deposited on the solid surface without any solvent.

Since the initial report [79], a large number of MW-accelerated solvent-free protocols have been illustrated for a wide variety of useful chemical transformations such as protection/deprotection (cleavage), condensation, rearrangement reactions, oxidation, reduction, and the synthesis of several heterocyclic compounds on mineral supports. These reactions have resulted in the synthesis of a range of industrially significant chemical precursors such as imines, enamines, enones, nitroalkenes, sulfur compounds, and heterocyclic compounds in a relatively more environmentally friendly manner. A vast majority of these solventless reactions have been performed using an unmodified household MW oven or commercial MW equipment usually operating at 2450 MHz in open glass containers with dry reactants. The general procedure involves simple mixing of neat reactants with the catalyst, their adsorption on mineral or "doped" supports, and exposure of the reaction mixture and the support to MW irradiation. Sequential workup and separation of expected products can be completed by extraction using reduced amounts of solvent compared to conventional solution-phase reactions. It is clear that this solventless approach addresses the problems associated with waste disposal of solvents that are used several-fold in chemical reactions, thus minimizing or avoiding the use of excess chemicals and solvents.

2.2.4.1
Protection–Deprotection (Cleavage) Reactions

Although inherently wasteful, protection–deprotection reaction sequences constitute an integral part of organic syntheses such as the preparation of monomers, fine chemicals, and precursors for pharmaceuticals. These reactions often involve the use of acidic, basic, or hazardous reagents, and toxic metal salts [80]. The solventless MW-accelerated cleavage of functional groups provides an attractive alternative to conventional deprotection reactions.

Acetal and dioxolane derivatives of aldehydes and ketones have been prepared using orthoformates, 1,2-ethanedithiol, or 2,2-dimethyl-1,3-dioxolane in acid-catalyzed reactions that proceed in the presence of *p*-toluenesulfonic acid (*p*-TsOH) or clay under solvent-free conditions (Scheme 2.2-21). The yields obtained with the MW method are better than those obtained using conventional heating in an oil bath [81].

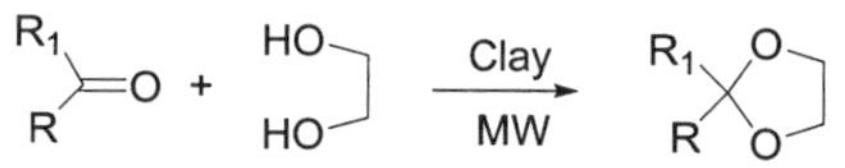

Scheme 2.2-21 Protection of carbonyls by ethanediol.

The acetals of 1-galactono-1,4-lactone have been prepared in excellent yields [82] to protect lactone by adsorbing the lactone and the aldehyde on montmorillonite K10 or KSF clay followed by heating the reaction mixture in a MW oven (Scheme 2.2-22).

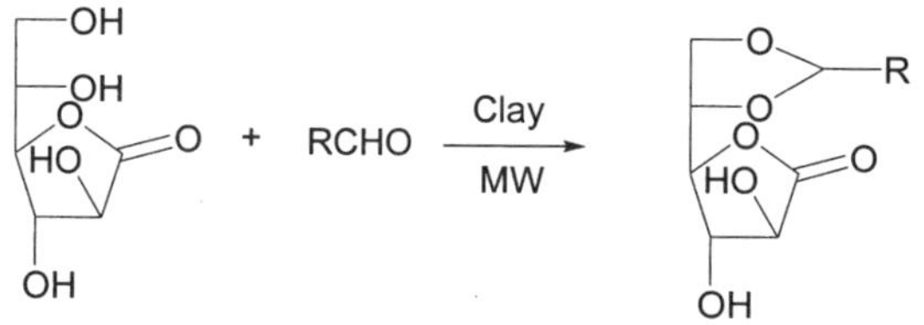

Scheme 2.2-22 Preparation of 1-galactano-1,4-lactone acetals.

The utility of recyclable alumina as a viable support surface for deacylation reactions is described by Varma and his colleagues [83] wherein the orthogonal deprotection of alcohols is possible under solvent-free conditions on a neutral alumina surface using MW irradiation (Scheme 2.2-23). Interestingly, chemoselectivity between alcoholic and phenolic groups in the same molecule has been achieved simply by varying the reaction time; phenolic acetates are deacetylated faster than alcoholic analogues.

In a very rapid reaction, the diacetate derivatives of aromatic aldehydes regenerate the aldehydes on MW irradiation on a neutral alumina surface (Scheme 2.2-24) [84]. The selectivity in these reactions is achieved by merely adjusting the time of irradiation. For example, the aldehyde diacetate is selectively removed in 30 s, whereas a 2-min period is required to cleave both the diacetate and ester groups. The protocol is applicable to compounds bearing olefinic moieties, such as cinnamaldehyde diacetate.

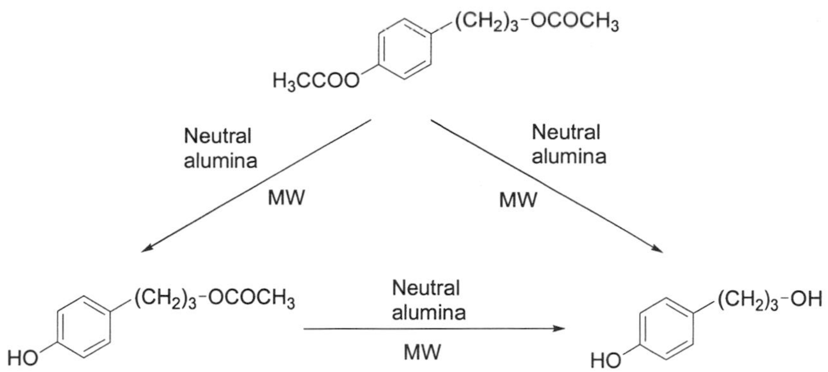

Scheme 2.2-23 Deprotection of alcohols and phenols on neutral alumina.

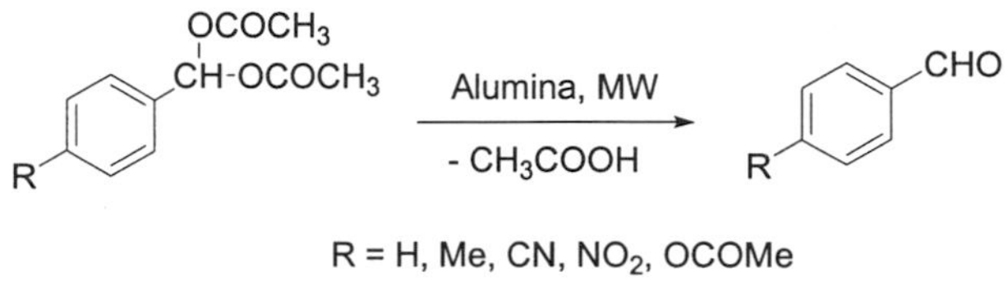

R = H, Me, CN, NO$_2$, OCOMe

Scheme 2.2-24 Regeneration of aldehydes from aldehyde diacetates.

An efficient solvent-free debenzylation process for the cleavage of carboxylic esters on an alumina surface has been developed by Varma and colleagues (Scheme 2.2-25) [85]. By changing the surface characteristics of the solid support from neutral to acidic, the cleavage of the 9-fluorenylmethoxycarbonyl (Fmoc) group and related protected amines can be achieved in a similar manner. The hydrolytic deprotection of carboxylic acids from their corresponding allyl esters, under "dry conditions", has also been reported on montmorillonite K10 clay [86].

Solvent-free deprotection of the *N-tert*-butoxycarbonyl (Boc) groups, a very commonly used protection group in organic synthesis, has been accomplished in the presence of neutral alumina that is "doped" with aluminum chloride (Scheme 2.2-26) [87]. This approach may find application in a typical peptide-bond-forming re-

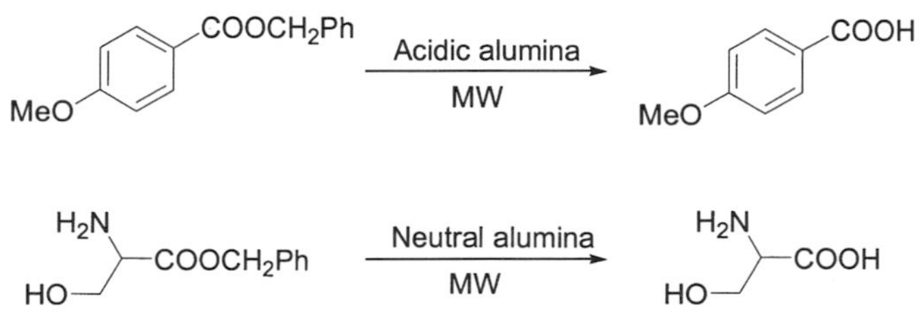

Scheme 2.2-25 Deprotection of benzyl ester on alumina.

action thus eliminating the use of irritating and corrosive chemicals such as trifluoroacetic acid and piperidine.

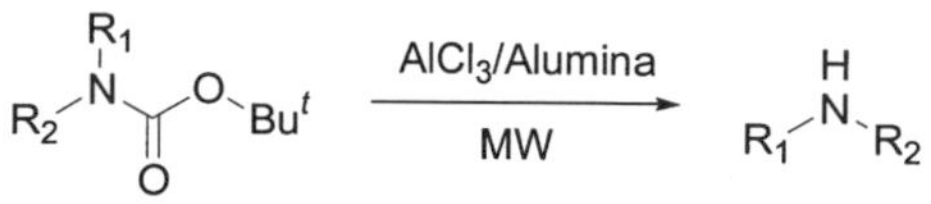

Scheme 2.2-26   Cleavage of *N-tert*-butoxycarbonyl groups.

tert-Butyldimethylsilyl (TBDMS) ether derivatives of alcohols can be rapidly cleaved to regenerate the corresponding hydroxy compounds on alumina using MW irradiation (Scheme 2.2-27) [88]. This approach circumvents the use of corrosive fluoride ions that is normally employed for cleaving such silyl protecting groups. Deprotection of trimethylsilyl ethers has also been accomplished on K10 clay [89] and oxidative cleavage in the presence of iron(III)/clay [90]

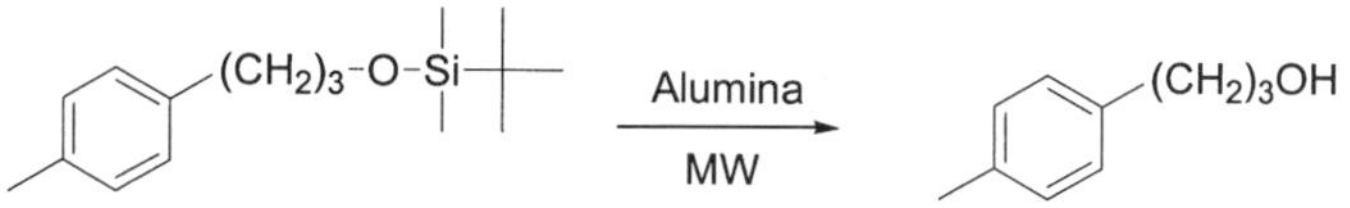

Scheme 2.2-27   Alumina-mediated cleavage of *t*-butyldimethylsilyl ether.

Thio acetals and ketals are key protecting groups employed in organic transformations, but the regeneration of carbonyl groups by cleavage of acid- and base-stable thioacetals and thioketals is a challenging task. Conventionally, the cleavage of thioacetals requires the use of toxic heavy metals such as Ti^{4+}, Cd^{2+}, Hg^{2+}, Tl^{3+}, or uncommon reagents such as benzeneseleninic anhydride [91]. A solid-state dethioacetalization reaction that proceeds in high yields has been reported by Varma et al. using montmorillonite K10 clay-supported iron(III) nitrate (clayfen) (Scheme 2.2-28), and this general reaction does not yield any byproducts except in the case of substrates bearing free phenolic groups, where ring nitration is observed. A report on the cleavage of thioacetals with clayan (80–89%) has also appeared subsequently [92]

The search for a solvent-free deprotection procedure has led to the use of the relatively benign reagent ammonium persulfate on silica, for regeneration of carbonyl compounds (Scheme 2.2-29) [93]. Neat oximes are simply admixed with the so-

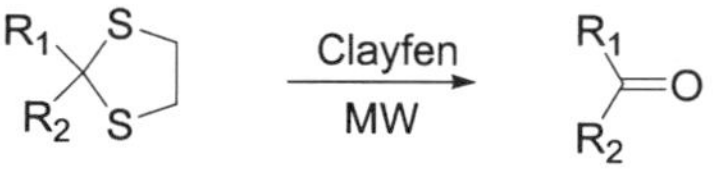

R_1 = Ph, Et, 4-$CH_3C_6H_4$,4-$NO_2C_6H_4$; R_2 = H, Et, Me;
   R_1-R_2 = isoflavanolyl, 2-methylcyclohexyl

Scheme 2.2-28   Solid-state dethioacetalization reaction on clayfen.

lid-supported reagent and then irradiated in a MW oven to regenerate free aldehydes or ketones; the process is applicable to both aldoximes and ketoximes. The critical role of the surface is apparent since the same reagent supported on a montmorillonite K10 clay surface delivers predominantly the Beckmann rearrangement products, the amides [94]

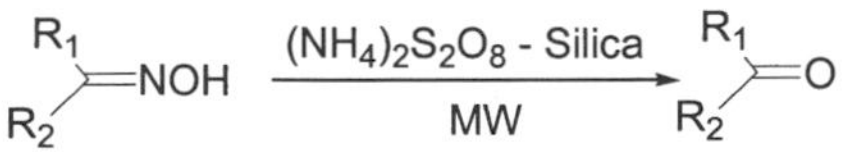

$$R_1R_2C{=}NOH \xrightarrow[\text{MW}]{(NH_4)_2S_2O_8 \text{ - Silica}} R_1R_2C{=}O$$

Scheme 2.2-29 Deoximation of carbonyl compounds on silica.

Dethiocarbonylation is the transformation of thiocarbonyl to carbonyl groups and has been accomplished with several reagents, namely trifluoroacetic anhydride, CuCl/MeOH/NaOH, tetrabutylammonium hydrogen sulfate/NaOH, clay/ferric nitrate, NOBF$_4$, bromate and iodide solutions, alkaline hydrogen peroxide, sodium peroxide, thiophosgene, trimethyloxonium fluoroborate, tellurium-based oxidants, dimethyl selenoxide, benzeneseleninic anhydride, benzoyl peroxide, and halogen-catalyzed alkoxides under phase-transfer conditions. These methods, unfortunately, have certain limitations, such as the use of stoichiometric amounts of oxidants that are often inherently toxic, long reaction times, or tedious procedures. Varma et al. have developed an efficient dethiocarbonylation process under solvent-free conditions using clayfen or clayan (Scheme 2.2-30) [95] that is accelerated by MW irradiation.

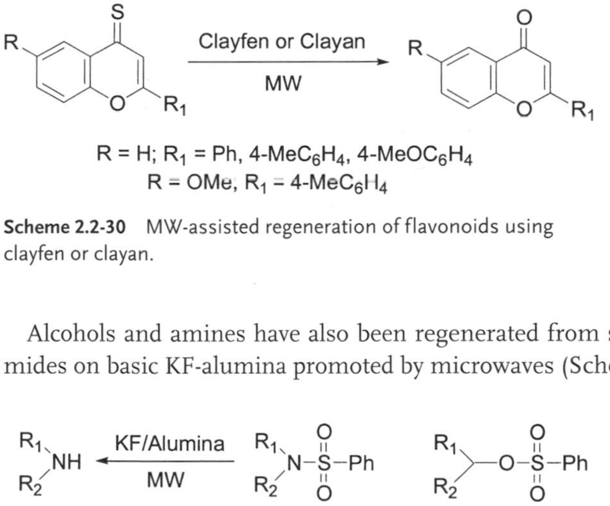

$$R = H; R_1 = Ph, 4\text{-MeC}_6H_4, 4\text{-MeOC}_6H_4$$
$$R = OMe, R_1 = 4\text{-MeC}_6H_4$$

Scheme 2.2-30 MW-assisted regeneration of flavonoids using clayfen or clayan.

Alcohols and amines have also been regenerated from sulfonates and sulfonamides on basic KF-alumina promoted by microwaves (Scheme 2.2-31) [96].

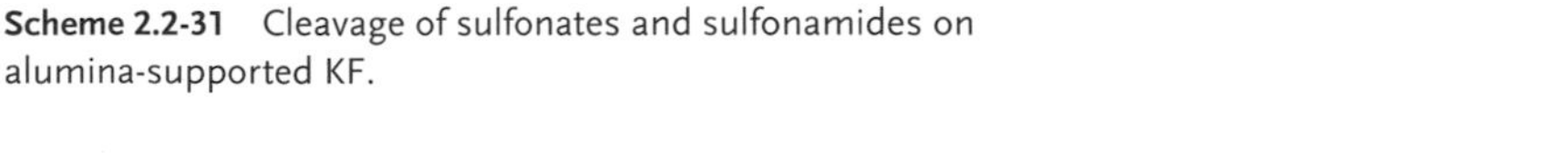

Scheme 2.2-31 Cleavage of sulfonates and sulfonamides on alumina-supported KF.

2.2.4.2
Condensation Reactions

Microwave-expedited condensation reactions using montmorillonite K10 clay or Envirocat reagent, EPZG®, have yielded a rapid synthesis of imines and enamines via the reactions of primary and secondary amines with aldehydes and ketones, respectively (Scheme 2.2-32) [97,98]. In these reactions, the generation of polar transition-state intermediates that readily couple to microwaves is probably responsible for the rapid imine or enamine formation. The use of a MW oven at lower power levels or intermittent heating has been used to prevent the loss of low-boiling reactants.

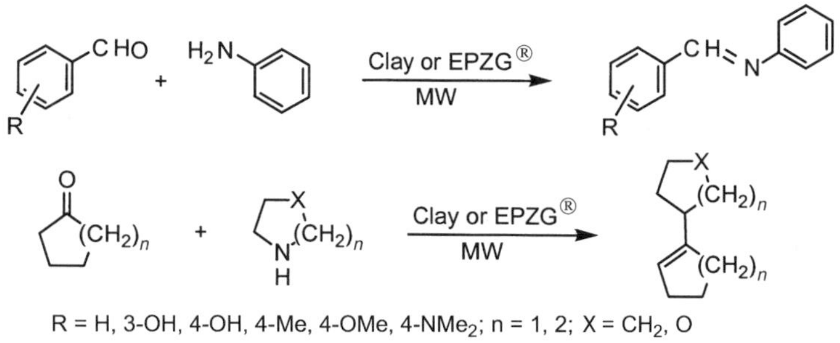
CHO
H_2N
+
Clay or EPZG®
MW
CH=N
R
O
$(CH_2)_n$
X
N
H
$(CH_2)_n$
R = H, 3-OH, 4-OH, 4-Me, 4-OMe, 4-NMe_2; n = 1, 2; X = CH_2, O

Scheme 2.2-32 Solventless MW synthesis of imines and enamines.

A class of previously unknown compounds, spiro[3*H*-indole-3,2'-[4*H*] pyrido [3,2-*e*]-1,3-thiazine]-2,4'-(1*H*) diones, can be synthesized by the reaction of in situ generated 3-indolylimine with 2-mercaptonicotinic acid under microwaves in the absence of solvent. Both neat reactions or reactions on solid supports such as silica, alumina etc., are effective in promoting the reaction, while reaction under thermal heating failed to proceed (Scheme 2.2-33) thus indicating some specific microwave effect [99].

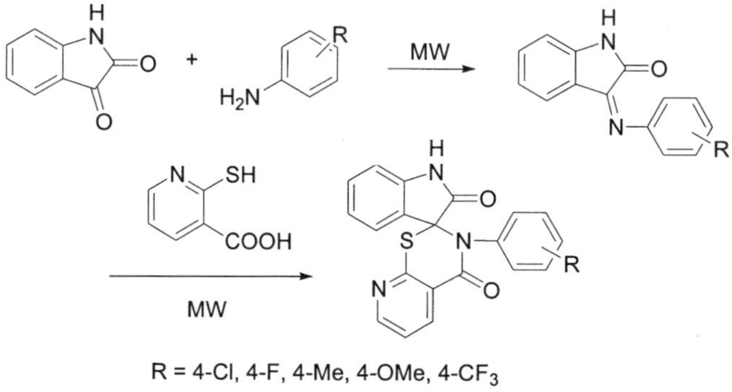
H
N
O
+
R
H_2N
MW
N
SH
COOH
S
R = 4-Cl, 4-F, 4-Me, 4-OMe, 4-CF_3

Scheme 2.2-33 MW-promoted synthesis of novel spiro[indole-pyrido thiazines].

2.2.4.3
Solventless Rearrangement Promoted by MW Irradiation

A solvent-free pinacol–pinacolone rearrangement was reported in 1989 using MW irradiation. The process involves irradiation of the gem-diols with Al(III)-montmorillonite K10 clay for 15 min to afford the rearrangement product in high yields (Scheme 2.2-34) [100]; the equivalent reaction performed under conventional heating in an oil bath requires a longer reaction time (15 h).

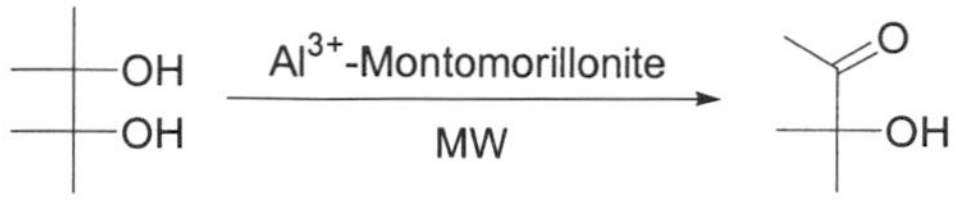

Scheme 2.2-34   Solvent-free pinacol-pinacolone rearrangement.

A facile ring-expansion reaction under solvent-free conditions accelerated by microwaves on an alumina surface has also been described (Scheme 2.2-35) which is preferable to the one conducted in methanol [101]

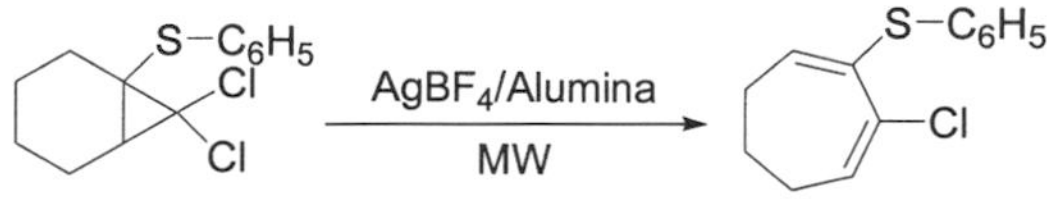

Scheme 2.2-35   MW-accelerated ring-expansion reaction.

The montmorillonite K10 clay surface is found to be superior among several acidic surfaces that have been used for the Beckmann rearrangement of oximes (Scheme 2.2-36) [94]. However, under the same conditions aldoximes dehydrate to the corresponding nitriles under solventless conditions.

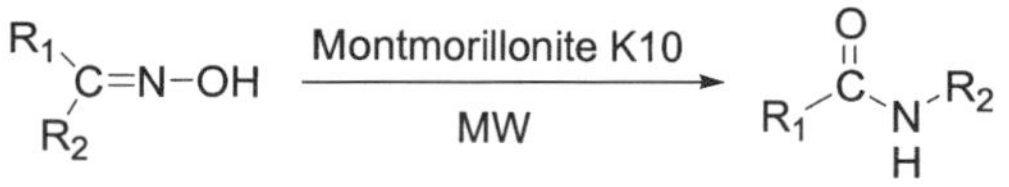

R$_1$ = Me, Ph; R$_2$ = Ph or substituted phenyl

Scheme 2.2-36   Beckmann rearrangement reaction on clay surface.

2.2.4.4
Oxidation Reactions – Oxidation of Alcohols and Sulfides

Limitations in the use of potassium permanganate and similar oxidizing agents [102–104] have already been mentioned in Section 2.2.2.4. The immobilization of metallic reagents on solid supports has addressed some of these limitations and

provides an attractive alternative in organic synthesis because of the enhanced selectivity and the containment of metals on the support surface, which precludes
them from leaching into the environment.

Silica-supported manganese dioxide (MnO_2) provides a rapid and high-yield route for the oxidation of alcohols to aldehydes and ketones. Benzyl alcohols are selectively oxidized to carbonyl compounds using 35% MnO_2 "doped" silica under MW
irradiation conditions (Scheme 2.2-37) [105]. Bentonite-type clay-supported MnO_2
has also been used for the oxidation of phenols to quinones (30–100%) [106] and
MnO_2 on silica effects the dehydrogenation of pyrrolidines (58–96%) [107]

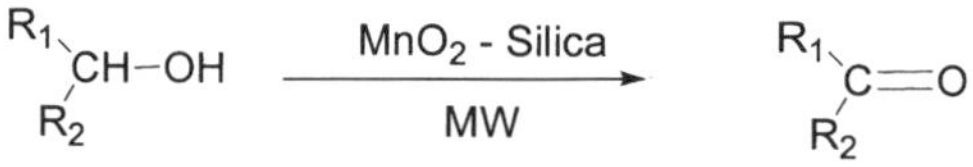

R$_1$\CH–OH $\xrightarrow[\text{MW}]{\text{MnO}_2 \text{ - Silica}}$ R$_1$\C=O

R_1 = H; R_2 = Ph, 4-MeC$_6$H$_4$, 4-MeOC$_6$H$_4$, PhCH=CH
R_1 = Et, Ph, PhCO; R_2 = Ph
R_1 = 4-MeOC$_6$H$_4$CO; R_2 = 4-MeOC$_6$H$_4$

Scheme 2.2-37 MW-assisted oxidation of alcohols.

The use of chromium(VI) reagents as oxidants is limited by their inherent toxicity, the need to prepare them in various complex forms (with acetic acid or pyridine), and complicated work-up procedures. Chromium trioxide (CrO_3) immobilized
on pre-moistened alumina affords efficient oxidation of benzyl alcohols to carbonyl compounds by simple mixing. Remarkably, neither the over-oxidation to carboxylic acids nor the usual formation of tar, a typical occurrence in many CrO_3 oxidations, is observed [108]. The reagent system is also used for the preparation of
acyclic α-nitro ketones by the oxidation of nitroalkanols under solvent-free conditions [109]

A rapid MW oxidation protocol for the oxidation of alcohols to carbonyl compounds has been reported by Varma et al. using clayfen under solvent-free conditions, which proceeds via the intermediacy of nitrosonium ions. Interestingly, no
carboxylic acids are formed in the oxidation of primary alcohols. The simple solvent-free experimental procedure involves mixing the neat substrates with clayfen
and a brief MW irradiation for 15–60 sec. This fast, manipulatively simple and selective protocol avoids excess use of solvents and toxic oxidants (Scheme 2.2-38)
[110]. The solid-state use of clayfen as an oxidant is more efficient than the method
of Laszlo et al. [111,112], affording higher yields and using half the amount of oxidant.

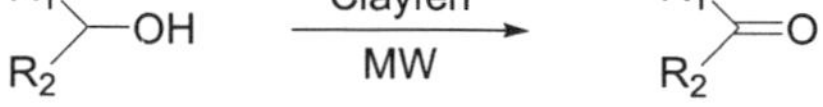

R$_1$>—OH $\xrightarrow[\text{MW}]{\text{Clayfen}}$ R$_1$>=O

R_1 = Alkyl, aryl; R_2 = H, alkyl, aryl

Scheme 2.2-38 Solventless oxidation of alcohols with clayfen.

Benzoins, both symmetrical and unsymmetrical, have been rapidly oxidized to benzils in high yields using the solid reagent copper(II) sulfate–alumina [113] or Oxone®–moist alumina [114,115] upon exposure to microwaves (Scheme 2.2-39). Normally, these oxidative transformations employ reagents such as nitric acid, Fehling's solution, thallium(III) nitrate, ytterbium(III) nitrate, ammonium chlorochromate–alumina and clayfen. In addition to extended reaction times, most of these processes suffer from the use of corrosive acids and toxic metallic compounds that generate undesirable waste products. The process, however, is applicable only to α-hydroxyketones as exemplified by mixed benzylic/aliphatic α-hydroxyketones or 2-hydroxypropiophenone, which delivers the corresponding vicinal diketone.

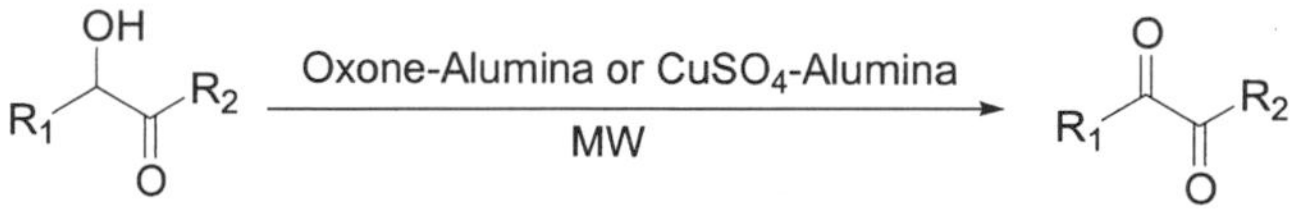

$R_1 = R_2 = Ph, 4\text{-}MeC_6H_4, 4\text{-}MeOC_6H_4, 4\text{-}ClC_6H_4$
$R_1 = Ph; R_2 = 4\text{-}MeC_6H_4, 4\text{-}MeOC_6H_4$
$R_1 = Me; R_2 = Ph$

Scheme 2.2-39 Oxidations of α-hydroxyketones on alumina.

Organohypervalent iodine reagents such as iodoxybenzene, *o*-iodoxybenzoic acid (IBX), bis(trifluoroacetoxy)iodobenzene (BTI), and Dess–Martin periodinane have been used for the oxidation of alcohols and phenols. Most of these reactions are conducted in high-boiling DMSO or relatively toxic acetonitrile, which increase the burden on the environment. Further, the use of inexpensive iodobenzene diacetate (IBD) as an oxidant has not been fully exploited. Varma et al. have reported the first use of supported iodobenzene diacetate as an oxidant. In this novel oxidative protocol, alumina-supported IBD under solvent-free conditions rapidly converts alcohols to the corresponding carbonyl compounds in almost quantitative yields. The use of alumina as a support improved the yields markedly as compared to neat IBD (Scheme 2.2-40). 1,2-Benzenedimethanol, however, undergoes cyclization to afford 1(3*H*)-isobenzofuranone [116]

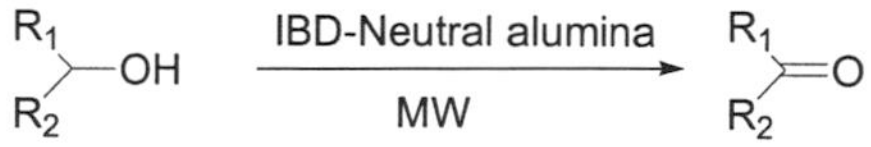

$R_1 = Ph$, Substituted Ph; $R_2 = H$, Et, COPh

Scheme 2.2-40 MW-assisted oxidation of alcohols by alumina-supported IBD.

The solid oxidizing reagent IBD–alumina has also been used for the rapid, high-yielding, and selective oxidation of alkyl, aryl, and cyclic sulfides to the corresponding sulfoxides upon MW irradiation [117].

The conventional reaction conditions for the oxidation of sulfides to the corresponding sulfoxides and sulfones are rather strenuous, requiring oxidants such as nitric acid, hydrogen peroxide, chromic acid, peracids, and periodate. The oxidation of sulfides to sulfoxides and sulfones is achieved in a selective manner using MW irradiation under solvent-free conditions with desired selectivity to either sulfoxides or sulfones over sodium periodate ($NaIO_4$) on silica (Scheme 2.2-41) [118]. A reduced amount of the active oxidizing agent, 20% $NaIO_4$ on silica, is employed, which is safer and easier to handle. Several refractory thiophenes can be oxidized under these conditions, benzothiophenes are oxidized to the corresponding sulfoxides and sulfones using ultrasonic and microwave irradiation, respectively, in the presence of $NaIO_4$–silica. A noteworthy feature of the protocol is its applicability to long-chain fatty sulfides that are insoluble in most solvents and are consequently difficult to oxidize.

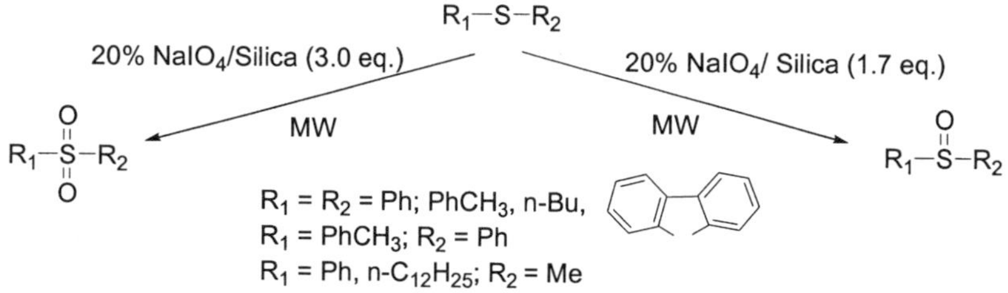

Scheme 2.2-41   Solventless oxidation of sulfides using sodium periodate–silica.

β,β-Disubstituted enamines have been successfully oxidized into carbonyl compounds under solvent-free conditions by alumina-supported $KMnO_4$ in both domestic and focused microwave ovens; the yields are better in the latter case. No carbonyl compounds are formed when the same reactions are conducted in an oil bath at 140°C, (Scheme 2.2-42) [119]. $KMnO_4$ supported on alumina oxidizes arenes to ketones within 10–30 min under solvent-free conditions using focused microwaves [120].

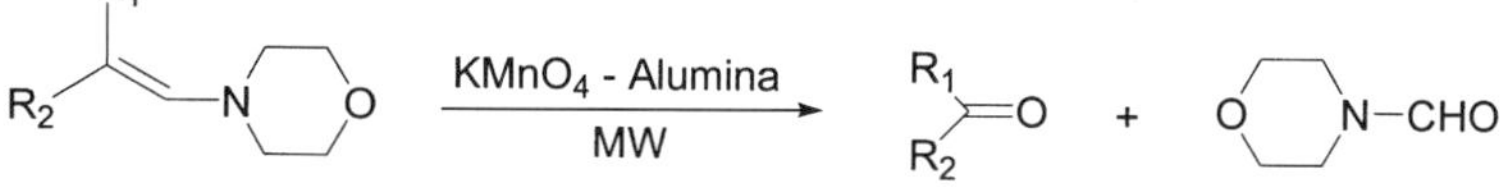

Scheme 2.2-42   Oxidation of enamines with $KMnO_4$ on alumina.

2.2.4.5
Reduction Reactions

Microwave-assisted reduction reactions are among the last to be studied. The kinetics of catalytic transfer hydrogenation of soybean oil under MW and thermal conditions have been examined and reaction rates are found to be eight times greater in microwaves than in conventional heating at the same temperature [121]. Ammonium formate has been used for allylic reduction [122] and the protocol has been subsequently adapted for the reduction of imines and in the dehalogenation of aryl halides in the presence of 10% Pd/C [123].

The relatively inexpensive and safe sodium borohydride ($NaBH_4$) has been extensively used as a reducing agent because of its compatibility with protic solvents. The solid-state reduction of carbonyl compounds has been accomplished by mixing them with $NaBH_4$ and storing the reaction mixture in a dry box for five days which is too long for it to be of any practical utility [124]. In a first example of its kind under solvent-free conditions to be accelerated by MW irradiation, Varma and coworkers have reported a simple method for the expeditious reduction of aldehydes and ketones that uses alumina-supported $NaBH_4$ and proceeds in the solid state [125]. The process involves simple mixing of carbonyl compounds with 10% $NaBH_4$ supported on alumina and exposure of the reaction mixture to microwaves in a household MW oven for 0.5–2 min (Scheme 2.2-43).The useful chemoselective feature of the reaction is apparent from the reduction of *trans*-cinnamaldehyde (cinnamaldehyde:$NaBH_4$–alumina, 1:1 mol equivalent); the olefinic moiety remains intact and only the aldehyde functionality is reduced in a facile reaction.

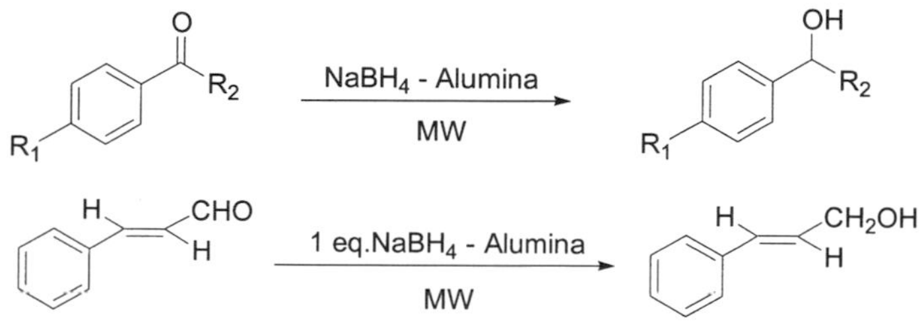

Scheme 2.2-43 Solventless reduction of carbonyls.

The reaction rate improves in the presence of moisture and the reaction does not proceed in the absence of alumina. The alumina support can be recycled and reused for subsequent reduction, repeatedly, by mixing with fresh borohydride without any loss in activity. In terms of safety, the air used for cooling the magnetron ventilates the microwave cavity, thus preventing any ensuing hydrogen from reaching explosive concentrations. The process has been nicely utilized for MW-enhanced solid-state deuteration reactions using sodium borodeuteride-impregnated alumina [126]. Subsequent extension of these studies to specific labeling has been explored [127] including deuterium exchange reactions for the preparation of reactive intermediates [128].

A solvent-free reductive amination protocol for carbonyl compounds using sodium borohydride supported on moist montmorillonite K10 clay is facilitated by MW irradiation (Scheme 2.2-44) [129]. Traditionally, sodium cyanoborohydride [130], sodium triacetoxyborohydride [131], and $NaBH_4$ coupled with sulfuric acid [132] are reagents used for the reductive amination of carbonyl compounds; the use of corrosive acids or cyanide-based reagents results in toxic-waste generation. The solid-state reductive amination of carbonyl compounds on alumina, clay, silica, and especially a K10 clay surface rapidly affords amines. Clay behaves as a Lewis acid and also provides water from its interlayers thus enhancing the reducing ability of $NaBH_4$. Thus practical applications of $NaBH_4$ reductions on mineral surfaces for in situ generated Schiff's bases have been successfully demonstrated.

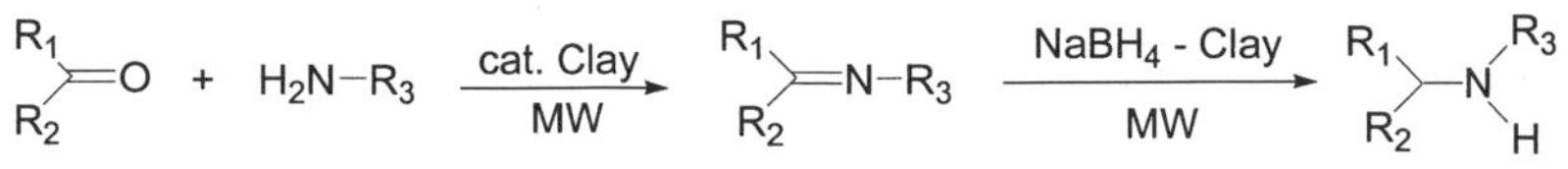

Scheme 2.2-44 Reductive amination of carbonyls.

2.2.4.6
Microwave-assisted Synthesis of Heterocyclic Compounds on Solid Supports

Heterocyclic chemistry has been a major beneficiary of MW-expedited solvent-free chemistry using mineral-supported reagents developed since the mid-1990s, and numerous reports have appeared that employ this technique to assemble a variety of heterocycles [133].

The estrogenic properties displayed by isoflav-3-ene derivatives have attracted the attention of medical chemists. Varma et al. have discovered a facile and general method for the MW-assisted synthesis of isoflav-3-enes substituted with basic moieties at the 2-position (Scheme 2.2-45) [134]. This convergent one-pot approach exploits the in situ generated enamine derivatives that are subsequently reacted with *o*-hydroxyaldehydes in the same pot.

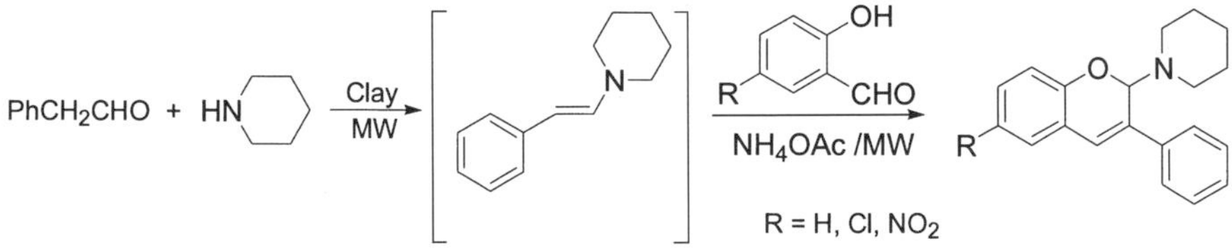

Scheme 2.2-45 Synthesis of 2-amino-substituted isoflav-3-enes from enamines.

The rapid MW-assisted synthesis of bridgehead nitrogen heterocycles has been achieved under solvent-free conditions as exemplified by the synthesis of pyrimidino[1,6-*a*]benzimidazoles [135] and 2,3-dihydroimidazo[1,2-*c*]pyrimidines [136] from *N*-acylimidates and activated 2-benzimidazoles or imidazoline ketene aminals respectively (Scheme 2.2-46).

The conventional preparations of thiazoles and 2-aroylbenzo[*b*]furans require the use of lachrymatory α-haloketones and thioureas (or thioamides). In a process that eliminates this problem, Varma et al. have synthesized various heterocycles via sim-

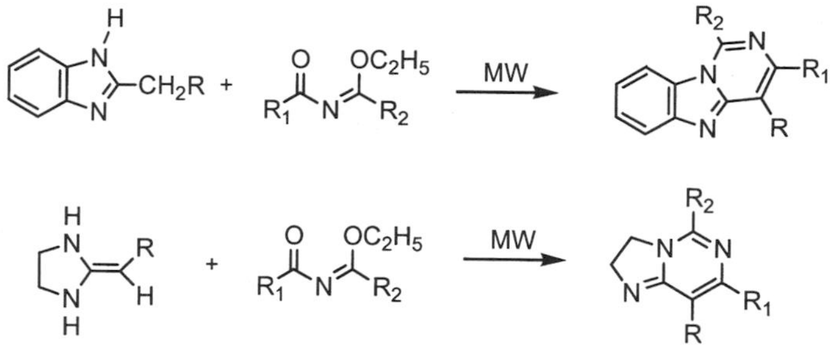

Scheme 2.2-46 MW-assisted solvent-free syntheses of bridge-he-aded heterocycles.

ple solvent-free reaction of thioamides, ethylenethioureas and salicylaldehydes with α-tosyloxyketones that are generated in situ from arylmethyl ketones and [hydroxy(tosyloxy)iodo]benzene (HTIB) under MW irradiation conditions (Scheme 2.2-47) [137].

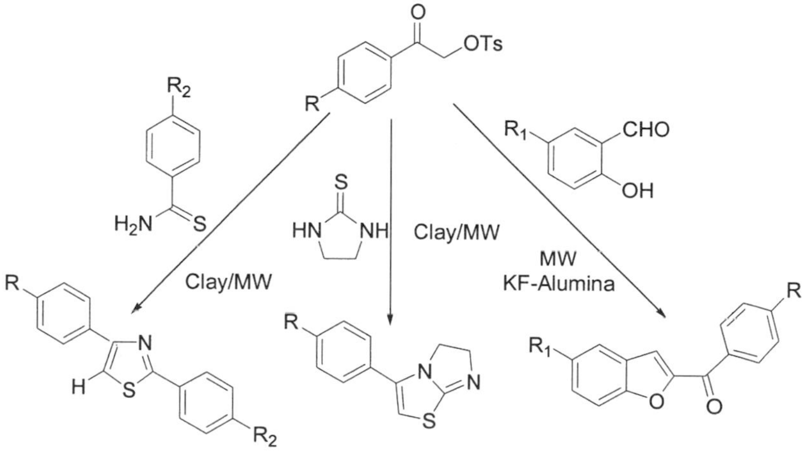

Scheme 2.2-47 Synthesis of heterocycles from in situ generated α-tosyloxyketones.

A multiple-component condensation (MCC) strategy enables the creation of a diverse array of molecules in a single step by changing the reacting components. The generation of small-molecule libraries is facilitated when the ease of reaction manipulation is coupled with efficient protocols, as exemplified by a library of imidazo[1,2-*a*]pyridines, imidazo[1,2-*a*]pyrazines and imidazo[1,2-*a*]pyrimidines which were readily obtained by varying the three components [138]. Experimentally, aldehydes and the corresponding 2-aminopyridine, pyrazine, or pyrimidine are admixed in the presence of a catalytic amount of clay (50 mg) to generate iminium intermediates to which isocyanides are subsequently added in the same reaction vessel.

The reactants are exposed to MW irradiation to afford the corresponding imidazo[1,2-*a*]pyridines, imidazo[1,2-*a*]pyrazines and imidazo[1,2-*a*]pyrimidines (Scheme 2.2-48). The process is general for all three components, i.e. aldehydes (aliphatic, aromatic, and vinylic), isocyanides (aliphatic, aromatic, and cyclic) and amines (2-aminopyridine, 2-aminopyrazine, and 2-aminopyrimidine).

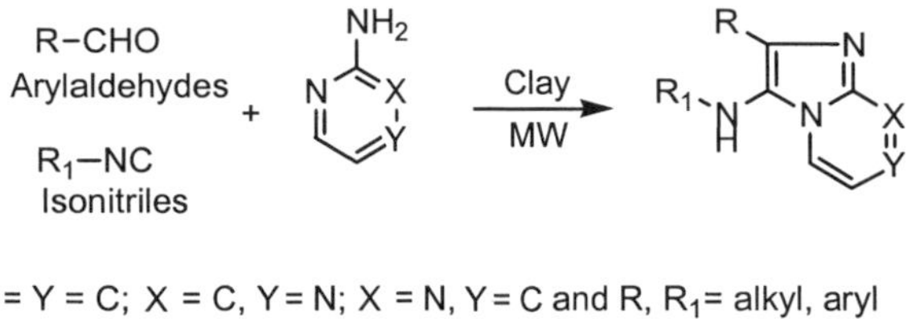

X = Y = C; X = C, Y = N; X = N, Y = C and R, R_1= alkyl, aryl

Scheme 2.2-48 MW-assisted three-component Ugi condensation reaction.

A similar MW strategy has been used to synthesize a set of pyrimidinones (65–95%) via the Biginelli condensation reaction in a household MW oven and has been successfully applied to combinatorial synthesis [139]. More recent examples describe a convenient synthesis of highly substituted pyrroles (60–72%) on silica gel using readily available α,β-unsaturated carbonyl compounds, amines, and nitroalkanes [140], and the use of neat reactants under solvent-free conditions to generate Biginelli and Hantzsch reaction products with enhanced yields and shortened reaction times [141].

2.2.5
Miscellaneous Reactions

2.2.5.1
Solvent-free Preparation of Ionic Liquids Using Microwaves

Room temperature ionic liquids (RTILs), comprising dialkylimidazolium cations and various anions, have received wide attention due to their potential in a variety of commercial applications such as electrochemistry, heavy metal ion extraction, phase-transfer catalysis, polymerization, catalyst supports, and as substitutes for traditional volatile organic solvents. Most of these polar ionic salts are good solvents for a wide range of organic and inorganic materials and consist of poorly coordinating ions, which provide a polar alternative for biphasic systems. Other significant attributes of these ionic liquids include barely measurable vapor pressure, potential for recycling, compatibility with various organic compounds and organometallic catalysts, and ease of separation of products from reaction mixtures. However, most of the initial preparations of ionic liquids involved heating the mixture of reactants for extended periods of time in refluxing solvents, the use of excess amount of alkyl halides to furnish the reaction, and large quantities of organic solvents for product purification. These aspects certainly diminished the true potential of RTILs as "green" solvents.

Ionic liquids, being polar and ionic in character, respond to MW irradiation very efficiently and therefore are ideal microwave-absorbing candidates for expediting chemical reactions. The first preparation of the 1,3-dialkylimidazolium halides via MW heating was described by Varma and coworkers (Scheme 2.2-49) [142,143]. This method reduces the reaction time from hours to minutes and avoids the use of a large excess of alkyl halides/organic solvents as the reaction medium. The efficiency of this approach has been extended to the preparation of other ionic salts bearing tetrafluoroborate anions [142,144], that involves exposing *N, N'*-dialkylimidazolium chloride and ammonium tetrafluoroborate salt to MW irradiation (Scheme 2.2-49).

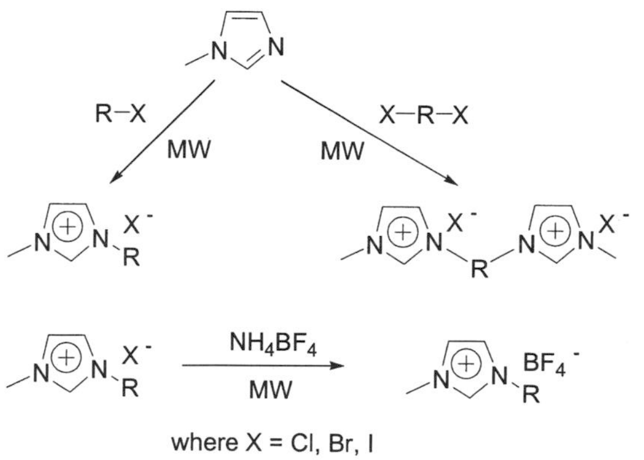

where X = Cl, Br, I

Scheme 2.2-49 MW-assisted solventless preparation of ionic liquids.

The surge of interest in this area continues, including the use of ultrasonic irradiation to prepare these solvents [142] and their use as catalysts for alkylation of isobutane with 2-butene [145] or for ruthenium-catalyzed tandem migration [146a] or silver-catalyzed coupling reactions [146b].

The company BASF is employing an ionic liquid in the manufacture of alkoxyphenylphosphines, the first commercial application of ionic liquids in an organic process wherein *N*-methylimidazole is used to scavenge acid generated in the process [147]. During the BASIL (biphasic acid scavenging utilizing ionic liquids) process, the ionic liquid separates as a clear liquid phase from the pure product and is recycled. Conventional scavengers such as trimethylamine form solids that are difficult to remove from the products and involve a large amount of organic solvents. In the BASIL process, liquids can be easily mixed and pumped without any plugging.

2.2.5.2
Enzyme-catalyzed Reactions

In traditional synthetic transformations, enzymes are normally used in aqueous or organic solvents at relatively low temperatures to preserve their activity. Therefore, it is not surprising that some of these reactions require long reaction times. In view of the newer developments that immobilize enzymes on solid supports [148], these reactions are now amenable to operation at higher temperature with adequate pH control. The application of MW irradiation has been investigated with Pseudomonas lipase dispersed in Hyflo Super Cell, which essentially consists of diatomaceous silica around pH 8.5–9.0 and commercially available SP 435 Novozym (Candida antarctica lipase grafted on an acrylic resin) [149].

The solvent-free resolution of racemic 1-phenylethanol has been achieved under MW-irradiation conditions by transesterification using the above enzymes (Scheme 2.2-50) [149]. A comparison of the MW-assisted reaction with that by conventional heating revealed enhanced enantioselectivity for the former, presumably due to the efficient removal of low molecular weight alcohols or water upon exposure to microwaves or alternatively an entropic effect due to dipolar polarization that induces a previous organization of the system. Thermostable enzymes such as crude homogenate of *Sulfolobus solfataricus* and recombinant β-glucosidase from *Pyrococcus furiosus* have been successfully applied to transglycosylation reactions where recycling of the biocatalysts is feasible [150].

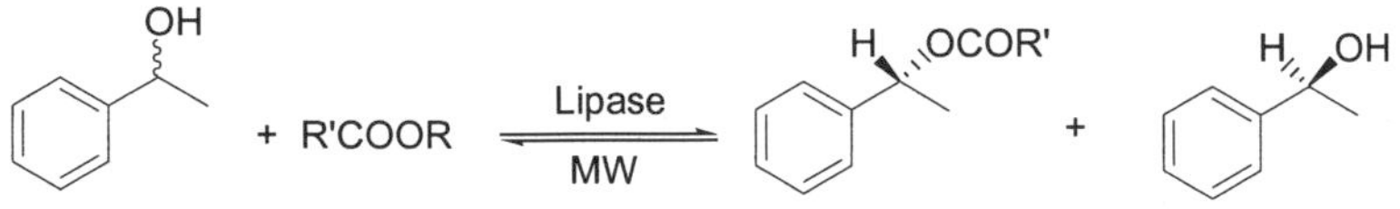

Scheme 2.2-50 MW-induced resolution of 1-phenylethanol via transesterification.

2.2.6
Conclusion

Solvent-free reactions are of immense value to the chemical community in view of their sustainable features, especially in instances where the final product is separated by simple filtration or phase separation without any use of solvent. Indeed, the best solvent is "no solvent" but in such cases the handling of materials and heat- and mass-transfer aspects need to be addressed in close cooperation with chemical engineers at the early stages of chemical process development for industrial success. These reactions are, in general, convenient to perform and have advantages over conventional heating protocols [14], especially when assisted by microwave irradiation [2, 3, 71–77]. The eco-friendly advantages of these reactions may be found in instances where catalytic amounts of reagents or supported agents are used, since they provide reduction or elimination of solvents thus preventing pollution "at source". Although not delineated completely, the reaction rate enhancements achieved in microwave-assisted methods may be ascribable to higher concentrations of reactants and non-thermal specific microwave effects. The ease in the preparation of catalysts by immobilization on solid supports and the acceleration of synthesis processes by microwave irradiation to shorten the reaction time and to eliminate side-product formation is already finding acceptance in the pharmaceutical industry (combinatorial chemistry) [151] and polymer syntheses [152–154] and may pave the way towards a greener and more sustainable approach to chemical syntheses and separations.

References

1 P. T. Anastas, J. C. Warner, *Green Chemistry: Theory and Practice*, Oxford University Press, New York, **1998**.

2 R. S. Varma, *Advances in Green Chemistry: Chemical Synthesis Using Microwave Irradiation*, AstraZeneca Research Foundation India, Bangalore, India, (free copy available from: azrefi@astrazeneca.com) **2002**.

3 R. S. Varma, *Tetrahedron*, **2002**, *58*, 1235–1255.

4 (a) P. Wasserscheid, T. Welton (Eds.), *Ionic Liquid in Synthesis*, Wiley-VCH Verlag, Weinheim, **2003**; (b) J. D. Holbrey, M. B. Turner, R. D. Rogers, *Ionic Liquids as Green Solvents*, ACS Symposium Series 856, American Chemical Society, Washington, DC, **2003**, pp. 2–12

5 (a) J. A. Darr, M. Poliakoff, *Chem. Rev.*, **1999**, *99*, 495–542; (b) A. A. Clifford, C. M. Rayer, *Tetrahedron Lett.*, **2001**, *42*, 323–326.

6 (a) W. Wei, C. C. K. Keh, C. J. Li, R. S. Varma, *Clean Tech. & Environ. Policy* **2005**, *7*, 62–69. (b) C. J. Li, T. H. Chen, *Organic Reactions in Aqueous Media*, J. Wiley & Sons, New York, **1997**.

7 Y. Ju, R. S. Varma, *Green Chem.*, **2004**, *6*, 219–221.

8 (a) M. Wende, J. A. Gladysz, *J. Am. Chem. Soc.*, **2003**, *125*, 5861–5872; (b) D. P. Curran, M. Amatore, D. Guthrie, M. Campbell, E. Go, Z. Y. Luo, *J. Org. Chem.*, **2003**, *68*, 4643–4647; (c) I. T. Horváth, *Acc. Chem. Res.*, **1998**, *31*, 641–650.

9 M. Freemantle, *Chem. Eng. News*, **1998**, *76*, 32–37.

10 C. M. Gordon, J. D. Hobrey, A. R. Kennedy, K. R. Seddon, *J. Mater. Chem.*, **1998**, *8*, 2627–2636.

11 K. R. Seddon, A. Stark, M. J. Torres, *Pure App. Chem.*, **2000**, *72*, 2275–2287.

12 D. Prajapati, M. Gohain, *Tetrahedron*, **2004**, *60*, 815–833.

13 P. Smaglik, *Nature*, **2000**, *406*, 807–808.

14 (a) A. Loupy, *Topics Curr. Chem.*, **1999**, *206*, 155–207; (b) A. Loupy, A. Petit, D. Bogdal, in *Microwaves in Organic Synthesis*, A. Loupy, ed., Wiley-VCH, Weinheim, **2002**, pp. 147–180; (c) K. Tanaka, F. Toda, *Chem. Rev.*, **2000**, *100*, 1025–1074; (d) K. Tanaka, *Solvent-free Organic Synthesis*, Wiley-VCH, Weinheim, **2003**.

15 G. Rothenberg, A. P. Downie, C. L. Raston, J. L. Scott, *J. Am. Chem. Soc.*, **2001**, *123*, 8701–8708.

16 G. W. V. Cave, C. L. Raston, J. L. Scott, *Chem. Commun.*, **2001**, 2159–2169.

17 D. Rajagopal, R. Narayanan, S. Swaminathan, *Proceedings Indian Academy of Sciences, Chemical Sciences*, **2001**, *113*, 197–213.

18 B. Kalluraya, G. Rai, *Synth. Commun.*, **2003**, *33*, 3589–3595.

19 F. Constabel, K. E. Geckeler, *Tetrahedron Lett.*, **2004**, *45*, 2071–2073.

20 (a) A. T. Nielsen, W. J. Houlihan, *Org. React.* **1968**, *16*, 1–438; (b) R. L. Reeves, in *Chemistry of Carbonyl Group*, S. Patai, ed., Wiley-Interscience, New York, **1988**, pp. 580–593; (c) H. O. House, *Modern Synthetic Reaction*, 2nd Edition, W. A. Benjamin, Menlo Park, CA, **1972**, pp. 629–682.

21 (a) C. L. Raston, J. L. Scott, *Green Chem.*, **2000**, *2*, 49–52; (b) G. W. V. Cave,

C. L. Raston, *Chem. Commun.*, **2000**, 2199–2200.

22 (a) E. Knoevenagel, *Chem. Ber.*, **1894**, *27*, 2345–2347; (b) B. M. Trost, *Comprehensive Organic Synthesis*, Pergamon Press, Oxford, **1991**, Vol. 2, pp. 341–355.

23 Z. Ren, W. Cao, W. Tong, *Synth. Commun.*, **2002**, *32*, 3475–3479.

24 J. L. Scott, C. L. Raston, *Green Chem.*, **2000**, *2*, 245–247.

25 R. S. Varma, K. P. Naicker, *Org. Lett.*, **1999**, *1*, 189–191.

26 S. Demirayak, K. Benkli, K. Guven, *Eur. J. Med. Chem.*, **2000**, *35*, 1037–1040.

27 D. Kumar, K. V. G. Chandra Sekhar, H. Dhillon, V. S. Rao, R. S. Varma, *Green Chem.*, **2004**, *6*, 156–157.

28 (a) A. Maercker, *Org. React.*, **1965**, *14*, 270–320; (b) B. E. Maryanoff, A. B. Reitz, *Chem. Rev.*, **1989**, *89*, 863–927.

29 F. Toda, H. Akai, *J. Org. Chem.*, **1990**, *55*, 3446–3447.

30 W. Liu, Q, Xu, Y. Ma, Y. Liang, N. Dong, D. Guan, *J. Organomet. Chem.*, **2001**, *625*, 128–131.

31 C. Xu, G. Chen, C. Fu, X. Huang, *Synth. Commun.*, **1995**, *25*, 2229–2230.

32 A. Spinella, T. Fortunati, A. Soriente, *Synlett*, **1997**, 93–95.

33 V. P. Balema, J. W. Wiench, M. Pruski, V. K. Pecharsky, *J. Am. Chem. Soc.*, **2002**, *124*, 6244–6245.

34 T. Thiemann, M. Watanabe, Y. Tanaka, S. Mataka, *New. J. Chem.*, **2004**, *28*, 578–584.

35 R. Gedye, F. Smith, K. Westaway, H. Ali, L. Baldisera, L. Laberge, J. Rousell, *Tetrahedron Lett.*, **1986**, *27*, 279–282.

36 A. Loupy, A. Petit, J. Hamelin, F. Texier-Boullet, P. Jacquault, D. Mathe, *Synthesis*, **1998**, 1213–1217.

37 C. Gabriel, S. Gabriel, E. H. Grant, B. S. J. Halstead, D. M. P. Mingos, *Chem. Soc. Rev.*, **1998**, *27*, 213–224.

38 C. R. Buffler, *Microwave Cooking and Processing*, Van Nostrand Reinhold, New York, **1993**, pp. 1–68.

39 T. E. Long, M. O. Hunt, *Solvent-free Polymerization and Process*, ACS Symposium Series 713, American Chemical Society, Washington DC, **1999**.

40 (a) M. Soukri, G. Guillaumet, T. Besson, D. Aziane, M. Aadil, E. M. Essassi, M. Akssira, *Tetrahedron Lett.*, **2000**, *41*, 5857–5860; (b) L. Domon, C. Le Coeur, A. Grelard, V. Thiéry, T. Besson, *Tetrahedron Lett.*, **2001**, *42*, 6671–6674.

41 (a) M. Ješelnik, R. S. Varma, S. Polanc, M. Kocevar, *Chem. Commun.*, **2001**, 1716–1717; (b) M. Ješelnik, R. S. Varma, S. Polanc, M. Kocevar, *Green Chem.*, **2002**, *4*, 35–38.

42 (a) Q. Buckingham, *Chem. Soc. Rev.*, **1969**, *23*, 37–56. (b) J. S. Clark, in *Comprehensive Organic Functional Group Transformations*, A. R. Katrizky, O. Meth-Cohn, C. W. Rees, eds., Pergamon, Oxford, **1995**, pp. 221–244.

43 C. F. Most, *Experimental Organic Chemistry*, J. Wiley & Sons, New York, **1998**, pp. 53–62.

44 S. Paul, M. Gupta, R. Gupta, A. Loupy, *Tetrahedron Lett.*, **2001**, *42*, 3827–3829.

45 P Biginelli, *Gazz. Chim. Ital.*, **1893**, *23*, 360–413.

46 G. Sabitha, G. S. K. K. Reddy, K. B. Reddy, J. S. Yadav, *Tetradedron Lett.*, **2003**, *44*, 6497–6499.

47 A. Shaabani, A. Bazgir, *Tetrahedron Lett.*, **2004**, *45*, 2575–2577.

48 J. H. M. Lange, P. C. Verveer, S. J. M. Osnabrug, G. M. Visser, *Tetrahedron Lett.*, **2001**, *42*, 1367–1369.

49 (a) D. Villemin, B. Martin, *Synth. Commun.*, **1995**, *25*, 3135–3136; (b) D. Villemin, B. Martin, *J. Chem. Res. (S)*, **1994**, 146–148.

50 D. Bogdal, *J. Chem. Res. (S)*, **1998**, 468–470.

51 J. J. Kiddle, *Tetrahedron Lett.*, **2000**, *41*, 1339–1341.

52 R. S. Varma, R. Dahiya, S. Kumar, *Tetrahedron Lett.*, **1997**, *38*, 5131–5133.

53 D. Villemin, F. Sauvaget, *Synlett*, **1994**, 435–436.

54 D. Barbry, P. Champagne, *Tetrahedron Lett.*, **1996**, *37*, 7725–7726.

55 B. M. Trost, I. Fleming, M. F. Semmelhack, in *Comprehensive Organic Synthesis*, Pergamon, New York, **1999**.

56 R. C. Larock, *Comprehensive Organic Transformations*, J. Wiley & Sons, New York, **1999**, pp. 1234–1249.

57 J. C. Lee, J. Y. Lee, S. J. Lee, *Tetrahedron Lett.*, **2004**, *45*, 4939–4941.

58 S. Narayan, T. Seelhammer, R. E. Gawley, *Tetrahedron Lett.*, **2004**, *45*, 757–759.

59 R. S. Varma, D. Kumar, *Org. Lett.*, **1999**, *1*, 697–700.

60 U. R. Pillai, E. Sahle-Demmessie, R. S. Varma, *J. Mat. Chem.*, **2002**, *12*, 3199–3221.

61 A. McKillop, K. W. Young, *Synthesis*, **1979**, 401–422, 481–500.

62 G. H. Posner, *Angew. Chem. Int. Ed. Engl.*, **1978**, *17*, 487–496.

63 A. Cornelis, P. Laszlo, *Synthesis*, **1985**, 909–918.

64 P. Laszlo, *Preparative Chemistry Using Supported Reagents*, Academic Press, San Diego, CA, **1987**.

65 K. Smith, *Solid Supports and Catalyst in Organic Synthesis*, Ellis Horwood, Chichester, **1992**.

66 M. Balogh, P. Laszlo, *Organic Chemistry Using Clays*, Springer-Verlag, Berlin, **1993**.

67 J. H. Clark, *Catalysis of Organic Reactions by Supported Inorganic Reagents*, VCH, New York, **1994**.

68 J. H. Clark, D. J. Macquarrie, *Chem. Commun.*, **1998**, 853–860.

69 G. W. Kabalka, R. M. Pagni, *Tetrahedron*, **1997**, *53*, 7999–8065.

70 R. J. Giguere, A. M. Namen, B. O. Lopez, A. Arepally, D. E. Ramos, G. Majetich, J. Defauw, *Tetrahedron Lett.*, **1987**, *28*, 6553–6556.

71 R. S. Varma, in *Microwaves: Theory and Application in Material Processing IV*, D. E. Clark, W. H. Sutton and D. A. Lewis, eds., American Ceramic Society, Westerville, OH, **1997**, pp. 357–365.

72 R. S. Varma, in *Green Chemical Syntheses and Processes*, ACS Symposium Series No. 767, P.T. Anastas, L. Heine, T. Williamson, eds., American Chemical Society, Washington DC, **2000**, pp. 292–313.

73 R.S. Varma, in *Green Chemistry: Challenging Perspectives*, P. Tundo, P.T. Anastas, eds., Oxford University Press, Oxford, **2000**, pp. 221–244.

74 R. S. Varma, *Green Chem.*, **1999**, 1, 43–55.

75 R. S. Varma, in *Microwaves in Organic Synthesis*, A. Loupy, ed., Wiley-VCH, Weinheim, **2002**, pp. 181–218.

76 R. S. Varma, *Pure Appl. Chem.*, **2001**, *73*, 193–198.

77 L. Perreux, A. Loupy, *Tetrahedron*, **2001**, *57*, 9199–9223.

78 A. Stadler, S. Pichler, G. Horeis, C. O. Kappe, *Tetrahedron*, **2002**, *58*, 3177–3183.

79 E. Gutiérrez, A. Loupy, G. Bram, E. Ruiz-Hitzky, *Tetrahedron Lett.*, **1989**, *30*, 945–948.

80 T. W Greene, P.G. M. Wuts, *Protective Groups in Organic Synthesis*, 2nd Edition, J. Wiley & Sons, New York, **1991**.

81 B. Perio, M. J. Dozias, P. Jacquault, J. Hamelin, *Tetrahedron Lett.*, **1997**, *38*, 7867–7870.

82 M. Csiba, J. Cleophax, A. Loupy, J. Malthete, S. D. Gero, *Tetrahedron Lett.*, **1993**, *34*, 1787–1790.

83 R. S. Varma, M. Varma, A. K. Chatterjee, *J. Chem. Soc, Perkin Trans. 1*, **1993**, 999–1001.

84 R. S. Varma, A. K. Chatterjee, M. Varma, *Tetrahedron Lett.*, **1993**, *34*, 3207–3210.

85 R. S. Varma, A. K. Chatterjee, M. Varma, *Tetrahedron Lett.*, **1993**, *34*, 4603–4606.

86 A. S. Gajare, N. S. Shaikh, B. K. Bonde, V. H. Deshpande, *J. Chem. Soc., Perkin Trans. 1*, **2000**, 639–640.

87 D. S. Bose, V. Lakshminarayana, *Tetrahedron Lett.*, **1998**, *39*, 5631–5634.

88 R. S. Varma, J. B. Lamture, M. Varma, *Tetrahedron Lett.*, **1993**, *34*, 3029–3032.

89 M. M. Mojtahedi, M. R. Saidi, M. M. Heravi, M. Bolourtchian, *Monatsh. Chem.*, **1999**, *130*, 1175–1178.

90 M. M. Mojtahedi, M. R. Saidi, M. Bolourtchian, M. M. Heravi, *Synth. Commun.*, **1999**, *29*, 3283–3284.

91 R. S. Varma, R. K. Saini, *Tetrahedron Lett.*, **1997**, *38*, 2623–2624.

92 H. M. Meshram, G. S. Reddy, G. Sumitra, J. S. Yadav, *Synth. Commun.*, **1999**, *29*, 1113–1115.

93 R. S. Varma, H. M. Meshram, *Tetrahedron Lett.*, **1997**, *38*, 5427–5428.

94 A. I. Bosch, P. de la Cruz, E. Diez-Barra, A. Loupy, F. Langa, *Synlett*, **1995**, 1259–1260.

95 R. S. Varma, D. Kumar, *Synth. Commun.*, **1999**, *29*, 1333–1335.

96 G. Sabitha, S. Abraham, B. V. S. Reddy, J. S. Yadav, *Synlett*, **1999**, 1745–1746.

97 R. S. Varma, R. Dahiya, S. Kumar, *Tetrahedron Lett.*, **1997**, 38, 2039–2042.

98 R. S. Varma, R. Dahiya, *Synlett*, **1997**, 1245–1246.

99 A. Dandia, K. Arya, M. Sati, S. Gautam, *Tetrahedron*, **2004**, *60*, 5253–5258.

100 E. Gutiérrez, A. Loupy, G. Bram, E. Ruiz-Hitzky, *Tetrahedron Lett.*, **1989**, *30*, 945–947.

101 D. Villemin, B. Labiad, *Synth. Commun.*, **1992**, *22*, 2043–2052.

102 I. E. Marko, P. R. Giles, M. S. Tsukazaki, M. Brown, C. J. Urch, *Science*, **1996**, *274*, 2044–2045.

103 A. J. Fatiadi, in *Organic Synthesis by Oxidation with Metal Compounds*, W. J. Mijs, C. R. H. I. DeJonge, eds., Plenum, New York, **1986**, pp. 119–260.

104 B. M. Trost, ed., *Comprehensive Organic Synthesis (Oxidation)*, Pergamon, New York, Vol. 7, **1991**.

105 R. S. Varma, R. K. Saini, R. Dahiya, *Tetrahedron Lett.*, **1997**, *38*, 7823–7825.

106 J. Gomez-Lara, R. Gutierrez-Perez, G. Penieres-Carillo, J. G. Lopez-Cortes, A. Escudero-Salas, C. Alvarez-Toledano, *Synth. Commun.*, **2000**, *30*, 2713–2715.

107 B. Oussaid, B. Garrigues, M. Soufiaoui, *Can. J. Chem.*, **1994**, *72*, 2483–2485.

108 R. S. Varma, R. K. Saini, *Tetrahedron Lett.*, **1998**, *39*, 1481–1843.

109 R. Ballini, G. Bosica, M. Parrini, *Tetrahedron Lett.*, **1998**, *39*, 7963–7965.

110 R. S. Varma, R. Dahiya, *Tetrahedron Lett.*, **1997**, *38*, 2043–2044.

111 P. Laszlo, *Acc. Chem. Res.*, **1986**, *19*, 121–127.

112 (a) A. Cornelius, P. Laszlo, *Synlett*, **1994**, 155–161; (b) P. Laszlo, in *Comprehensive Organic Synthesis*, Vol. 7, B. M. Trost, I. Fleming, eds., Pergamon, Oxford, **1991**, pp. 839–848.

113 R. S. Varma, D. Kumar, R. Dahiya, *J. Chem. Res. (S)*, **1998**, 324–325.

114 R. S. Varma, R. Dahiya, D. Kumar, *Molecules Online*, **1998**, *2*, 82–85.

115 R. S. Varma, K. P. Naicker, D. Kumar, R. Dahiya, P. J. Liesen, *J. Microwave Power Electromagn. Energy*, **1999**, *34*, 113–123.

116 R. S. Varma, R. Dahiya, R. K. Saini, *Tetrahedron Lett.*, **1997**, *38*, 7029–7031.

117 R. S. Varma, R. K. Saini, R. Dahiya, *J. Chem. Res (S)*, **1998**, 120–121.

118 R. S. Varma, R. K. Saini, H. M. Meshram, *Tetrahedron Lett.*, **1997**, *38*, 6525–6528.

119 H. Benhaliliba, A. Derdour, J.-P. Bazureau, F. Texier-Boullet, J. Hamelin, *Tetrahedron Lett.*, **1998**, *39*, 541–542.

120 A. Oussaid, A. Loupy, *J. Chem. Res. (S)*, **1997**, 342–343.

121 S. Leskovsek, A. Smidovnik and T. Koloini, *J. Org. Chem.*, **1994**, *59*, 7433–7436.

122 A. K. Bose, M. S. Manhas, B. K. Banik, E. W. Robb, *Res. Chem. Intermed.*, **1994**, *20*, 1–12.

123 B. K. Banik, K. J. Barakat, D. R. Wagle, M. S. Manhas, A. K. Bose, *J. Org. Chem.*, **1999**, *64*, 5746–5753.

124 F. Toda, K. Kiyoshige, M. Yogi, *Angew. Chem., Int. Ed. Engl.*, **1989**, *28*, 320–321.

125 R. S. Varma, R. K. Saini, *Tetrahedron Lett.*, **1997**, *38*, 4337–4338.

126 W. T. Erb, J. R. Jones, S.-Y. Lu, *J. Chem. Res (S)*, **1999**, 728–730.

127 N. Elander, J. R. Jones, S. Y. Lu, S. Stone-Elander, *Chem. Soc. Rev.*, **2000**, *29*, 239–250.

128 K. Fodor-Csorba, G. Galli, S. Holly, E. Gacs-Baitz, *Tetrahedron Lett.*, **2002**, *43*, 4337–4339.

129 R. S. Varma, R. Dahiya, *Tetrahedron*, **1998**, *54*, 6293.

130 R. F. Borch, M. D. Berstein, H. D. Durst, *J. Am. Chem. Soc.*, **1971**, *93*, 289–300.

131 A. F. Abdel-Magid, K. G. Carson, B. D. Harris, C. A. Maryanoff, R. D. Shah, *J. Org. Chem.*, **1996**, *61*, 3849–3862.

132 V. Giancarlo, A. G. Giumanini, P. Strazzolini, M. Poiana, *Synthesis*, **1993**, 121–124.

133 R. S. Varma, *J. Heterocycl. Chem.*, **1999**, *36*, 1565–1571.

134 R. S. Varma, R. Dahiya, *J. Org. Chem.*, **1998**, *63*, 8038–8041.

135 M. Rahmouni, A. Derdour, J.-P. Bazureau, J. Hamelin, *Tetrahedron Lett.*, **1994**, *35*, 4563–4565.

136 M. Rahmouni, A. Derdour, J.-P. Bazureau, J. Hamelin, *Synth. Commun.*, **1996**, *26*, 453–454.

137 R. S. Varma, D. Kumar, P. J. Liesen, *J. Chem. Res., Perkin Trans. 1*, **1998**, 4093–4096.

138 R. S. Varma, D. Kumar, *Tetrahedron Lett.*, **1999**, *40*, 7665–7668.

139 C. O. Kappe, D. Kumar. R. S. Varma, *Synthesis*, **1999**, 1799–1803.

140 B. C. Ranu, A. Hajra, U. Jana, *Synlett*, **2000**, 75–76.

141 M. Kidwai, S. Saxena, R. Mohan, R. Venkataraman, *J. Chem. Soc., Perkin Trans. 1* **2002**, 1845–1846.

142 R. S. Varma, in *Ionic Liquids as Green Solvents. Progress and Prospects*, ACS Symposium Series 856, R. Rogers, K. R. Seddon, eds., American Chemical Society, Washington, DC, **2003**. pp. 82–92.

143 (a) R. S. Varma, V. V. Namboodiri, *Chem. Commun.*, **2001**, 643–644; (b) R. S. Varma, V. V. Namboodiri, *Pure Appl. Chem.*, **2001**, *73*, 1309–1313.

144 V. V. Namboodiri, R. S. Varma, *Tetrahedron Lett.*, **2002**, *43*, 5381–5383.

145 K. Yoo, V. V. Namboodiri, P. G. Smirniotis, R. S. Varma, *J. Catal.*, **2004**, *222*, 511–519.

146 (a) X.-F. Yang, M. Wang, R. S. Varma, C.-J. Li, *J. Mol. Cat. A, Chemical*, **2004**, *214*, 147–154; (b) Z., Li, C. Wei, L. Chen, R. S. Varma, C.-J. Li, *Tetrahedron Lett.*, **2004**, *45*, 2443–2446.

147 (a) M. Freemantle, *Chem. Eng. News*, **2003**, *81*, 9; (b) R. D. Rogers, K. R. Seddon, *Science*, **2003**, *302*, 792–793.

148 E. Guibe-Jampel, G. Rousseau, *Tetrahedron Lett.*, **1987**, *28*, 3563–3564.

149 J.-R. Carrillo-Munoz, D. Bouvet, E. Guibe-Jampel, A. Loupy, A. Petit, *J. Org. Chem.*, **1996**, *61*, 7746–7749.

150 M. Gelo-Pujic, E. Guibe-Jampel, A. Loupy, A. Trincone, *J. Chem. Soc. Perkin Trans. 1*, **1997**, 1001–1002.

151 (a) F.-R. Alexandre, L. Domon, S. Frère, A. Testard, V. Thiéry, T. Besson, *Molecular Diversity*, **2003**, *7*, 273–280; (b) G. A. Strohmeier, C. O. Kappe, *J. Comb. Chem.*, **2002**, *4*, 154–161.

152 X. Fang, R. Hutchenon, D. A. Scola, *J. Polym. Sci., Part A: Poly. Chem.*, **2000**, *38*, 1379–1390.

153 S. Velmathi, R. Nagahata, J. Sugiyama, K. Takeuchi, in *Microwave 2004, Proceedings of International Symposium on Microwave Science and Its Application to Related Science*, Takamatsu, Japan, **2004**, pp. 91–93.

154 W. Zhang, J. Gao, C. Wu, *Macromolecules*, **1997**, *30*, 6388–6390.

2.3
Combinatorial Chemistry on Solid Phases

Mazaahir Kidwai and Richa Mohan

2.3.1
Introduction

The initial report on solid-phase synthesis appeared in 1963, when Merrifield [1] developed a way to make peptides by solid-phase synthesis. Since then the general field of solid-phase organic synthesis (SPOS) has grown enormously. The fundamental technique is based on polymeric resin beads to which a reactant is covalently bonded (Fig. 2.3-1).

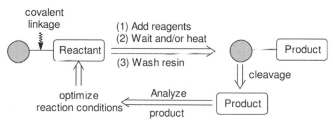

Fig. 2.3-1 The solid-phase synthesis process on insoluble polymer beads.

This immobilized group serves as the starting unit for the linear assembly of additional covalently bonded units. When the desired sequence of molecular building blocks is completed, the original covalent bond to the resin is cleaved to generate the new molecule as a separate entity. The original support used by Merrifield was a copolymer of styrene and divinyl benzene. Currently a wide range of solid supports is available. The principal advantage of SPOS is that the product is always associated physically with the resin beads. Thus the reactions can be driven to completion by using excess reagents, and products are readily separated from reactants by simple filtration.

Several advances have been made in the field of SPOS, notable among them the concepts of parallel synthesis and divergent synthesis [2]. These strategies permit the rapid production of large collections of compounds. Also new technologies ha-

Green Separation Processes. Edited by C. A. M. Afonso and J. G. Crespo
Copyright © 2005 WILEY-VCH Verlag GmbH & Co. KGaA, Weinheim
ISBN 3-527-30985-3

ve been developed based on miniaturization, automation, and robotics. Combinatorial chemistry is the result of these advances in strategies and instrumentation.

Combinatorial chemistry is a technology for creating molecules in bulk and testing them rapidly for desirable properties. Compared with conventional one-molecule-at-a-time discovery strategies, combinatorial synthesis has the ability to generate a large number of chemical compounds very quickly. In its early days, combinatorial chemistry was seen as a brute-force method in which very large collections of compounds (>100 000) could be synthesized. It was more focused on using existing synthetic protocols than on developing new synthetic processes. However, it became obvious that this approach did not yield the desired results and, consequently, much more effort was made in rational design of new molecules for combinatorial synthesis. This subsequently resulted in more extensive chemistry development, which has been coupled with the development and implementation of computational concepts to aid in the design of smaller and more diverse libraries.

2.3.2
Theory

The synthesis of collections of different molecules by varying the combinations of molecular building blocks in each synthetic step is the core of the combinatorial chemistry strategy [3], which was in part enabled by the advent of SPOS. This approach has been explored within the chemistry community since the mid-1990s, with applications ranging from drug discovery to catalyst design to materials science. The compound collections (or libraries) are prepared either in parallel or via a split-pool approach on solid supports. The split-pool approach allows significantly larger libraries to be prepared with less synthetic manipulation than parallel reactions, but requires extra post-synthesis deconvolution/decoding steps to determine the structure of each library member. The purification benefits of having a large molecule covalently bound to an insoluble polymeric support and the ability to push reactions to completion using excess of reagents are both features that make SPOS an attractive platform for numerous combinatorial applications.

2.3.3
Combinatorial Chemistry Applications on a Solid Phase (CCSP)

CCSP can be used in numerous types of organic reactions. Some of the applications are outlined below.

(a) The parallel solid-phase synthesis of an 18-member library of 2-substituted pyrimidines using a chlorogermane-functionalized resin **1** has been reported [4].

Immobilization of the required dioxolane-protected 4-lithioacetophenone, deprotection with pyridinium *p*-toluene sulphonate (PPTS), and condensation with Bredereck's reagent gave resin-bound enaminone **2**. The suitability of resin **2** for library synthesis was verified by a trial condensation with acetamidine hydrochlori-

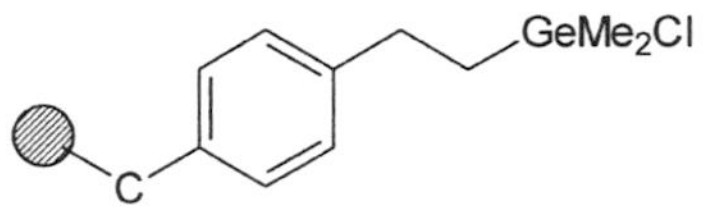

GeMe$_2$Cl

(1)

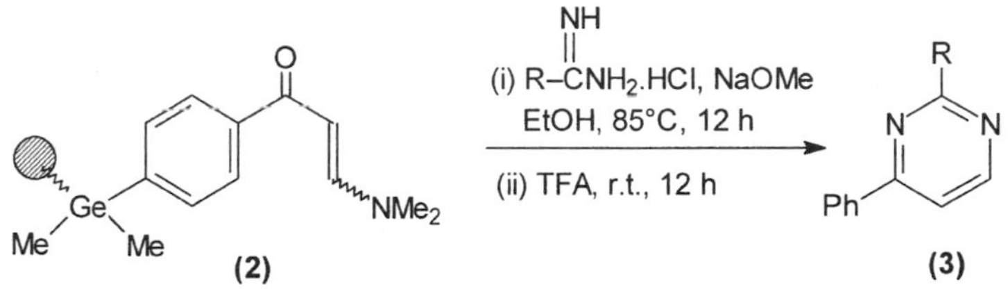

NH
||
(i) R–CNH$_2$.HCl, NaOMe
EtOH, 85°C, 12 h

(ii) TFA, r.t., 12 h

Ge

Me   Me

NMe$_2$

(2)

R

N   N

Ph

(3)

Quadragel^{TM} (0.7 mmolg^{-1})

Scheme 2.3-1

de followed by electrophilic ipso-degermylative cleavage using trifluoroacetic acid (TFA). This gave 2-methyl-pyrimidine **3** in >98% crude purity. Thus an array of 16 alkyl-, aryl-, and heteroaryl-substituted amidine hydrochlorides was then employed for the parallel SPS of a library of 2-substituted-4-phenyl pyrimidines on a ~ 0.02-mmol scale. The resulting pyrimidines were released from the resin in a traceless fashion using TFA.

(b) R.J. Simon et al. [5] described the development of oligomeric N-substituted glycines as a motif for the generation of chemically diverse libraries. They referred to these oligomers as "peptoids". 4-[2',4'-Dimethoxyphenyl(9-fluorenyl-methoxy carbonylaminomethyl)phenoxy] resin (Rink amide resin) was used for the preparation of C-terminal amides. Standard solid-phase peptide synthesis techniques were used with in situ activation by either benzotriazol-1-yloxytris (pyrrolidino) phosphonium hexafluorophosphate or bromotris (pyrrolidino) phosphonium hexafluorophosphate. The standard synthesis cycle involved swelling the resin in either dichloromethane or *N,N*-dimethyl formamide, deprotection with 20% (v/v) piperidine in dimethyl formamide (DMF), thorough washing with the reaction solvent, reaction with the activated amino acid for either a single or double cycle, and in most cases, capping unreacted amino groups with acetic anhydride and pyridine in dichloromethane.

The methods for oligomerization of the peptoid monomers are broadly based on the techniques developed for solid-phase peptide synthesis. A number of modifications are necessary, since difficulties are often encountered with N-alkyl amino acids; for example, proline and other N-alkylamino acids often couple poorly under standard conditions. Currently a wide range of functionalized polystyrenes and polystyrene copolymers is available. The immobilized functional group typically is an amino, carboxy, halo, hydroxy, or phosphino group. The polymer is joined to the immobilized functional group by a connecting section known as a *linker*. The linker provides the covalent bond that is cleaved in the final step. The selection of polymer support and linker is dictated by the choice of reagents needed to deblock protecting groups, effect the couplings, and cleave the linker bond, involving stability and lability to acids or bases.

(c) A small combinatorial library of furans [6] has been generated via split synthesis using eight carboxylic acids, two malonyl chlorides, and two acetylenes as shown in Scheme 2.3-2.

The removal of resin was done by washing with CH_2Cl_2 followed by washing with water.

(d) A library of 192 structurally diverse 1,4-benzodiazepine derivatives containing a variety of chemical functionalities including amides, carboxylic acids, amines, phenols, and indoles has been constructed in a combinatorial fashion from three components, 2-aminobenzophenones, amino acids, and alkylating agents, by employing Geysen's pin apparatus [7] (Scheme 2.3-3).

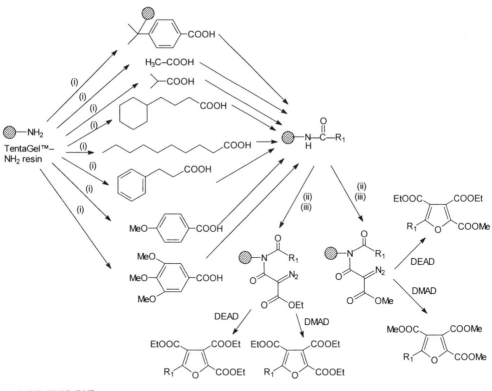

(i) DIC, DMAP, DMF

(ii) Cl $\overset{O}{\text{—}}$ $\overset{O}{\text{—}}$ OEt/OMe, Benzene, 60°C, 1.5 hrs

(iii) TsN$_3$, Et$_3$N, CH$_2$Cl$_2$, 18 hrs.

Scheme 2.3-2

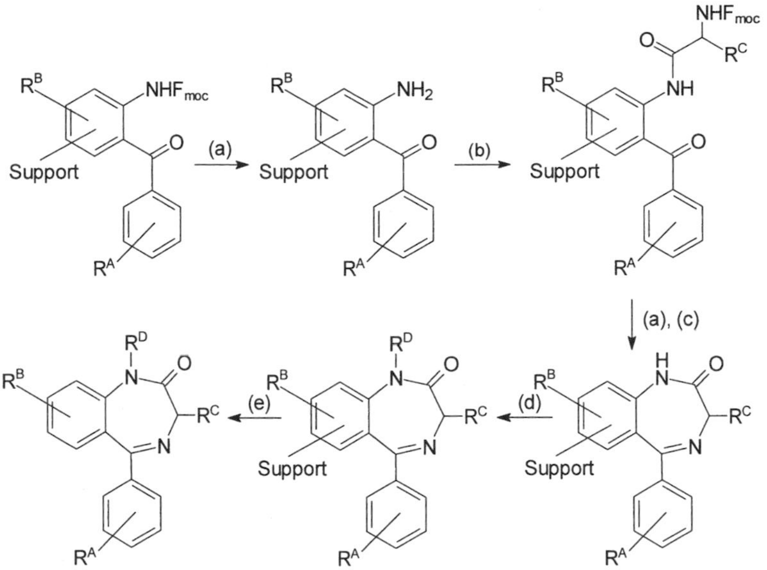

Scheme 2.3-3 1,4-Benzodiazepine synthesis on a solid support, (a) 20% piperidine in DMF; (b) *N*-Fmoc-aminoacid fluoride in CH$_2$Cl$_2$; (c) 5% acetic acid in DMF; (d) lithia-

ted 5-phenylmethyl-2-oxazolidinone in DMF/THF, 1:10 (vol/vol), followed by alkylating agent in DMF; (e) trifluoroacetic acid/H$_2$O/Me$_2$S, 95:5:10 (vol/vol).

The pins were removed from the glove bag, rinsed with DMF, DMF/H$_2$O, MeOH, (air-dried) and CH$_2$Cl$_2$ and cleaved from the support by immersion in 85:10:5 (v/v) TFA/dimethylsulfide/H$_2$O for 24 h. For benzodiazepine derivatives incorporating tryptophan, 85:5:5:5 (v/v) TFA/dimethyl sulfide/H$_2$O/1,2-ethanediol was employed as the cleavage mixture.

(e) Synthesis of highly substituted imidazole libraries on a solid support using an aldehyde, an amine, and a 1,2-dione in the presence of NH$_4$OAc has been reported [8] (Scheme 2.3-4). The synthesis was accomplished by attaching the alde-

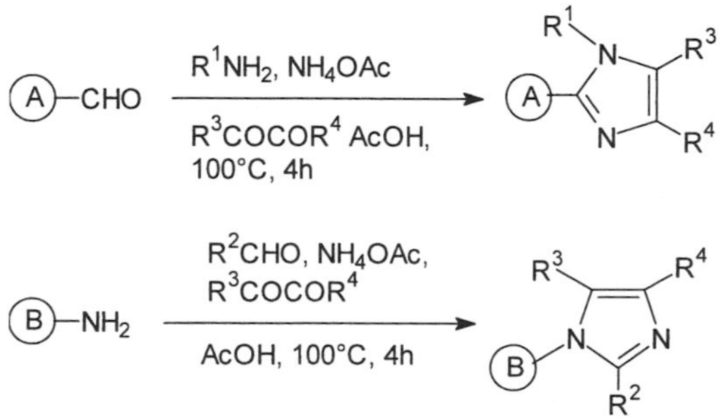

Scheme 2.3-4 (i) DIC, DMAP, THF, 23°C, 48 h, carboxybenzaldehyde (ii) *N*-fmoc-6-aminohexanoic acid, DIC, DMAP, CH$_2$Cl$_2$, 23°C, 24 h.

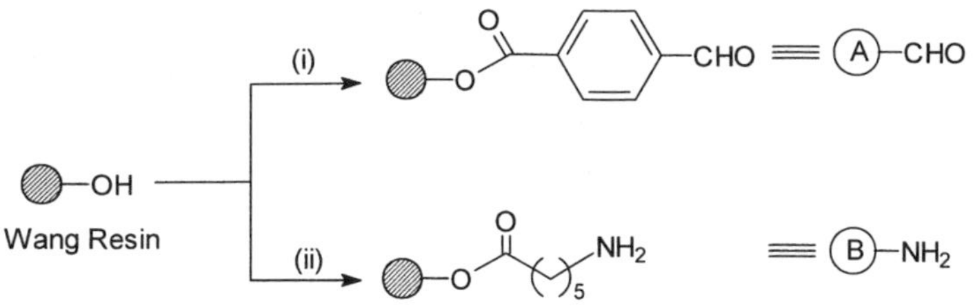

Scheme 2.3-5 Preparation of imidazoles on a solid support.

hyde or amine component to Wang resin via an ester or ether linkages (Scheme 2.3-5).

The imidazoles were removed by washing (CH_2Cl_2/CH_3OH) and treatment with 20% TFA in CH_2Cl_2.

(f) The recently developed radiofrequency encoded combinatorial (REC) chemistry strategy has been applied to the design and synthesis of a taxoid library of general formula **4** [9]. Taxol template **5** was reacted with excess 2-chlorotrityl resin (10 mequiv) via standard trityl ester formation to afford the polymer-bound compound **6**. The loaded resin **6** was then distributed into 400 microreactors (20 mg each) and the microreactors were then subjected to various reactions (Scheme 2.3-6).

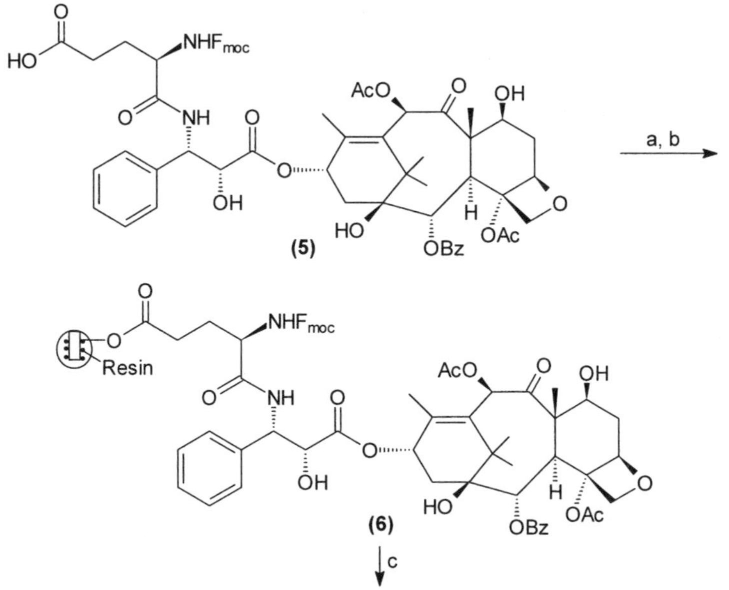

Scheme 2.3-6 (a) 2-Chlorotrityl resin (8.0 g, 10 mequiv), DIEA, CH_2Cl_2, r.t., 3 h, MeOH, r.t., 0.5 h; (b) distribution of resin into 400 microreactors; (c) 5% piperidine in DMF, r.t., 0.5 h; (d) microreactors split into 20 equal pools, and each pool treated with carboxylic acid (R^1), DIEA, PyBOP, DMF, r.t., 4 h; (e) microreactors resplit into 20 new pools and treated with carboxylic acid (R^2), DIC, DMAP, CH_2Cl_2, r.t., 48 h; (f) microreactors were decoded and distributed into 400 glass vials; (g) the final product was cleaved from resin by treatment with AcOH, CH_2Cl_2, CF_3CH_2OH (2:7:1, v/v/v), r.t. 4 h.

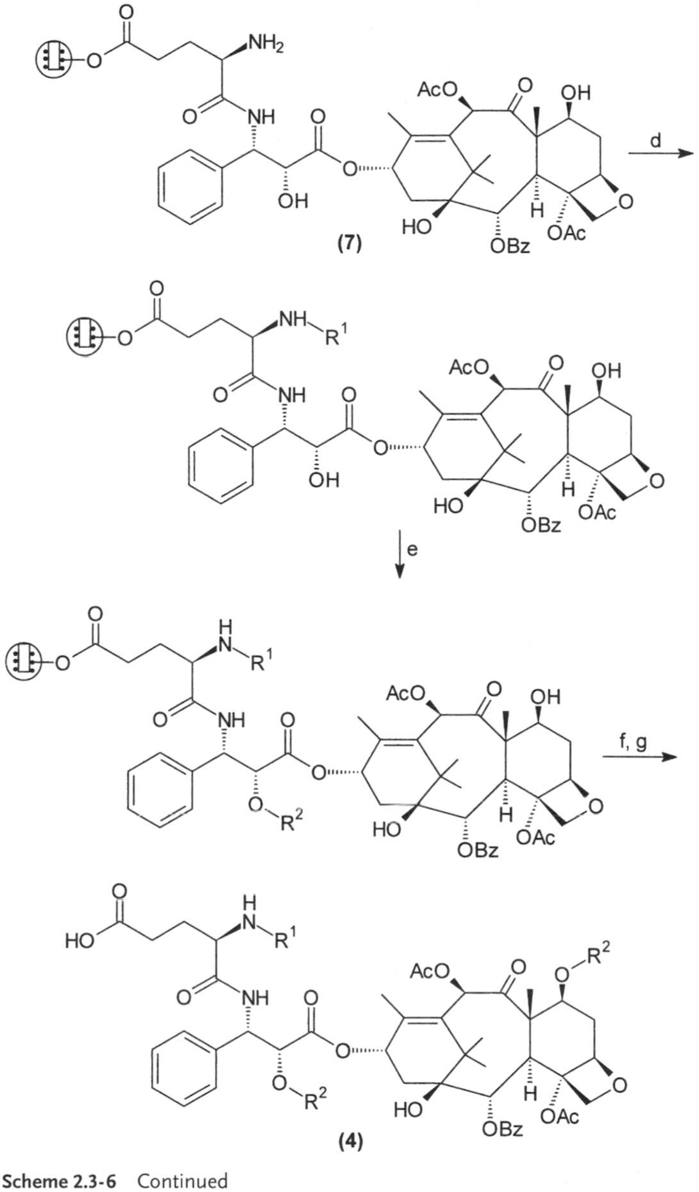

Scheme 2.3-6 Continued

(g) Multicomponent condensations have proven to be very simple and effective methods for combinatorial synthesis. In 1996, Gordeev and co-workers [10] developed a solid-phase synthesis of 1,4-dihydropyridines (Scheme 2.3-7). In this a β-dicarbonyl compound reacted with a polystyrene-based acid-cleavable resin **8** to afford the enamine **9** on the solid support. Next **9** was treated with either a preformed α-aryl methylene-β-dicarbonyl compound or directly with an aromatic or heteroaromatic aldehyde and a β-dicarbonyl compound to afford **10** which underwent imine–enamine tautomerism to produce **11**. Acidic cleavage from the resin produced **12**, which then completed the cyclo-condensation, yielding the 1,4-dihydropyridines **13**.

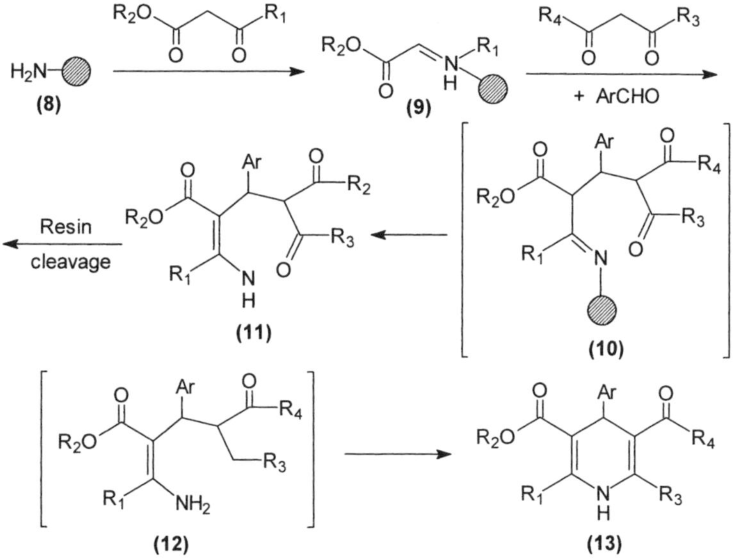

Scheme 2.3-7   Solid-phase split-and-mix combinatorial 1,4-dihydropyridine synthesis by Gordeev et al. [10].

(h) More recently, the Biginelli reaction has proven to be very applicable to combinatorial chemistry, and many diverse dihydropyrimidone compound libraries have been synthesized for high-throughput screening. One of the first solid-phase modifications of the Biginelli reaction was reported in 1995 by Wipf and Cunningham [11] (Scheme 2.3-8). The synthesis required the attachment of a γ-aminobutyric acid-derived urea to Wang resin, providing **14**. Following this, the acid-catalyzed Biginelli reaction was performed in which four equivalents of β-ketoester **16** and aryl aldehyde **15** was used. After cleavage of the resultant dihydropyrimidine **17** from the resin with TFA, further derivatization of the resulting acid **18** was possible.

It is clear from the reaction conditions given for these examples that solid-phase reactions are heterogeneous and often take considerably longer than their homogeneous solution-phase counterparts (frequently up to 10 times longer). Reagent diffusion into the polymer matrix is invariably a slow process, especially for large macrobeads. As a result, solid-phase reactions often require more "forcing" reaction conditions than their solution-phase counterparts, such as high heat and prolonged reaction-times, both of which can generate unwanted reaction by-products. Moreover, solid-phase reactions require extra steps to attach and detach the com-

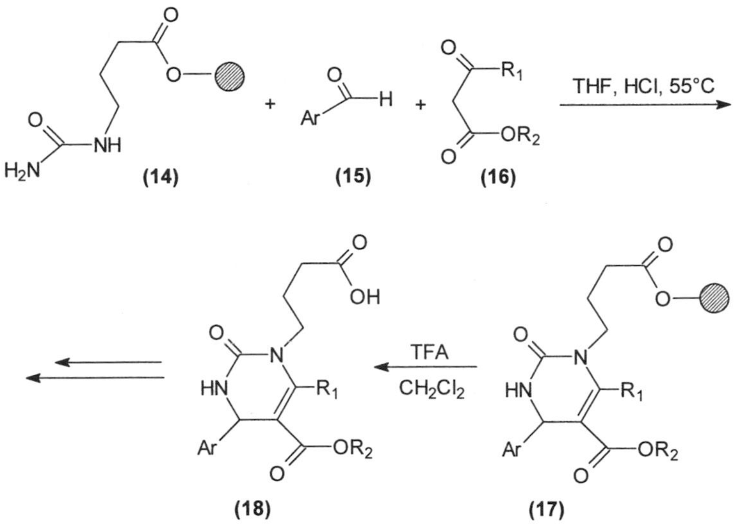

Scheme 2.3-8 Solid-phase synthesis of dihydropyrimidones by Wipf and Cunningham.

pound to and from the resin and lack a qualitative method to quickly monitor their progress.

2.3.4
Microwave-assisted Solid-phase Synthesis

For combinatorial chemistry to deliver on its promise, general methods need to be discovered to accelerate solid-phase reactions to the point at which they are equivalent in rate to, or faster than, homogeneous reactions. Microwave-assisted organic synthesis has had considerable success in achieving this goal. Since 1990, many organic transformations have been accelerated by subjecting them to microwave (MW) irradiation. In most cases reaction times of hours to days have been reduced to seconds to minutes.

Thermally demanding reactions, such as the Diels–Alder reaction, are often completed in hours in solution but when performed on a solid phase can take several days. If such solid-phase reactions are performed with MW irradiation, the reaction times can be reduced from days to minutes.

For example, Yu et al. showed that polystyrene-bound peptides could be hydrolyzed in 7 min in a domestic MW oven, a process normally taking 24 h. Furthermore, traditional solid-phase peptide couplings were achieved in 4 min in 99–100% conversion with no detected racemization. A broad range of solid-phase reactions was found to undergo substantial rate acceleration, including Claisen and Knoevenagel condensations, nucleophilic substitutions, succinimide and hydantoin formation, and Suzuki couplings.

2.3.4.1
Microwave-assisted Combinatorial Synthesis on Solid Phases [12]

Given the success of several MW-assisted solid-phase reactions, the next critical step is to extend this enabling methodology to solid-phase combinatorial library synthesis. Since the late 1990s, several parallel libraries have been prepared on polymeric supports using MW-assisted reactions as a key step. Although each library has been small (ranging from 5 to 96 members), these efforts have laid the foundation for the generation of larger libraries in the future.

Some of the notable examples of these libraries generated via a MW-assisted solid phase are given below. Multicomponent reactions (MCRs), in which three or more reactants combine to give a single product, have received much attention owing to their elegance, simplicity, and overall efficiency in comparison to multistep syntheses.

2.3.4.2
Microwave-assisted Polymer-supported Library Synthesis

One example of a library synthesized via MW-assisted solid-phase MCRs is that of Hoel and Nielson [13] (Scheme 2.3-9). The authors performed Ugi-4CC (four component condensation) reactions in a monomodal MW reactor between polymer (TentaGel)-bound amines and various aldehydes, carboxylic acids, and isocyanides to yield a "mini library" of 18α-acylaminoamides in just 15 minutes per compound. The reaction time was reduced by a factor of three. The yields were variable, but the authors reported highly pure products (>95%).

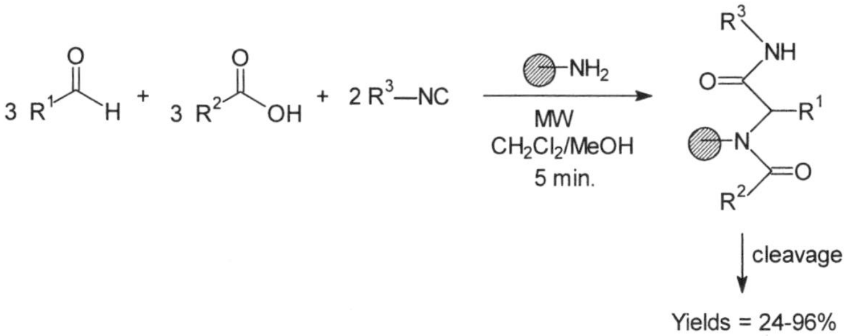

Scheme 2.3-9 Parallel library generated via Ugi-4CC reactions on amino TentaGel resin.

In two elegant examples (shown in Schemes 2.3-10 and 2.3-11), Ley and coworkers [14] employed MW irradiation in reactions mediated by their solid-supported reagents, a polymer-supported thionating agent and a supported reagent for the conversion of isothiocyanates to isocyanides, respectively. Using a monomodal MW oven, both reactions gave highly pure products in excellent yields in a fraction of the time required with traditional heating (15–150 min compared to ~ 30 h). The newly formed isocyanides were further processed in an Ugi-4CC to yield 18 novel bicyclic products.

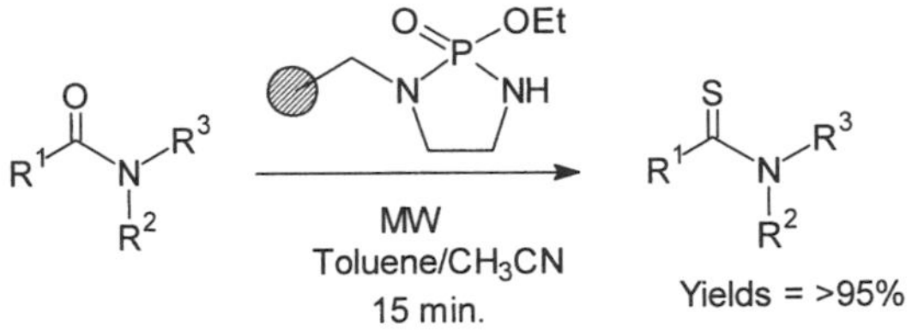

Scheme 2.3-10 Polystyrene-supported thionating reagents for the conversion of amides to thioamides.

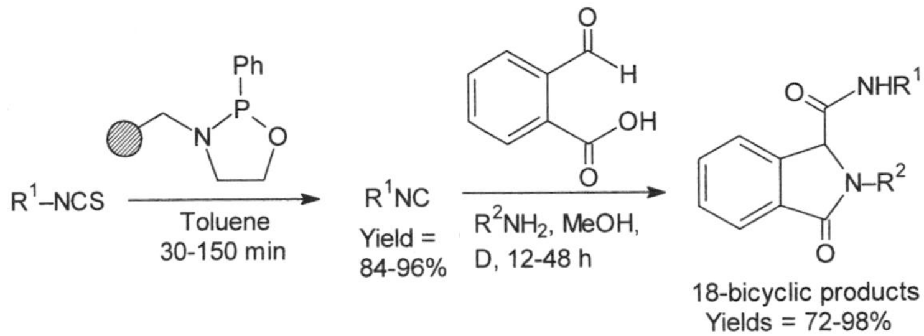

Scheme 2.3-11 Polystyrene supported [1,3,2]oxazaphospholidene for the conversion of isothiocyanates to isocyanides.

2.3.4.3
Microwave-assisted Solvent-free Library Synthesis

All the MW-assisted reactions presented above have been conducted in polar organic solvents, the improper handling of which can be a significant safety hazard. Numerous groups have explored solvent-free MW-assisted organic syntheses in which reagents are either adsorbed onto a polar, inorganic support (e.g. alumina, silica, clay, or zeolites) or mixed and subjected to irradiation neat. The reduced use of organic solvents makes this methodology more environmentally benign.

Recently this approach has been extended to parallel combinatorial library synthesis. One example is the solvent-free synthesis of a 96-member library of substituted pyridines via a one-step Hantzsch-3CC conducted in 96-well micro titer filter plates (Scheme 2.3-12). Here the β-ketoester and aldehyde reagents were impregnated onto a 5:1 bentonite clay–ammonium nitrate mixture. Irradiation for 5 min in a domestic MW oven, followed by washing the product off the support into a receiver daughter plate gave substituted-pyridine products in >70% purity overall.

Several additional examples of solvent-free MW-accelerated processes for parallel synthesis are known (Scheme 2.3-13).

In most of the parallel reactions, solvent is rarely used and the starting materials or reagents are either supported on mineral solid supports or these supports are

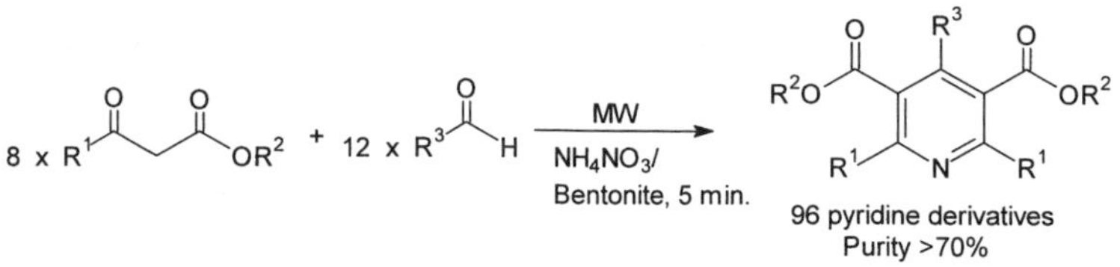

Scheme 2.3-12 Parallel library generated via Hantzsch-3CC reactions under solvent-free conditions.

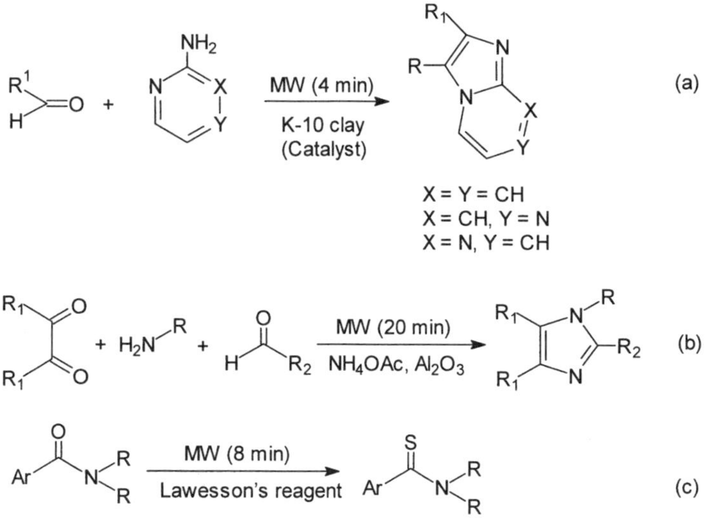

(a)

X = Y = CH
X = CH, Y = N
X = N, Y = CH

(b)

(c)

Scheme 2.3-13 (a) Synthesis of imidazo-annulated pyridines, pyrazines and pyrimidines by a Ugi-type 3CC reaction; (b) synthesis of 1,2,4,5-substituted imidazoles on acidic alumina [15]; (c) synthesis of thioamide library via thionation using Lawesson's reagent.

used in catalytic amounts. The desired products are extracted by organic solvents from the solid support and are devoid of any starting material.

2.3.4.5
Microwave-assisted Parallel Library Synthesis on Planar Supports

The application of MW irradiation to solid-phase chemistry has not been restricted to spherical polymer beads. Recently large arrays of compounds have been synthesized on planar solid supports including cellulose, polypropylene, and SiO_2 TLC plates.

Williams [16] synthesized an array of N-substituted aryl piperazines by first spotting the neat reagents onto a glass-backed TLC plate and subjecting the plate to MW irradiation for 5 min in a domestic oven (Fig. 2.3-2). After cooling, the reaction array was then eluted and visualized using UV light and chemical stains. No starting material was observed and all reactions cleanly gave one product.

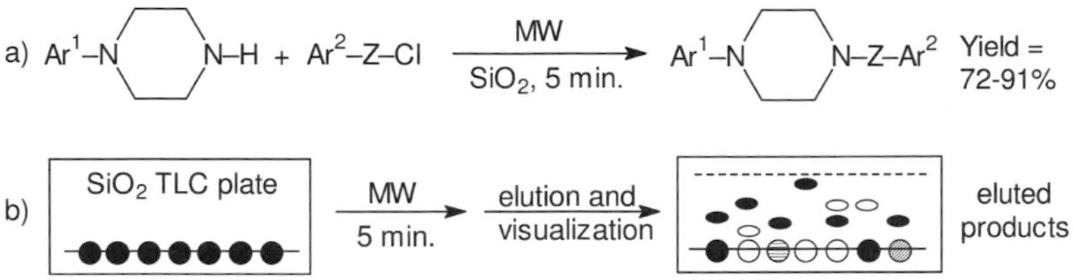

a) Ar^1-N⟨piperazine⟩$N-H$ + Ar^2-Z-Cl $\xrightarrow[\text{SiO}_2, \text{5 min.}]{\text{MW}}$ Ar^1-N⟨piperazine⟩$N-Z-Ar^2$ Yield = 72-91%

b)

| SiO₂ TLC plate |
| ● ● ● ● ● ● ● |

$\xrightarrow[\text{5 min.}]{\text{MW}}$ elution and visualization $\longrightarrow$

eluted products

Fig. 2.3-2 (a) Solvent-free synthesis of N-substituted aryl piperazines on SiO₂. (b) Schematic diagram of thin-layer chromatography (TLC) as a tool for parallel screening in MW-assisted synthesis.

2.3.5
Conclusion

In comparison to solid-phase syntheses, conventional solution-phase syntheses require fewer steps, but purification of intermediates is much more difficult. In solution-phase syntheses exact stoichiometric conditions have to be employed to avoid tedious separations. In contrast, the solid-phase approach allows easy separations, sometimes simply by filtration as mentioned. The application of MW irradiation to expedite solid-phase organic reactions could be the tool that allows combinatorial chemistry to deliver on its promise of providing rapid access to large collections of diverse small molecules. With new therapeutic targets emerging from genomic and proteomic research efforts, there is an urgent need to develop methods to synthesize small-molecule modulators efficiently – solid-phase combinatorial chemistry is poised to help chemists meet this challenge.

References

1 R.B. Merrifield, *J. Am. Chem. Soc.*, **1963**, *85*, 2149–2154.

2 F.D. Zaragoza, *Organic Synthesis on Solid Phase*, Wiley-VCH, Weinheim, **2000**.

3 S. Borman, *Chem. Eng.*, **1998**, April 6, 47–67.

4 A.C. Spivey, R. Srikaran, C.M. Diaper and D.J. Turner, *Org. Biomol. Chem.*, **2003**, *1*, 1638–1640. Reproduced by permission of The Royal Society of Chemistry.

5 R.J. Simon, R.S. Kania, R.N. Zuckermann, V.D. Huebner, D.A. Jewell, S. Banville, S. Ng, L. Wang, S. Rosenberg, C.K. Marlowe, D.C. Spellmeyer, R. Tan, A.D. Frankel, D.V. Santi, F.E. Cohen and P.A. Bartlett, *Proc. Natl. Acad. Sci. USA*, **1992**, *89*, 9367–9371.

6 M.R. Gowravaram and M.A. Gallop, *Tetrahedron Lett.*, **1997**, *38(40)*, 6973–6976.

7 B.A. Bunin, M.J. Plunkett and J.A. Ellman, *Proc. Natl. Acad. Sci. USA*, **1994**, *91*, 4708–4712. Copyright (1994) National Academy of Sciences, USA.

8 S. Sarshar, D. Siev and A.M.M. Mjalli, *Tetrahedron Lett.*, **1996**, *37(6)*, 835–838.

9 Reprinted with permission from X.-Y. Xiao, Z. Parandoosh and M.P. Nova, *J.* *Org. Chem.*, **1997**, *62*, 6029–6033. Copyright (1997) American Chemical Society.

10 M.F. Gordeev, D.V. Patel and E.M. Gordon, *J. Org. Chem.*, **1996**, *61*, 924–928.

11 P. Wipf and A. Cunningham, *Tetrahedron Lett.*, **1995**, *36*, 7819–7822.

12 For two recent reports on microwave-assisted combinatorial chemistry, see: (a) H.E. Blackwell, *Org. Biomol. Chem.*, **2003**, *1*, 1251–1255. Reproduced by permission of The Royal Society of Chemistry; (b) A. Lew, P.O. Krutzik, M.E. Hart and A.R. Chamberlin, *J. Comb. Chem.*, **2002**, *4*, 95–105.

13 A.M.L. Hoel and J. Nielson, *Tetrahedron Lett.*, **1999**, *40*, 3941–3944.

14 (a) S.V. Ley, A.G. Leach and R.I. Storer, *J. Chem. Soc. Perkin Trans. 1*, **2001**, 358–361; (b) S.V. Ley and S.J. Taylor, *Bioorg. Med. Chem. Lett.*, **2002**, *12*, 1813–1816.

15 Reprinted from A.Y. Usyatinsky and Y.L. Khmelnitsky, *Tetrahedron Lett.*, **2000**, *41*, 5031–5034. Copyright (2000), with permission from Elsevier.

16 L. Williams, *Chem. Commun.*, **2000**, 435–436.

Part 3
New Developments in Separation Processes

Green Separation Processes. Edited by C. A. M. Afonso and J. G. Crespo
Copyright © 2005 WILEY-VCH Verlag GmbH & Co. KGaA, Weinheim
ISBN 3-527-30985-3

3.1
Overview of "Green" Separation Processes

Richard D. Noble

3.1.1
Background

Separation processes are any set of operations that separate solutions of two or more components into two or more products that differ in composition. These may either remove a single component from a mixture or separate a solution into its almost pure components. This is achieved by exploiting chemical and physical property differences between the substances through the use of a separating agent (mass or energy).

Separation processes are used for three primary functions: purification, concentration, and fractionation. Purification is the removal of undesired components in a feed mixture from the desired species. For example, acid gases, such as sulfur dioxide and nitrogen oxides, must be removed from power-plant combustion gas effluents before they are discharged into the atmosphere. Concentration is performed to obtain a higher concentration of desired components that are initially dilute in a feed stream. An example is the concentration of metals present in an electroplating process by removal of water. This separation allows metals to be recycled back to the electroplating process rather than discharged to the environment. Lastly, in fractionation, a feed stream of two or more components is segregated into product streams of different components, typically relatively pure streams of each component. The separation of radioactive wastes with short half-lives from those having much longer half-lives facilitates proper handling and storage.

Separation processes can be placed into two fundamental categories – equilibrium processes and rate processes – that are analyzed differently. These separation categories are designed using thermodynamic equilibrium relationships between phases and the rate of transfer of a species from one phase into another, respectively. The choice of which analysis to apply is governed by which is the limiting step. If mass transfer is rapid, such that equilibrium is quickly approached, then the separation is equilibrium limited. On the other hand, if mass transfer is slow, such that equilibrium is not quickly approached, the separation is mass-transfer li-

Green Separation Processes. Edited by C. A. M. Afonso and J. G. Crespo
Copyright © 2005 WILEY-VCH Verlag GmbH & Co. KGaA, Weinheim
ISBN 3-527-30985-3

mited. In some separations, the choice of analysis depends upon the type of process equipment used.

Equilibrium processes are those in which cascades of individual units, called stages, are operated, with two streams typically flowing countercurrent to each other. The degree of separation in each stage is governed by a thermodynamic equilibrium relationship between the phases. One example is distillation, in which a different temperature at each stage alters the vapor-phase equilibrium between the components of, typically, a binary mixture. The driving force for separation is the tendency to establish a new equilibrium between the two phases at the temperature of each stage. The end result is the separation of two liquids with dissimilar boiling temperatures. Other equilibrium-based processes that will be covered in this text include extraction and solid extraction, or leaching. Extraction is the removal of a species from a liquid in which it is dissolved by means of another liquid for which it has a higher affinity, and leaching is the removal of a species from a solid phase by means of a liquid for which it has stronger affinity.

Rate processes, on the other hand, are limited by the rate of mass transfer of individual components from one phase into another under the influence of physical stimuli. Concentration gradients are the most common stimuli, but temperature, pressure, or external force fields can also cause mass transfer. One mass-transfer-based process is gas absorption, a process by which a vapor is removed from its mixture with an inert gas by means of a liquid in which it is soluble. Desorption, or stripping, on the other hand, is the removal of a volatile gas from a liquid by means of a gas in which it is soluble. Adsorption consists of the removal of a species from a fluid stream by means of a solid adsorbent with which it has a higher affinity. Ion exchange is similar to adsorption, except that the species removed from solution is replaced with a species from the solid resin matrix so that electroneutrality is maintained. Lastly, membrane separations are based upon differences in permeability (transport through the membrane) due to size and chemical selectivity for the membrane material between components of a feed stream.

3.1.2
Pollution Sources

The National Research Council released a report [1] that states:

"The expanding world population is having a tremendous impact on our ecosystem, since the environment must ultimately accommodate all human derived waste materials. The industries that provide us with food, energy and shelter also introduce pollutants into the air, water, and land. The potential for an increasing environmental impact will inevitably result in society's setting even lower allowable levels for pollutants."

The report further concludes: "In the future, separation processes will be critical for environmental remediation and protection." Chemical separations are used to reduce the quantity of potentially toxic or hazardous materials discharged to the environment. In addition, separations that lead to recovery, recycle, or reuse of materials also prevent discharge.

Sources of pollution vary from small-scale businesses, such as dry cleaners and gas stations, to very large-scale operations, such as power plants and petrochemical facilities. The effluent streams of industry are particularly noticeable because of their large volumes [1]. Sources include both point source and non-point source pollution. Point source pollution can be traced directly to single outlet points, such as a pipe releasing into a waterway. Non-point source pollutants, on the other hand, such as agricultural run-off, cannot be traced to a single definite source. The emissions from both span a wide range of gas, liquid, and solid compounds.

A large majority of air polluting emissions comes from mobile sources. The automobile is an obvious example, but other vehicles, such as semi-trucks, trains, and aircraft also contribute. Emissions from mobile sources include CO_2, volatile organic compounds (VOCs), NO_x, and particulates. The latter may also have heavy metals, such as lead or mercury, or hazardous organics attached. Stationary sources typically burn or produce fossil fuels – coal, gasolines, and natural gas. These produce gaseous sulfur compounds (H_2S, SO_2, etc.), nitrogen oxides (NO_x), CO_2, and particulates. Fuel producers and distributors also typically produce VOCs. Most of these pose human health concerns and many contribute to the acid rain problem and global warming effect.

Water pollution also comes from a variety of sources. Agricultural chemicals (fertilizers, pesticides, herbicides) find their way into groundwater and surface water through water run-off from farming areas. For example, agricultural drainage water with high concentrations of selenium threatens the Kesterson National Wildlife Refuge in California. Chemical discharge from sources ranging from household releases (lawn fertilizers, detergents, motor oil) to industrial releases into surface or groundwater supplies is an obvious problem. Industrial discharges can occur from leaking storage facilities as well as process effluent. Municipal water treatment effluent is another prevalent source. MTBE, a gasoline additive used until recently to reduce air pollution, has been identified as a source of water pollution, demonstrating that the solution to one environmental concern can create a problem elsewhere. Isolation and recovery of these and other water pollutants will be a challenge for innovative separation techniques.

Pollution of soils also occurs through a variety of sources. Municipal and industrial waste has been buried in landfills, which sometimes leak even if lined with durable impermeable materials. Periodic news accounts of hazardous chemicals migrating through soil to threaten water supplies and homes are reminders of this issue. Chemical discharge directly onto surface soil from periodic equipment cleaning, accidental discharges (spills), abandoned process facilities, or disposal sites is another environmental challenge. Subsurface contamination can also occur as a result of leaking underground storage tanks.

In addition to air, water, and soil pollution, large quantities of solid and liquid wastes generated by both industry and domestic use must be remediated, recycled, or contained. Industrial wastes include overburden and tailings from mining, milling, and refining, as well as residues from coal-fired steam plants and the wastes from many manufacturing processes. The nuclear and medical industries generate radioactive solid wastes that must be carefully handled and isolated. Effective ways of fractionating long-lived radioactive isotopes from short-lived ones are needed because the long-lived ones require more expensive handling and storage. The environmental problems of residential wastes are increasing as the population grows. It is important to segregate and recycle useful materials from these wastes. In many places, there are no effective options for dealing with toxic liquid wastes. Landfill and surface impoundment are being phased out. There is a strong incentive toward source reduction and recycling, which creates a need for separations technology [1].

All of the above separation needs are oriented primarily toward removal and isolation of hazardous material from effluent or waste streams. Pollutants are frequently present in only trace quantities, such that highly resolving separation systems will be required for detection and removal. The problem of removing pollutants from extremely dilute solutions is becoming more important as allowable release levels for pollutants are lowered. For example, proposed standards for the release of arsenic prescribe levels at or below the current limit of detection. Another example is pollution of water with trace quantities of dioxin. In research being carried out at Dow Chemical U.S.A., concentrations of adsorbed dioxin at the part-per-quadrillion (10^{15}) level have been successfully removed from aqueous effluents. That technology has now been scaled up, such that dioxin removals to less than ten parts per quadrillion are being achieved on a continuous basis on the 20 million gallon per day waste water effluent stream from Dow's Midland, Michigan, manufacturing facility.

3.1.3
Environmental Separations

Based upon sources of pollution and the nature of polluted sites (air, land, or water), environmental separations can be categorized in the following manner:

1. Clean-up of existing pollution problems
   Examples:
   - surface water contamination (organics, metals, etc.)
   - ground water contamination (organics, metals, etc.)
   - airborne pollutants (SO_x, NO_x, CO, etc.)
   - soil clean-up (solvent contamination, heavy metals, etc.)
   - continuing discharges to the environment
     automobiles
     industries: chemical, nuclear, electronics.
2. Pollution prevention
   Examples:
   - chemically benign processing
     hybrid processing

A unit operation is any single step in an overall process that can be isolated and that also tends to appear frequently in other processes. For example, a car's carburetor is a single unit operation of the engine, just as the heart is a unit operation of the human body. The concept of a unit operation is based on the idea that general analysis will be the same for all systems because individual operations have common techniques and are based on the same scientific principles. In separations, a unit operation is any process that uses the same separation mechanism. For example, adsorption is a technique in which a solid sorbent material removes specific components, called solutes, from either gas or liquid feed streams because the solute has a higher affinity for the solid sorbent than it does for the fluid. The mathematical characterization of any adsorption column is the same regardless of which solutes are being removed from a given fluid by a given sorbent or the amount of fluid processed in a given time. Hence, the design and analysis of a particular separation method will be the same regardless of the species and quantities to be separated. Once it is understood how to evaluate an adsorption column to separate a binary component feed stream, the same principles can be applied to any binary mixture. An important aspect of design and analysis is the scale (or size) of a process. Those in which separations technology is required span several orders of magnitude in terms of their throughput. For example, industrial separation of radioisotopes occurs at a production rate on the order of 10^{-6} to 10^{-3} kg h^{-1}, while coal-cleaning plants operate at a production rate greater than 10^6 kg h^{-1}. If new design criteria had to be developed for a given separation technique each time that the scale of the process was changed, the analysis would be of very limited value and one would have to write a book for each separation technology to cover all the potential process sizes. The concept of a unit operation, therefore, allows us to apply the same design criteria and analysis for a given separation technology, irrespective of the size. This very important element in evaluation allows one to scale-up or scale-down a process based upon results obtained on a different-size piece of equipment. This is the basis for conducting tests on bench or pilot-scale equipment and using the results for the design of the full-scale process. In addition, bench or pilot-scale test results can be used to determine the effect of a single separation step or other unit operation on an overall process. The configuration and flow patterns of any single step can affect the entire process and are usually determined experimentally.

3.1.6
Separation Mechanisms

Separation processes rely on various mechanisms, implemented via a unit operation, to perform the separation. The mechanism is chosen to exploit some property difference between the components. They fall into two basic categories: the partitioning of the feed stream between phases and the relative motion of various chemical species within a single phase. These two categories are often referred to as equilibrium and mass transfer rate processes, respectively. Separation processes

can often be analyzed with either equilibrium or mass transfer models. However, one of these two mechanisms will be the limiting, or controlling, factor in the separation and is, therefore, the design mechanism.

For a separation to occur, there must be a difference in either a chemical or a physical property between the various components of the feed stream. This difference is the driving force basis for the separation. Some examples of exploitable properties are listed in the Table 3.1-1. Separation processes generally use one of these differences as their primary mechanism.

Table 3.1-1 Exploitable properties for separations.

Equilibrium properties	Rate properties
Vapor pressure	Diffusivity
Partitioning between phases	Ionic mobility
Solubility of a gas in a liquid	Molecular size and shape
Sorption of a solute in a fluid onto a sorbent	
Chemical reaction equilibrium	
Electric charge	
Phase change	
solid/liquid	
liquid/gas	

The following factors are important considerations in the choice of a property difference:

- Prior experience. The reliability and "comfort" factor go up if there has been prior *positive* experience in the use of a certain property difference for certain applications.
- The property itself. How simple will it be to implement? How much prior experience has there been in the use of this property for separations?
- The magnitude of the property difference. Obviously, the larger the difference between the components to be separated, the easier the separation. How large is good enough? This answer is based on experience and the value of the components to be separated. Preliminary calculations for various separation processes (using various property differences) can be very useful.
- Chemical behavior under process conditions. Will the process fluids chemically attack the separation equipment (corrosion, morphological changes such as swelling, etc.) and/or react themselves (polymerize, oxidize, etc.)? This is a very important consideration as it affects the lifetime and reliability of the process.
- Quantities and phases that need to be processed. This is an economic consideration as the cost of implementing various property differences (energy input as heat, for example) is a function of the size of the process. The phases (gas versus liquid, for example) also affect the equipment size and material handling.

• Separation criteria. What are the concentrations of the various components in the feed stream? What purity and recovery are needed? How many components in the feed need to be separated?

Table 3.1-2 contains a list of several common separation processes, their primary separation mechanisms, and the separating agent used. The separating agent concept will be explained in some detail in a later section of this chapter.

Table 3.1-2 Separation mechanisms and agents associated with some separation processes.

Separation Process	Separation Mechanism	Separating Agent
Distillation	Vapor pressure	Heat
Extraction	Partitioning between phases	Immiscible liquid
Adsorption	Partitioning between phases	Solid sorbent
Absorption	Partitioning between phases	Nonvolatile liquid
Filtration	Molecular size and shape	Membrane
Ion exchange	Chemical reaction equilibrium	Solid ion-exchanger
Gas separation	Diffusivity and phase partitioning	Membrane
Electrodialysis	Electric charge and ionic mobility	Charged membrane/ electric field

3.1.7
Equilibrium Processes

In an equilibrium process, two phases (vapor, liquid, or solid) are brought into contact with each other, mixed thoroughly, then separated with a redistribution of the components between phases. Often multiple contacts are made in a series of cascading steps in which the two phases flow counter-current to each other. At each contact the phases are allowed to approach thermodynamic equilibrium. Once equilibrium is reached, there can be no more separation without a change in the operating parameters of the system that affect the equilibrium relationship. The next stage in the cascade, therefore, has at least one process change that alters the equilibrium relationship to establish a new equilibrium relationship. The cascade should be designed such that conditions are altered at each stage to move closer towards the desired separation. For example, distillation is a fractionating separation in which a binary (or greater) feed stream is separated into two (or more) product streams based upon their differences in boiling point. One type of distillation column has a series of cascading contact trays such that the temperature increases from the top tray, which is just above the boiling point of the lower boiling point component, to the bottom tray, which is just below the boiling point of the higher boiling point component. Thus, the lower boiling point component is enriched in the gas phase, while the higher boiling point component is enriched in the liquid phase. Each tray from the top to the bottom of the column, then, operates at a higher temperature such that a new equilibrium is established down the length of the column. As the temperature increases down the column, the lower boiling point species tends to vaporize more and move up the column as a gas stream, while the higher boiling point component continues down the column as a liquid. The final result is a vapor stream leaving the top of the column which is almost pure in the lower boiling point species and a liquid stream exiting the bottom of the column that is almost pure in the higher boiling point component.

For phase partitioning (equilibrium), the variable of interest is the solute concentration in the first phase that would be in equilibrium with the solute concentration in a second phase. For example, in the distillation example above, each component is partitioned between the vapor and liquid phases. The mathematical description of the equilibrium relationship is usually given as the concentration in one phase as a function of the concentration in the second phase as well as other parameters. One example is the Henry's law relation for the mole fraction of a solute in a liquid as a function of the mole fraction of the solute in the gas phase which contacts the liquid.

$$y = mx \tag{1}$$

where y is the mole fraction in the gas phase, x is the mole fraction in the liquid phase, and m is the Henry's law constant.

A second example is the Langmuir isotherm characterizing adsorption, which relates the equilibrium amount of a solute sorbed onto a solid to the concentration in the fluid phase in contact with the solid.

$$X = \frac{K_s bC}{1 + bC} \tag{2}$$

where X is the amount of solute sorbed per weight of sorbate, C is the solute concentration in the fluid phase, and K_s and b are constants.

An important factor in the use of phase partitioning for separations is the degree of change in composition between the two phases. In the limit where the composition in each phase is identical, separation by this mechanism is futile. For vapor–liquid equilibrium, this condition is called an azeotrope. Irrespective of the phases, the condition corresponds to a partition coefficient of unity.

3.1.8
Rate Processes

Rate processes are those in which one component of a feed stream is transferred from the feed phase into a second phase owing to a gradient in a physical property. Gradients in pressure or concentration are the most common. Other gradients include temperature, electric fields, and gravity. The limiting step upon which design is based is the rate of transfer of the particular component from the feed material to the second phase. For relative motion of the various chemical species (rate), the mathematical description relates the rate of transfer of a particular component across a boundary to a driving force. One example is Fick's law, which relates the flux of a component, N_A, across a layer (fluid or solid) to the concentration gradient within the layer:

$$N_A = -D_A \frac{dC_A}{dx} \tag{3}$$

where D_A is the diffusion coefficient of A in the medium (a physical property found in many handbooks) and

$\frac{dC_A}{dx}$ is the concentration gradient of A in the direction of interest.

A second example is the use of a mass-transfer coefficient to relate the flux across a fluid boundary layer (fluid region over which the solute concentration changes from the bulk phase value) to the concentration difference across the layer:

$$N_A = k(C_{A1} - C_{A2}) \tag{4}$$

where N_A is the mass flux of A across the fluid layer, k is the mass-transfer coefficient, and C_A is the concentration of component A.

In choosing between these two models, one needs to consider the specific process. The use of mass-transfer coefficients represents a lumped, more global view of the many process parameters that contribute to the rate of transfer of a species from one phase to another, while diffusion coefficients are part of a more detailed model. The first gives a macroscopic view, while the latter gives a more microscopic view of a specific part of a process. For this reason, the second flux equation is a more engineering representation of a system. In addition, most separation processes involve complicated flow patterns, limiting the use of Fick's Law.

3.1.9
Countercurrent Operation

The analysis of equilibrium stage operations is normally performed on the basis of countercurrent flow between two phases. Because most separation processes, whether described in terms of equilibrium or mass-transfer rates, operate in this flow scheme, it is useful to compare countercurrent to cocurrent flow. Figure 3.1-2 illustrates cocurrent and countercurrent operation. Assuming mass transfer across a barrier between the two fluid phases, generic concentration profiles can be drawn for each case (Fig. 3.1-3).

A few points become obvious. First, in each case, the concentration difference across the barrier changes with axial position x. So, the flux (or rate) will change with position. Second, for cocurrent flow, the concentration difference (the driving force for mass transfer) becomes very small as the flow moves axially away from the entrance ($x = 0$). So, the separation becomes less efficient as the barrier beco-

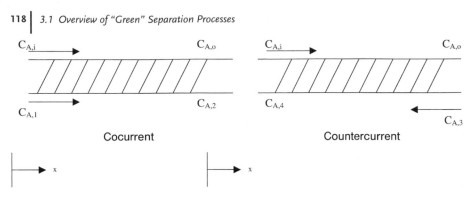

Fig. 3.1-2   Cocurrent versus countercurrent operation.

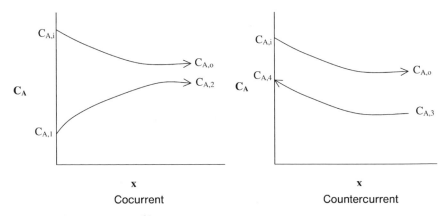

Fig. 3.1-3   Concentration profiles.

mes longer in the axial direction. For countercurrent operation, the driving force is maintained at a larger value along the barrier and $C_{A,4}$ can be larger than $C_{A,o}$. Therefore, countercurrent operation is usually the preferred method.

To account for the variation in driving force, a log-mean driving force is used instead of a linear one:

$$\Delta C_{lm} = \frac{\Delta C_1 - \Delta C_2}{ln\left(\dfrac{\Delta C_1}{\Delta C_2}\right)} \qquad (5)$$

For countercurrent flow:
$\Delta C_1 = C_{A,i} - C_{A,4f}$ and $\Delta C_2 = C_{A,o} - C_{A,3}$
For cocurrent flow:
$\Delta C_1 = C_{A,i} - C_{A,1}$ and $\Delta C_2 = C_{A,o} - C_{A,2}$
ΔC_1 and ΔC_2 are the concentration differences on each end of the barrier.

3.1.10
Productivity and Selectivity

In the evaluation of a separation process, there are two primary considerations: productivity and selectivity. The productivity, or throughput, of a process is the measure of the amount of material which can be treated by this process in a given amount of time. This quantity is usually specified by the feed flowrate to the process and/or the amount of a product stream. The selectivity of the process is the measure of the effectiveness of the process to separate the feed mixture. Selectivity is usually given by a separation factor (α_{ij}), which is a ratio of compositions in the product streams for an equilibrium process or rates of mass transfer for a rate process:

$$\alpha_{ij} = \frac{x_{i1}/x_{j1}}{x_{i2}/x_{j2}} \tag{6}$$

where x_{i1} is the fraction of component i in stream 1, x_{j1} is the fraction of component j in stream 1, x_{i2} is the fraction of component i in stream 2, and x_{j2} is the fraction of component j in stream 2.

Various terms will be used to represent these quantities, depending on the process. A target for the composition of one or more product streams usually dictates the separation requirement for a particular process.

There are two types of separation factor commonly used; the ideal and the actual separation factor. The ideal separation factor is based on the equilibrium concentrations or transport rates due to the fundamental physical and/or chemical phenomena that dictate the separation. This is the separation factor that would be obtained without regard to the effects of the configuration, flow characteristics, or efficiency of the separation device. This value can be calculated from basic thermodynamic or transport data, if available, or obtained from small-scale laboratory experiments. For an equilibrium-based separation, the ideal separation factor would be calculated based on composition values for complete equilibrium between phases. For a rate-based separation, this factor is calculated as the ratio of transport coefficients, such as diffusion coefficients, without accounting for competing or interactive effects. Each component is assumed to move independently through the separation device.

3.1.10.1
Equilibrium Processes

The separation factor, α_{ij}, is the ratio of the concentration of components i and j (mole fractions, for example) in product stream 1 divided by the ratio in product stream 2. For example, in distillation α_{ij} is defined in terms of vapor and liquid mole fractions (Fig. 3.1-4).

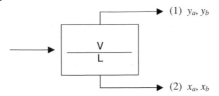

Fig. 3.1-4 Distillation product streams.

$$\alpha_{ij} = \frac{x_{i1}/x_{j1}}{x_{i2}/x_{j2}} = \frac{y_a/y_b}{x_a/x_b} \tag{7}$$

where y_a, y_b are the mole fractions of components a and b in the overhead stream and x_a, x_b are the mole fractions of a and b in the bottoms stream.

[Note: not the feed conditions]

Example: $x_a = x_b = 0.5$, $y_a = 0.6$, $y_b = 0.4$

$$\alpha_{ab} = \frac{0.6/0.4}{0.5/0.5} = \frac{1.5}{1.0} = 1.5$$

Ideal: The calculation is based simply on the vapor pressure of each component.

Actual: The calculation would take into account the non-idealities in solution (fugacity versus pressure, for example).

3.1.10.2
Rate Processes

Component flowrates can be used for rate processes, as shown in Fig. 3.1-5.

Mass Separating Agent

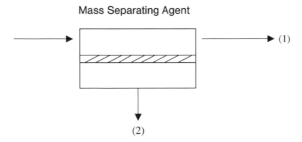

(1)

(2)

Fig. 3.1-5 Membrane separation product streams.

3.1.10.3
Membrane Separation

In computing the separation factor one must use appropriate physical parameters, such as operating conditions and equipment size (membrane area in this case), to relate the flux to a driving force. The compositions of streams 1 and 2 may be used; however, it is better to use the ratio of permeabilities, transport coefficients, or other measures of the inherent separating ability of the device. One can think of α as a flux ratio scaled for a unit driving force.

$$\alpha_{ab} = \frac{(flux / driving\ force)_a}{(flux / driving\ force)_b} = \frac{Q_a}{Q_b} \tag{8}$$

Ideal: The calculation is based on single component measurements. Normally, it does not account for the configuration or flow characteristics of the separation device.

Actual: The calculation includes any competitive effects, interactive effects, and effects arising from the device itself.

The separation factor is usually given as a value of one or greater. The selectivity of the separation is improved as the value of this ratio is increased.

When one determines the separation for an actual feed mixture in a separation device, an actual separation factor is obtained. This value is usually obtained by measurements on the device and is usually less than the ideal value ($\alpha_{actual} < \alpha_{ideal}$).

3.1.11
Separating Agents

Many separation processes are based on the formation of an additional phase that has a different composition from the feed stream(s). One possible way of forming another phase is the addition of energy (energy separating agent) to convert a liquid stream to a vapor stream. Distillation exploits this idea to separate mixtures of liquids that boil at different temperatures. Crystallization processes use energy to separate liquid mixtures with components that solidify at different temperatures. The temperature is lowered until the species with the higher solidification temperature crystallizes out of solution. Evaporation and drying are other processes in which energy addition promotes the separation by formation of a new phase.

Another large class of separations makes use of a change in solute distribution between two phases in the presence of mass not originally present in the feed stream. This mass separating agent, MSA, is added as another process input to cause a change in solute distribution. The MSA can alter the original phase equilibrium or facilitate the formation of a second phase with a concentration of components different from that in the original phase. One of the components of the original feed solution must have higher affinity for the MSA than for the original solution. This solute will then preferentially transfer from the original feed solution to the MSA phase. Once the MSA has been used to facilitate a separation, it must normally be removed from the products and recovered for recycle in the process. Hence, use of an MSA requires two separation steps; one to remove a solute from a feed stream and a second to recover the solute from the MSA.

MSAs can consist of solids, liquids, or gases. Figure 3.1-6 shows various combinations of feed phase and MSA phase with examples of various separation processes involved. In almost all cases, the use of an MSA involves two steps. The initial step is the use of the MSA in the separation process and the second step is the regeneration of the MSA.

	Solid	Liquid	Gas
Feed Solid		Leaching	Steam Stripping
Liquid	Ion Exchange	Solvent Extraction	Stripping
Gas	Adsorption	Absorption	

Fig. 3.1-6 Examples of separation processes using mass separating agents.

A separation involving an energy separating agent (ESA) can involve input and removal steps as in distillation, for example, where there is a reboiler for energy input and a condenser for energy removal. In other cases, such as evaporation, the vapor can be discharged without the need for a heat removal step. An energy separating agent has the advantage that no additional material is introduced into the system. While heat is the most common energy separating agent used, examples of separation processes using gravitational, electric and magnetic fields have been reported.

Some general statements regarding the use of separating agents can now be made:

- **Separation processes use mass and/or energy separating agents to perform the separation.** A mass separating agent can be a solid, a liquid, or a gas (see Fig. 3.1-6). Heat is the most common energy separating agent, used in distillation. External fields, such as magnetic and electric, are sometimes used as energy separating agents.
- **A different component distribution between two phases is obtained.** This distribution change can be accomplished in two ways:

   1. The original phase equilibrium is altered. Two phases are originally present and the role of the separating agent is to change the composition in each phase relative to the initial values.

Examples:
mass separating agent: one approach is the addition of a selective complexing agent to a liquid phase to increase the solubility of a solute. For example, a selective chelating agent can be added to an organic phase to form a complex with a metal ion in an aqueous phase when the two phases are in contact.
energy separating agent: heat input to change the temperature of a gas/liquid system. The most common example is distillation.

2. A second phase is generated, with a different component distribution.
Examples:
mass separating agent: addition of a solid sorbent that is selective for a solute in a feed stream. An example is the removal of VOCs from an air stream using activated carbon.
energy separating agent: heat input to a liquid phase to change the solute solubility in that phase. An example is evaporation to remove water from a waste stream and concentrate the stream prior to disposal.

- **Separating agents employ four methods to generate selectivity**:

1. Modification of phase equilibrium
Examples:
mass separating agent: an ion-exchange resin used to selectively partition ions into the resin.
energy separating agent: heat removal to precipitate salts from an aqueous stream.

2. Geometry differences
Examples:
mass separating agent: a membrane to filter suspended solids based on size.
energy separating agent: gravity settling to separate particles by size.

3. Kinetics (rate of exchange) between phases
Examples:
mass separating agent: the use of amines in gas sorption to change the rate of acid gas (CO_2, H_2S) uptake into the liquid phase.
energy separating agent: heat input into the amine solution to accelerate the rate of gas desorption, which regenerates the solution.

4. Rate of mass transfer within a phase.
Examples:
mass separating agent: intra-particle diffusion in a porous sorbent
energy separating agent: application of an electric field across a liquid phase to accelerate charged particles relative to uncharged ones.

Mass separating agents are generally characterized by their capacity to incorporate the desired solute (sometimes called loading) and their ability to discriminate between solutes (selectivity). Energy separating agents are usually described by the amount required to achieve a certain throughput (productivity) and selectivity for a given process. These values relate directly to the equipment size needed for a given separation.

3.1.12
Selection of a Separation Process

This section is included to give the reader some "food for thought" in deciding how they might use the material in the subsequent chapters. For additional perspectives on this topic, consult Refs [5–14]. The following should be taken as a heuristic or guide.

1. Assess the feasibility. What are the property differences that you plan to exploit for the separation(s)? Which processes use this property difference as their primary separating mechanism? What operating conditions are associated with the feed stream (flowrate, temperature, pressure, pH, reactive components, etc.)? Are these conditions "extreme" relative to normal operating conditions for a given separation process?

2. Determine the target separation criteria. What purity and recovery are needed for the various components in the feed stream? For a feasible separation process, what is the separation factor based on the property difference chosen? For an equilibrium-based process, this would be the separation factor for one stage.

There are various molecular properties which are important in determining the value of the separation factor for various separation processes:

– *Molecular weight* Usually, the higher the molecular weight of a compound is, the lower the vapor pressure. Molecular weight is also related to molecular size, which affects diffusion rates and access to the interior of porous materials.
– *Molecular volume* This is a measure of density since there is an inverse relationship between density and volume. As will be seen in the analysis of various separation processes, density can be a significant variable.
– *Molecular shape* The molecular shape can certainly affect the access of certain molecules to pores and chemical binding sites. The shape also will affect how the molecules order in a liquid or solid phase.
– *Intermolecular forces* The strength of these forces can affect the vapor pressure and solubility in certain solvents. One property is the dipole moment, which is a measure of the permanent charge separation within a molecule (polarity). Another property is the polarizability which is a measure of a second molecule's ability to induce a dipole in the molecule of interest. The dielectric constant is a physical property that can be used as a measure of both the dipole moment and the polarizability.
– *Electrical charge* The ability of a molecule to move in response to an electric field is a function of the electrical charge.
– *Chemical complexation* Separations involving selective chemical reactions can impart higher selectivity (separation factors) than the use of a physical property difference alone.

The relative ability of the various compounds in the feed stream to react will directly affect the separation selectivity.

3. Do the easiest separation first. This may seem obvious, but often the starting place is non-obvious when one is faced with a complex feed mixture to separate. What are some examples of "easy?"
   - The separation technique itself is simple. If particles are present in a fluid phase, filtration can do the separation of the two phases. Separation of components in each phase can then be done in a subsequent step, if needed.
   - The technique has a high reliability. Consider separation steps that have been used for long periods of time with positive results (high "comfort" factor) ahead of newer approaches. Newer approaches are easier to implement further downstream, when their impact on the entire process is less than if they were the first step.
   - Remove the component with the highest mole fraction first. If the separation factors for each component in the feed mixture are approximately equal, the component with the highest mole fraction is usually the easiest to remove first. An alternative is to remove the component with the highest volatility first if distillation is feasible.
   - When more than one step is required for the separation sequence:
     - (i) Recover the mass separating agent and/or dissolved products immediately after the process step involving a mass separating agent.
     - (ii) Do not use a mass separating agent to remove or recover a mass separating agent.

3.1.13
A Unified View of Separations

Having discussed the concept of unit operations in the context of separation, it is useful at this point to reiterate what is gained by this approach. King [15] lists several major gains in understanding, insight, capability, and efficiency that come from a unified view of separations. The first such gain recognized historically is that methods of analyzing the degrees of separation achieved in different separation processes are similar. Hence, a comprehensive knowledge of separations enables one to transfer the uses of separations to different scales of operation, ranging from analytical separations to large-process separations. In addition, the interactions of mass transfer and phase equilibria and their resultant effects are similar to related types of contacting equipment. The concepts used for stage efficiencies (fraction of equilibrium attained) are also common.

Developments of powerful computational algorithms have provided immense gains in computing capacity and the widespread use of personal computers has meant that much less attention has to be given to methods of calculating degrees of separation. This allows for much more emphasis on process selection, synthesis, and improvement. However, while many simulators exist to model separation processes, it is critical that the designer understand the fundamental principles prior to using them. A unified view makes it possible to identify and select among candidate separation processes for a given task on a basis of knowledge. An understanding of patterns of stage-to-stage changes in composition in countercurrent separations and the causes of particular patterns is useful across the board for improvement of design and operating conditions to gain a greater degree of separation and/or lesser flows and equipment cost. In addition, an understanding of the factors governing energy consumption allows greater insight into reducing consumption and/or achieving an optimal combination of equipment and operating costs.

A general understanding of separations facilitates the development of entirely new methods of separation. Insight into the capabilities of a variety of methods helps us to identify when the ability to separate will pose a major process limit. An understanding of solution and complexation chemistry makes it possible to identify and select among potential mass separating agents for different applications and to transfer the use of particular agents and chemical functionalities among different types of separation processes.

Acknowledgement

Portions extracted from *Principles of Chemical Separations with Environmental Applications*, Richard D. Noble and Patricia A. Terry, Cambridge University Press, Cambridge, 2004.

References

1 National Research Council, *Separation & Purification: Critical Needs and Opportunities*, National Academy Press, **1987**.

2 Chemical Manufacturers Association **1993**, *Designing Pollution Prevention into the Process*, © CMA XXXX.

3 K. L. Mulholland and J. A. Dyer **2000**, Reduce Waste and Make Money, *Chem. Eng. Prog.* January.

4 E. B. Cowling **1982**, Acid Rain in Historical Perspective, *Environ. Sci. Technol.* 16 (2).

5 G. Keller **1987**, *Separations: New Directions for an Old Field*, AIChE Monograph Series, Vol. 83, no. 17, AIChE, New York.

6 H.R. Null **1986**, in *Handbook of Separation Processes*, ed. R.W. Rousseau, Wiley-Interscience, New York, Ch. 22.

7 P.C. Wankat **1988**, *Equilibrium Staged Separations*, Prentice-Hall.

8 P.C. Wankat **1990**, *Rate-Controlled Separations*, Elsevier.

9 D.H. Belhatecke **1995**, Choose appropriate wastewater treatment technologies, *Chem. Eng. Prog.* August, 32–51.

10 C.A. Zinkus, W.D. Byers, W.W. Doerr **1998**, Identify appropriate water reclamation technologies, *Chem. Eng. Prog.* May, 19–31.

11 B. Fitch **1974**, Choosing a separation technique, *Chem. Eng. Prog.*, December, 33–37.

12 C.J. King **1986**, in *Handbook of Separation Processes*, ed. R.W. Rousseau, Wiley-Interscience, New York, Ch. 15.

13 M. Hyman, L. Bagaasen **1997**, Select a Site Cleanup Technology, *Chem. Eng. Prog.* August, 22–43.

14 C.J. King **1971**, *Separation Processes*, McGraw-Hill.

15 C.J. King **2000**, From unit operations to separation processes, *Sep. Pur. Methods* 29 (2), 233–245.

3.2
Distillation

Sven Steinigeweg and Jürgen Gmehling

3.2.1
Introduction

A typical chemical process can be roughly divided in preparation, reaction, and separation steps. Although the reaction is the heart of the chemical process, the separation step often accounts for about 60–80% of the total costs.

The separation step most commonly used in industrial practice is distillation. Although distillation and evaporation are rather mature technologies, which have been used by humankind for about 2000 years, recent developments have contributed to enhanced distillation processes with regard to consumption of resources.

The principle of distillation is the use of differences in volatilities of the components to be separated. Distillation processes are usually carried out in countercurrent mode in multistage units. The differences that can be obtained in concentrations of the components in the vapor and liquid phases are determined by the vapor–liquid equilibrium (VLE). Until the 1970s reliable data for vapor–liquid equilibria could only be obtained by measurement, which, for a mixture containing more than two components, required a large number of time-consuming measurements. Advances in chemical thermodynamics have resulted in methods activity coefficient models (g^E models or equations of state) for the calculation of the phase-equilibrium behavior of multicomponent mixtures on the basis of binary subsystems. In the case that no information about the binary subsystems is available, predictive methods (group contribution methods) are available to allow estimation of the required phase equilibria.

According to the second law of thermodynamics, separating a mixture is a less favored process than mixing. The main costs of distillation units are usually energy costs. Hence, the most important focus for designing sustainable distillation processes is the minimization of energy consumption. The most important advantage of distillation in comparison with several other thermal separation technologies (e.g. absorption, adsorption, extraction) is the use of energy as the agent for separation, since energy can easily be added or be removed from the system. Other techniques, such as extraction, use a mass separating agent. This requires

Green Separation Processes. Edited by C. A. M. Afonso and J. G. Crespo
Copyright © 2005 WILEY-VCH Verlag GmbH & Co. KGaA, Weinheim
ISBN 3-527-30985-3

the addition of a further component (e.g. a solvent), which has to be regenerated. A further advantage of distillation is that the density of the involved fluid phases are very different. Nowadays, synthesis and design of separation processes are usually carried out not only according to economic factors but also taking environmental considerations into account (Kheawhom and Hirao, 2004).

3.2.2
Phase Equilibria

According to the twelve principles of Green Chemistry introduced by Anastas et al. (1998) minimizing waste and optimizing energy requirements are essential for a sustainable development of chemical processes. This leads to a demand for a distinct knowledge and reliable calculation of the VLE. The principle of distillation and evaporation is that the most volatile component is enriched in the vapor phase. Reliable calculation of the VLE behavior is therefore a crucial step for the design and optimization of distillation processes. Calculating the phase equilibrium means to answer the question: What are the concentrations and the pressure in the vapor phase at given liquid phase concentrations and temperature under the condition that both phases are in equilibrium? The number of phases is not limited to two. In cases of hetero-azeotropic mixtures there will be two liquid phases besides the vapor phase.

3.2.2.1
Calculation of Vapor-liquid Equilibria

Calculation of phase equilibria is based on thermodynamics of mixtures (Gmehling and Kolbe, 1992; Gmehling and Brehm, 1996; Gmehling et al., 2004). According to Gibbs, two or more phases are in equilibrium when the chemical potentials are the same in all phases. Lewis proposed to use fugacities instead of chemical potentials. Fugacities correspond to partial pressures for non-associating mixtures at low pressures. Phase equilibrium exists when the fugacities of the different components are the same in the different phases. This means that for VLE the following conditions have to be fulfilled:

$$f_i^V = f_i^L \tag{1}$$

$$T^V = T^L \tag{2}$$

$$P^V = P^L \tag{3}$$

For the calculation of VLE it is necessary to relate the fugacity to measurable parameters, such as concentration, temperature and pressure. Therefore auxiliary quantities are introduced: the fugacity coefficient φ_i and the activity coefficient γ_i which are defined for liquid and vapor phase as follows:

$$\varphi_i^L \equiv \frac{f_i^L}{x_i P} \tag{4}$$

$$\varphi_i^V \equiv \frac{f_i^V}{\gamma_i P} = \frac{f_i^V}{p_i} \tag{5}$$

The activity coefficient is defined as:

$$\gamma_i \equiv \frac{f_i}{x_i f_i^0} \tag{6}$$

f_i^0 is the standard fugacity which can be chosen arbitrarily. Combining the definitions of the activity coefficient and the fugacity coefficient with the phase-equilibrium condition leads to two routes for the calculation of VLE:

Route A $\quad x_i \varphi_i^L = \gamma_i \varphi_i^V \tag{7}$

Route B $\quad x_i \gamma_i f_i^0 = \gamma_i \varphi_i^V P \tag{8}$

Experimental data necessary to describe this behavior are available in large computerized data bases (e.g. Dortmund Data Bank, DDB). A small part of the data is also published in data collections (Gmehling et al., 1977; Sørensen et al., 1979; Gmehling et al., 1986; Gmehling et al., 1988; Gmehling et al., 2004;). Both routes allow the calculation of VLE (see Chapter 3.2.2.1, Sections 3.2.2.1.1 and 3.2.2.1.2) for multicomponent systems when the behavior of the binary subsystems is known.

Route A requires an equation of state and sophisticated mixing rules for calculating the fugacity coefficient for both the vapor and the liquid phase. The advantage of using equations of state is that other information (e.g. molar heat capacities, densities, enthalpies, heats of vaporization), which is necessary for designing and optimizing a sustainable distillation process, is also obtained at the same time.

Besides the standard fugacity, Route B needs a model for calculating the activity coefficient. The fugacity of the pure liquid at system pressure and system temperature is usually chosen as the standard fugacity. Therefore, standard fugacity is defined as

$$f_i^0 = f_i^s = \varphi_i^s \cdot P_i^s \tag{9}$$

This requires a knowledge of the saturation vapor pressure, which is usually calculated from the Antoine equation with the Antoine constants A, B and C and the absolute temperature T:

$$\log P^s = A - \frac{B}{T + C} \tag{10}$$

Antoine constants (A, B, C) for a few thousand compounds can be found in comprehensive databases (e.g. the DDB). In order to account for the pressure dependence of the standard fugacity and the resulting deviations between system pressure and vapor pressure the Poynting correction Poy_i is introduced:

$$f_i^0(P) = \varphi_i^s P_i^s \exp \frac{v_i^L(P - P_i^s)}{RT} = \varphi_i^s P_i^s \text{Poy}_i \tag{11}$$

Hence the VLE is given by:

$$x_i \gamma_i \varphi_i^s P_i^s \text{Poy}_i = y_i \varphi_i^V P \tag{12}$$

Under the condition that system pressure and saturation pressure are of the same order of magnitude the Poynting factor Poy_i is close to unity and can be neglected. Furthermore, for compounds that do not associate strongly the fugacity coefficient in the vapor phase is nearly identical with the saturation fugacity coefficient.

This leads to the simplified equation for VLE calculations:

$$x_i \gamma_i P_i^s = y_i P \tag{13}$$

Neglecting the activity coefficient and hence assuming ideal behavior of the mixture leads to Raoult's law.

According to Eqs. (7), (8) and (13), partition coefficients K_i and separation factors α_{ij} can be calculated from the following equations:

Route A $\qquad K_i \equiv \dfrac{y_i}{x_i} = \dfrac{\varphi_i^L}{\varphi_i^V} \qquad \alpha_{ij} \equiv \dfrac{K_i}{K_j} = \dfrac{y_i/x_i}{y_j/x_j} = \dfrac{\varphi_i^L \varphi_j^V}{\varphi_i^V \varphi_j^L} \tag{14}$

Route B $\qquad K_i = \dfrac{\gamma_i P_i^s}{P} \qquad \alpha_{ij} \equiv \dfrac{K_i}{K_j} = \dfrac{y_i/x_i}{y_j/x_j} = \dfrac{\gamma_i P_i^s}{\gamma_j P_j^s} \tag{15}$

3.2.2.1.1 Using Activity-Coefficient Models

In addition to calculation of the saturation vapor pressure, a model for calculating the activity coefficient is required. The activity coefficient depends on concentration, and also on pressure and temperature. These dependences can be related to partial molar excess enthalpies and partial molar excess volumes:

$$\left(\frac{\partial \ln \gamma_i}{\partial (1/T)} \right)_{P,x} = \frac{h_i^E}{R} \tag{16}$$

$$\left(\frac{\partial \ln \gamma_i}{\partial P} \right)_{T,x} = \frac{v_i^E}{RT} \tag{17}$$

For distillation processes the pressure dependence can usually be neglected whereas the temperature dependence has to be taken into account to develop and optimize sustainable and resource-saving processes.

It is nearly impossible to find experimental VLE data for multicomponent systems. Most published data relates to binary systems. Therefore, it is most important to calculate the activity coefficient for multicomponent systems from knowledge of the binary subsystems. The concept of local composition introduced by Wilson (Wilson, 1964) leads to a number of models, which allow the calculation of activity coefficients for multicomponent systems using only binary parameters. The most commonly used methods are the Wilson (Wilson, 1964), NRTL (Renon and Prausnitz, 1968), or UNIQUAC (Abrams and Prausnitz, 1975) models. All these models only need two binary interaction parameters, which are obtained by fitting them to experimental VLE data. The NRTL equation uses a non-randomness parameter as a further adjustable parameter. In addition, Wilson and UNIQUAC equations require pure component data (respectively molar volumes and relative van der Waals properties). In contrast to the Wilson model, UNIQUAC and NRTL methods are also applicable to liquid–liquid equilibria. Details about the equations can be found in literature (Gmehling and Brehm, 1996; Gmehling and Kolbe, 1992). If no experimental data are available, the required activity coefficients can be predicted by so-called group contribution methods. The great advantage of these methods is that the number of different functional groups is much smaller than the number of different molecules. In group contribution methods a molecule is divided into several functional groups as can be seen in Fig. 3.2-1 (e.g. ethanol can be subdivided into a CH_3 group, an OH group and a CH_2 group).

Group interaction parameters are determined by fitting them to a large number of experimental phase equilibrium data. The required activity coefficients can then

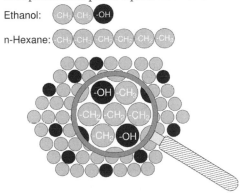

Fig. 3.2-1 Group contribution concept.

be predicted by dividing the molecules into their functional groups and calculating the required activity coefficients via group interaction parameters. Examples for group contribution methods, which are applied widely are ASOG (Kojima and Tochigi, 1979) or UNIFAC (Fredenslund et al., 1977). However, for highly diluted mixtures, asymmetric systems and heats of mixing UNIFAC did not show satisfactory results. This is not surprising since the group interaction parameters are obtained by fitting experimental VLE data in a concentration range between 5 and 95%. To overcome these limitations, temperature-dependent group interaction parameters and a modified combinatorial part was introduced leading to modified UNIFAC (Weidlich and Gmehling, 1987; Gmehling et al., 2002). Parameters for modified

UNIFAC are obtained by fitting them simultaneously to all consistent data stored in the DDB database (VLE, h^E, γ^∞, ...).

3.2.2.1.2 Using Equations of State (EOS) for VLE Calculations

As mentioned above, the use of equations of state for VLE calculations has distinct advantages compared to the alternative route. However, an equation of state has to be used that is able to describe the *PVT* behavior for both the liquid and the vapor phase.

Most commonly used are further developments of the cubic van der Waals EOS. This EOS for the first time allowed the description of different phenomena, such as condensation, vaporization, occurrence of the two-phase region and critical phenomena, using only two parameters a and b which take into account the interaction forces between the molecules and the volume of the molecules. The introduction of a further parameter, the acentric factor ω, which can be derived from vapor pressure data, leads to a more reliable description of the saturation vapor pressures.

In practice, the most commonly used EOSs are the Soave–Redlich–Kwong equation and the Peng–Robinson equation. These equations were developed for pure components only. Applying these models to multicomponent systems requires mixing rules for the calculation of the parameters a and b in the mixture. These parameters have to be calculated from the pure component parameters a_{ii} and b_i.

Simple empirical mixing and combination rules often used are given in Eqs. (18)–(20):

$$b = \sum z_i b_i \tag{18}$$

$$a = \sum_i \sum_j z_i z_j a_{ij} \tag{19}$$

$$\text{with } a_{ij} = \sqrt{a_{ii} a_{jj}} \left(1 - k_{ij}\right) \tag{20}$$

The required parameters k_{ij} are fitted to experimental binary VLE data. However, problems with this empirical mixing rules arise for highly polar or associating mixtures. g^E mixing rules as introduced by Huron and Vidal (1979) lead to an improved description of these systems. These types of mixing rules include g^E, which can be calculated from a g^E model like UNIQUAC.

The group contribution concept was combined with cubic EOSs leading to a group contribution equation of state which is capable of predicting the VLE behavior of systems with sub- and supercritical compounds. The PSRK model (predictive Soave–Redlich–Kwong) developed by Holderbaum and Gmehling (1991) uses a g^E mixing rule with original UNIFAC for calculating the required g^E-values. Although PSRK provides good results, some weaknesses caused by the Soave–Redlich–Kwong equation and original UNIFAC are left. The introduction of a volume-translated group contribution equation of state based on the Peng–Robinson model (VTPR) in combination with g^E mixing rules based on modified UNIFAC leads to a highly sophisticated model which has been successfully applied to a wide variety of systems including polymers and electrolytes (Ahlers and Gmehling, 2001). Figure 3.2-2 shows a comparison of experimental and predicted data for alkane–ketone systems for different phase equilibrium data and excess properties (VLE, SLE, h^E, azeotropic data, γ^∞) using VTPR and modified UNIFAC.

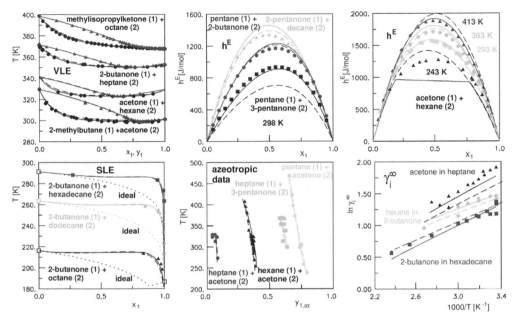

Fig. 3.2-2 Typical results for modified UNIFAC (- - - -) and the VTPR- model (——) for different alkane–ketone systems.

3.2.2.1.3 Azeotropy

Owing to the non-ideality of binary or multicomponent mixtures, the liquid phase composition is often identical with the vapor phase composition. This point is called an azeotrope and the corresponding composition is called the azeotropic composition. An azeotrope can not be circumvented by conventional distillation since no enrichment of the low-boiling component can be achieved in the vapor phase. Separating azeotropic mixtures therefore requires special processes, e.g. azeotropic or extractive distillation or pressure swing distillation. Azeotropic information is available in literature (Gmehling et al., 2004).

For a better understanding of the separation of multicomponent mixtures, the application of residue curves is helpful. Residue curves were introduced by Schreinemakers in 1901 (Schreinemakers, 1901a, 1901b) and are applied to distillation processes by Doherty and coworkers (Doherty and Malone, 2001). Residue curves describe the change of composition in the reboiler over time for open vaporization.

They can be calculated by integration of the differential equations taking into account the vapor–liquid equilibrium behavior. Details can be found in Doherty and Malone (2001). Residue curves can be related to the composition profile along a distillation column working at infinite reflux and can therefore be used to estimate if the desired purities can be achieved. Residue curve maps can also be used to identify different bottom and top products depending on the feed composition, and hence are valuable tools for designing separation processes. Figure 3.2-3 shows the residue curve maps for two ternary systems. Whereas the system benzene–N-methyl pyrrolidone (NMP)–cyclohexane shows only one binary and no ternary azeotrope the system chloroform–acetone–methanol shows strongly non-ideal behavior with four azeotropes, three binary and one ternary, and separation boundaries that cannot be circumvented by ordinary distillation. The direction of the residue curves, indicated by the arrows, points to the heavy boiler (the bottom product of a distillation column).

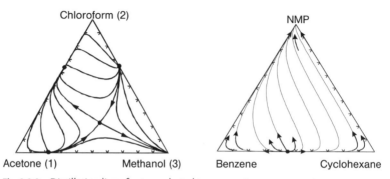

Fig. 3.2-3 Distillation lines for two selected ternary systems.

3.2.2.2
Calculation of Distillation Processes

Developing distillation processes that require a minimum in energy, provide a maximum in safety and lead to products with high purity, which is mandatory for a green process, requires detailed calculations of the distillation columns.

Industrial-scale distillation processes are carried out in distillation columns. The main parts of the columns are "internals", contacting devices that ensure an intense contact between liquid and vapor phase. In practice two different contacting devices are used: trays and packings. Usually one separation stage as realized in an evaporation process is not enough to reach the desired purity. Distillation columns combine several separation stages. On each theoretical stage of a distillation column VLE is reached. The vapor phase will be condensed on the stage above. The best separation will be realized when liquid and vapor phase flow countercurrently inside the column. Therefore a part of the distillate has to be recycled into the column (reflux) to ensure a liquid flow inside the column. The reflux ratio is defined as the ratio of the amount of liquid recycled and the amount of distillate:

$$v = \frac{\dot{L}_R}{\dot{D}} \tag{21}$$

The VLE (K-factor as the. separation factor) has the most important influence on the number of separation stages required.

Calculating the profiles in a distillation column means to solve the so-called MESH equations. These equations combine the material balance, equilibrium condition, summation condition and heat balance which have to be solved numerically:

$$M_{i,j} = \dot{L}_{j+1}x_{i,j+1} + \dot{V}_{j-1}y_{i,j-1} + \dot{F}_j z_{i,j} - (\dot{L}_j + \dot{S}_j^L)x_{i,j} - (\dot{V} - \dot{S}_j^V)y_{i,j} \tag{22}$$

$$E_{i,j} = y_{i,j} - K_{i,j}x_{i,j} \tag{23}$$

$$S_{y,j} = \sum y_{i,j} - 1 \tag{24}$$

$$S_{x,j} = \sum x_{i,j} - 1 \tag{25}$$

$$H_j = \dot{L}_{j+1}h_{j+1}^L + \dot{V}_{j-1}h_{j-1}^V + \dot{F}_j h_{F,j} - (\dot{L}_j + \dot{S}_j^L)h_j^L - (\dot{V}_j + \dot{S}_j^V)h_j^V + \dot{Q}_j \tag{26}$$

In the case of reactive distillation, the MESH equations also have to account for chemical reaction (heat of reaction, change of the mole numbers by chemical reaction).

By solving the MESH equations the required quantities (compositions in the liquid and vapor phases, temperatures, amount of liquid and vapor flow) for every theoretical stage can be calculated.

Owing to the availability of high-speed computers, short cut methods for designing distillation processes (e.g. McCabe–Thiele and Ponchon–Savarit for binary systems or the equations of Fenske, Underwood and Gilliland for multicomponent mixtures, see Gmehling and Brehm, 1996 and Sattler, 2001 for details) are no longer required.

Modern process simulators (e.g. Aspen-Plus from AspenTech or ChemCad from Chemstations) simultaneously solve the MESH equations using algorithms based on Newton–Raphson methods (Gmehling and Brehm, 1996). However, for highly non-ideal or complex systems, modifications have been developed to enhance convergence behavior.

Nowadays, modern computers enable the process engineer to design and optimize separation processes with rigorous models. Often, the main bottleneck remaining is the availability of parameters for calculating the VLE behavior reliably. The importance of reliable parameters cannot be overemphasized. According to a recent analysis performed by Kister (2002) one of the main reasons for errors between simulations during process development and the real behavior of the distillation column is the usage of wrong or inaccurate models and parameters for the description of the VLE behavior.

The main assumption of the concept is that phase equilibrium will be reached on each separation stage. However, in distillation columns the residence time of the components is not sufficient to reach an enrichment as predicted by the VLE calculation. Kinetic aspects have to be taken into account. This can be achieved by introducing empirical factors like the Murphree efficiency (Sattler, 2001). Since the Murphree efficiency may depend on the operating conditions, a more sophisticated model should be used if transport effects cannot be neglected. A rate-based approach allows for differences in transport coefficients for the different components. This requires detailed knowledge about the diffusivities, which can only be estimated roughly for multicomponent mixtures (Taylor and Krishna, 1993). The resulting model for calculating the column is much more complex. A rate-based approach is often used when packings are applied as contacting devices. Packings are continuous contacting devices and separation takes place along the packing height. Nevertheless, it is possible to model these columns assuming a number of theoretical stages. One meter of packing height is equivalent to a certain number of theoretical stages (NTSM value: number of theoretical stages per meter). The NTSM value depends on the operating conditions and on the components to be separated. The packing vendors usually supply this information, which is necessary for column design.

3.2.3
Distillation Processes

3.2.3.1
Separating Azeotropic Mixtures

The most widely used technologies for separating azeotropic mixtures are extractive distillation, azeotropic distillation and pressure swing distillation.

Pressure swing distillation takes advantage of the pressure dependence of the azeotropic composition. Two columns operating at different pressures are used. It is readily apparent that applying pressure swing distillation demands in-depth knowledge of the VLE of the system to be separated. Since azeotropic behavior is directly related to the non-ideality of the mixture, a sophisticated model for calculating activity or fugacity coefficients as well as consistent experimental data are required to describe the pressure dependence of the azeotropic compositions. But the separation factor is often still close to unity indicating that a large number of theoretical stages are necessary and the need for both vacuum and higher pressure will demand high expenditure from the equipment point of view. Since pressure swing distillation requires a strongly pressure-dependent azeotropic composition, the temperature dependence of the saturation vapor pressures of the components should be different as is the case of systems with organic compounds and water. The temperature dependence of the saturation vapor pressure can be calculated using the Clausius–Clapeyron equation:

$$\frac{dP^s}{dT} = \frac{\Delta h_V}{T(v^V - v^L)} \tag{27}$$

In a case of azeotropic distillation an additional component is introduced that forms a lower boiling binary or ternary azeotrope (e.g. a hetero-azeotrope), which is easier to separate than the original azeotrope.

Extractive distillation uses a selective solvent (entrainer). Here the entrainer influences the ratio of the activity coefficients of the components in order to alter the separation factor far from unity. Often about 70% of the liquid phase inside the column consist of the entrainer. A typical extractive distillation process for separating aromatics (benzene) from aliphatics (cyclohexane) is show in Fig. 3.2-4.

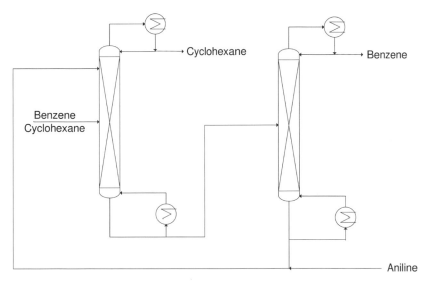

Fig. 3.2-4 Extractive distillation process for the separation of benzene from cyclohexane using aniline as an entrainer.

A good indication of whether or not a certain component is a suitable entrainer from the thermodynamic point of view is the selectivity at infinite dilution. On the other hand, from the point of view of green separation processes, the amount of entrainer needed has to be minimized. Hence, the entrainer should combine a high selectivity with a high capacity. To classify selectivity and capacity of an entrainer the selectivity and capacity at infinite dilution, S_{ij}^{∞} and k_i are used. The selectivity at infinite dilution S_{ij}^{∞} is defined as the ratio of the activity coefficients at infinite dilution γ_i^{∞}. It can be shown that the capacity k_i is also related to γ_i^{∞}:

$$S_{i,j}^{\infty} = \frac{\gamma_i^{\infty}}{\gamma_j^{\infty}} \tag{28}$$

$$k_i = \frac{1}{\gamma_i^\infty}$$ (29)

Entrainers with high selectivities usually suffer from a low capacity and vice versa, so, mixtures of entrainers are often used in industrial practice. Adding a small amount of water to *N*-methyl pyrrolidone (NMP) for separating aromatics from aliphatics increases the performance of the entrainer. Water shows a good selectivity but a poor capacity whereas NMP combines a moderate selectivity with a moderate capacity.

Extractive distillation is most widely used in industrial scale for separating aromatics from aliphatics. These processes typically use nitrogen-heterocycles as entrainers, such as NMP or *N*-formylmorpholine (NFM). Finding a suitable entrainer is a very difficult task since the influence of the entrainer on the activity coefficients of the components to be separated has to be calculated accurately. For a rough estimation of suitable entrainers some rules of thumb have been published (Matsuyama und Nishimura, 1977). Nowadays, computer programs connected to a large electronic database and group contribution methods are available to calculate the influence of the entrainer on the activity coefficients for a large number of potential entrainers within minutes (Gmehling and Möllmann, 1998).

Ionic liquids are a class of novel solvents with a melting point below 100°C and a negligible vapor pressure, which are interesting entrainers for extractive distillation. Examples of ionic liquids that have been investigated with respect to their potential as entrainers are 1-R-3-methyl-imidazolium-bis(trifluoromethyl-sulfonyl)-imides ($[RMIM]^+[CF_3SO_2]_2N^-$).

This class of solvents can be regarded as designer solvents since the anions and cations can be chosen nearly independently of each other. Therefore the best properties can be combined. Unfortunately, so far no thermodynamic model is available that is capable of describing the properties reliably enough. Experimental data needed to develop these models are scarce. Nevertheless, various ionic liquids have been investigated and interesting results have been published (Krummen et al., 2002).

Figure 3.2-5 shows separation factors at infinite dilution α_{ij}^∞ with NMP and NMP–water mixtures and with various ionic liquids as possible entrainers for the separation of benzene–cyclohexane mixtures. It can be seen that the ionic liquid shows much better performance than NMP. Hence less theoretical stages are needed for the separation, leading to lower energy consumption and lower demand in equipment. Conventional extractive distillation processes require an additional column for regenerating the entrainer.

Except for pressure swing distillation, which makes use of the pressure dependence of the azeotropic composition, all technologies require an entrainer to separate azeotropic mixtures. For some systems the azeotrope vanishes at certain pressures. This means that ordinary distillation at a different pressure (pressure or vacuum distillation) may circumvent the azeotrope. Although pressure swing distillation does not require an additional component in the process, it is not necessarily environmentally advantageous when compared to the alternatives, since additional

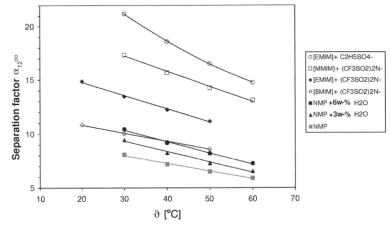

Fig. 3.2-5 Separation factors at infinite dilution for the mixture cyclohexane(1)–benzene(2) in different ionic liquids, NMP and NMP + water.

demands in energy and technical requirements may lead to a tremendous consumption of resources.

3.2.3.2
Coupled Columns

Separating a multicomponent mixture consisting of n different components usu ally requires $(n - 1)$ distillation columns. Minimizing the consumption of energy and resources means to decrease the number of columns necessary. This is possible if more than one (or from the last column two) pure component can be obtained from one column. If this is not possible, at least the energy consumption can be minimized by using waste heat from one column to heat the other columns. This is called indirect column coupling. In direct coupling, columns are combined by means of their product streams.

In the small column shown in Fig. 3.2-6a the mixture abc is separated into two binary mixtures – ab at the top and bc at the bottom. Without heating or cooling, ab and bc are fed at different locations into the second column. Pure a can be obtained at the top, pure c can be obtained from the bottom stream while pure b can be obtained from the side stream. Direct coupling can be used to separate a similar ternary mixture in a single distillation column. The principle is shown in Fig. 3.2-6b. The intermediate boiler b can be withdrawn from the middle of the column. In a conventional distillation column without a dividing wall, mixing on the feed tray would lead to large amounts of a and c in the side stream. The intermediate boiler, which will be present in the middle of the column with high purity would be diluted by the feed mixture. This can be avoided if the two columns of Fig. 3.2-6a are integrated into one column by means of a dividing vertical wall in the middle of the column. This so-called dividing wall column is an example of a direct cou-

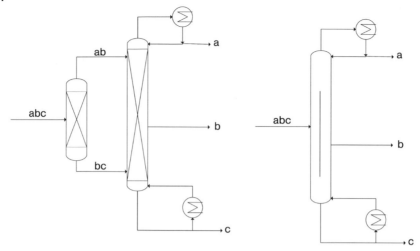

Fig. 3.2-6 Direct coupling of distillation columns.

pled column. It is thus possible to split a ternary mixture into pure components in a single distillation column by applying the dividing wall concept.

Indirect or thermal coupling involves using waste heat from one column to heat another column. The condenser of the first column is the reboiler of the other column. An example of the indirect coupling of columns is pressure swing distillation for separating azeotropic mixtures as mentioned in Section 3.2.3.1. Even if no azeotropes are present in a mixture it is often beneficial to operate columns at different pressure in order to thermally integrate them. This may lead to a lower consumption of energy since cooling and heating steps can be avoided. In the case of separating binary mixtures with two distillation columns it is possible to thermally couple the columns as can be seen in Fig. 3.2-7 for the separation of tetrahydrofuran (THF) from water. The feed enters the first column operating at 1 bar and the heavy boiler separates, forming the bottom stream. The azeotropic mixture, which is present at the top of the column, is condensed and fed into the second column. The condenser of the lower column section is the reboiler of the top column section. Using the right equipment it is also possible to establish a pressure swing distillation process within a single column consisting of two sections operating at different pressures. This can be seen on the right-hand side of Fig. 3.2-7.

Thermally integrating distillation columns is a chance to overcome one of the major drawbacks of this separation principle – the huge demand in energy.

Stichlmair and Fair (1998) give a detailed description of direct and indirect coupling of distillation columns.

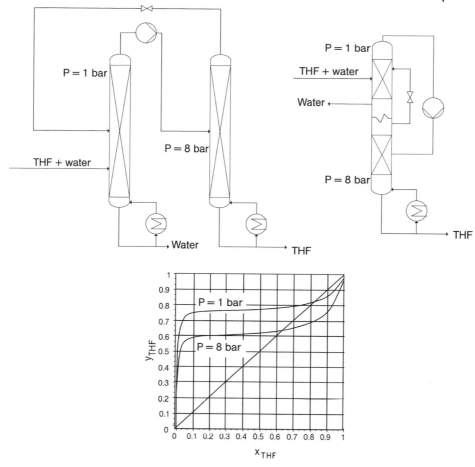

Fig. 3.2-7 Indirect coupling of distillation columns for the separation of the system THF–water, which forms a strongly pressure-dependent azeotrope (feed composition left from the azeotrope).

3.2.3.3
Reactive Distillation

Combining reaction and separation within one distillation column often offers distinct advantages when compared to the conventional chemical plant where reaction and separation are performed subsequently. The combination of reaction and distillation within one apparatus is called reactive distillation.

Especially for equilibrium limited and consecutive reactions, reactive distillation offers advantages. Higher selectivities in the case of consecutive reactions can be achieved by low local product concentrations if the product is removed from the reaction zone directly by distillation. Conversions higher than equilibrium conversions (100% in the ideal case) are obtained if the product is removed from the reactive section by distillation. Furthermore, high concentrations of reactants in the reactive section lead to a significant increase in reaction rate. In cases of exothermic reactions the heat of reaction can directly be used for distillation. Azeotropes with the reactants can be circumvented by reaction, and the temperature in the reactive section will not exceed the boiling point of the mixture, which is determined by phase equilibria and can be controlled by column pressure, leading to safer processes. Reactive distillation has been widely investigated for esterification, trans-esterification and etherification reactions (Sundmacher and Hoffmann, 1996; Pöpken et al., 2001; Steinigeweg and Gmehling, 2002, 2003a, 2003b). The application of reactive distillation to further reaction types has also been discussed (e.g. isomerizations, dimerizations, chlorinations, hydrations, alkylations) (Ruiz et al., 1995). The application to inorganic processes such as the production of silane (SiH_4) (Müller et al., 2002) is also interesting.

For a suitable reaction, reactive distillation is a highly attractive process alternative. Waste and energy consumption will be minimized. Reflux streams and equipment such as additional distillation towers or reactors can be avoided.

An example that demonstrates the advantages of reactive distillation is the synthesis of methyl acetate by esterification of acetic acid with methanol. A chemical plant consisting of two reactors and nine distillation columns can be replaced by just one reactive distillation column. Fig. 3.2-8 shows the conventional process for

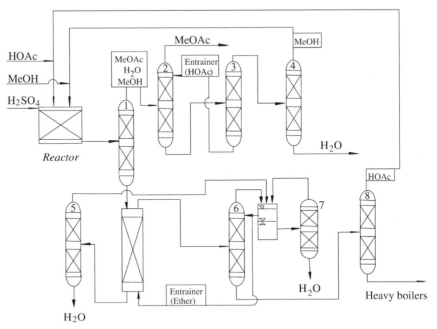

Fig. 3.2-8 Classical setup for the production of methyl acetate (MeOAc) by esterification of acetic acid (HOAc) with methanol (MeOH).

the production of methyl acetate. In this case the azeotrope methyl acetate–methanol can be circumvented by reaction. Acetic acid, which is one of the reactants, is an entrainer for the separation of the second azeotrope present in the methyl acetate–water system. Therefore acetic acid is fed into the column a few stages above the reactive section to make use of this extractive effect. The reactive distillation column for the synthesis of methyl acetate hence combines three columns in one: conventional distillation above the extractive section, an extractive distillation unit above the reactive section and a reactive distillation section. The additional columns shown in Fig. 3.2-9 are required for separating impurities which are present in the feedstock.

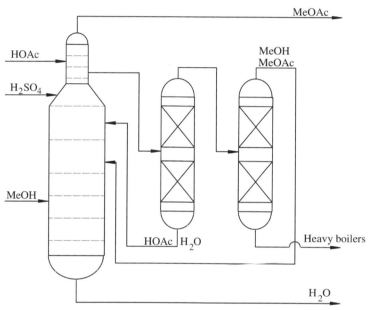

Fig. 3.2-9 Reactive distillation process for the esterification of acetic acid (HOAc) with methanol (MeOH) for the production of methyl acetate (MeOAc).

A further process for which reactive distillation is commonly used in industrial practice is the synthesis of methyl-*tert*-butyl ether (MTBE) which is an additive for gasoline. MTBE is produced by etherification of isobutene with methanol. The process based on reactive distillation leads to conversions of 99%. Isobutene and methanol are fed into the pre-reactor where the equilibrium conversion is obtained. The stream leaving the pre-reactor is fed into the reactive distillation column where MTBE is obtained as the bottom product.

Rigorous simulations performed by process simulators are absolutely necessary for successful process development. The development of reactive distillation processes can be divided into the following steps (Steinigeweg, 2003):

- Thermodynamic analysis of the system.
- Determination and modeling of reaction kinetics.
- Modeling and process simulation of the reactive distillation process and experimental validation in a pilot-plant scale.
- Evaluation of different process alternatives using simulation and comparison with conventional process alternatives.
- Scale-up and development of control strategies.

3.2.3.3.1 **Thermodynamic Properties**

For conventional distillation, reliable and consistent data about the VLE behavior are absolutely necessary. This becomes even more important for reactive distillation processes because the phase equilibrium determines the local reactant concentrations and therefore the local product concentrations, the reaction rate and the conversion. Although VLE data for reactive systems are most important for reactive distillation, experimental data is hard to find in the literature.

Determining experimental data is difficult for reactive systems because the composition of the mixture changes with time. Special equipment and apparatus are necessary for the measurement of the VLE of reactive systems. Measurements at low temperatures are suitable for slow reactions like esterification or etherification. Heat of mixing data can then be used to describe reliably the temperature dependence of the activity coefficients and allow extrapolation to higher temperatures.

As an alternative, group contribution methods can be applied to predict the required activity coefficients. Methods like UNIFAC or modified UNIFAC are based on the functional groups of the components and group interaction parameters, which are fitted to a large number of experimental data. The calculated activity coefficients therefore have a high accuracy (Gmehling et al., 2002) and can be used for reactive systems. Especially for very fast reactions, group contribution methods are often recommended.

The question whether or not reactive distillation is a favorable, or at least possible, process alternative can be answered with the help of reactive residual curves. Reactive residual curves, like conventional residual curves, describe the change of composition in the reboiler over time during an open vaporization. Besides information about the VLE behavior, the chemical equilibrium constant or the reaction kinetics are necessary for the simulation. For very fast reactions simplifying chemical equilibria can be assumed. According to Gibbs' phase rule (Gmehling and Kolbe, 1992), the dimensions of the system decrease by one if chemical equilibrium is assumed. This can be used for a transformation of coordinates leading to a plot in two-dimensional space which enhances the visibility of the system's behavior. For certain systems a new type of azeotrope, a so-called reactive azeotrope, has been found. Reactive azeotropes are formed when the change of composition caused by the reaction in the liquid phase is revoked by the composition change caused by vaporization. No change in the overall composition can be achieved.

Reactive residual curves can be used to estimate the potential products that can be obtained from a reactive distillation column. The main drawback of this concept is that reactive residual curves reflect the composition profile along the column only for columns operating at infinite reflux. Working with finite reflux ratios leads to small deviations between residual curve and column profile. Introducing a second feed location leads to column profiles totally different from the residual curves. Therefore the application range of reactive residual curves is limited to processes with only one feed location. Since reactions usually have more than one reactant the resulting reactive distillation process usually has more than one feed location. For these processes, analysis of reactive residual curves does not lead to correct results.

Heterogeneously catalyzed reactive distillation offers several advantages when compared to the homogeneously catalyzed process alternative. The size and location of the reactive section can be chosen independently of thermodynamic constraints, and an additional process step for separating the catalyst from the product can be avoided. In order to fix the catalyst in the reactive section of the column, heterogeneously catalyzed reactive distillation processes require special internals, which will be discussed in Section 3.2.4.

3.2.3.3.2 Reaction Kinetics and Modeling

For modeling reactive distillation processes both the reaction and the separation by distillation can be described using different approaches. These models can be used and combined into a resulting model of different complexity. For the reaction part either a kinetic model can be applied or an infinitely fast chemical reaction can be assumed, leading to a chemical equilibrium on every reactive stage. As mentioned above, two models can be used for modeling separation by distillation. The equilibrium-stage model assumes to reach vapor–liquid equilibrium on each theoretical stage. A rate-based model takes into account the transport limitations.

This leads to different models for describing reactive distillation processes:

1. Chemical equilibrium + equilibrium-stage model   low complexity
2. Chemical equilibrium + rate-based approach   medium complexity
3. Reaction kinetics + equilibrium-stage model   medium complexity
4. Reaction kinetics + rate-based approach   high complexity

It has been shown (Pöpken et al., 2001; Beßling et al., 1998) that the first model usually provides the highest deviations between calculation results and the behavior of real reactive distillation columns. The assumption that chemical equilibrium is reached is not adequate for most of the chemical reactions of commercial interest. Changes in composition and heat are taken into account by using the equilibrium constant K and the heat of reaction. The second model cannot be recommended because chemical reactions are slower than the time needed to reach VLE. Therefore, it makes no sense to assume kinetic limitations for the distillation part but to neglect the reaction kinetics.

For calculations on reactive distillation columns, a modification of the MESH equations is necessary. A term for the composition change by reaction has to be added to the material balance ($r_{i,j}H_j^L$) and a term for the heat of reaction ($-r_{i,j}H_j^L\Delta h_{R,j}^L$)

has to be introduced into the heat balance (H_j^L is the liquid holdup on stage j). Since chemical equilibrium is assumed for the first and second models the reaction rate ($r_{i,j}$) can be substituted by the changes in molar composition according to the chemical equilibrium composition.

It has been shown (Pöpken et al., 2001; Steinigeweg and Gmehling, 2002, 2003a, 2003b; Beßling et al., 1998) that this model is widely applicable, at least for columns in pilot plants, and it forms the basis for process optimization and evaluation.

More detailed models (fourth model) also account for the transport properties of the components, leading to a rate-based approach. The resulting model shows a high complexity but also leads to good results in the comparison to calculated with experimental results (Noeres et al., 2002).

Investigations carried out in pilot plants indicate that an equilibrium-stage model is sufficient for the synthesis and design of reactive distillation processes.

Scale-up of reactive distillation processes strongly depends on the internals used. For Katapak several studies have been performed and published (Moritz, 1999).

As can be seen from reactive distillation, developing sustainable processes means the integration of several different process steps within a small number of pieces of apparatus in order to benefit from synergistic effects. This concept, called process intensification, can also be applied to other processes. Reactive distillation itself can be combined with further separation steps in order to increase the degree of process intensification. A simple example is the esterification of acetic acid with methanol, in which a reactive distillation process and an extractive distillation process are combined within one column. Integrations of this type have also be applied to other reactions (Steinigeweg, 2003). Reactive distillation has also been combined with pervaporation (Steinigeweg and Gmehling, 2003b).

Process intensification does have tremendous additional potential for optimized processes with respect to energy consumption and is therefore most important for developing sustainable chemical processes.

3.2.3.4
Combination of Distillation with Other Unit Operations

For difficult separations such as those of closely boiling or azeotropic mixtures, the combination of distillation with other unit operations such as crystallization, membrane separation processes or extraction offers a potential alternative. Design and development of these so called hybrid separation processes is ambitious. Like reactive distillation, a process combining different unit operations shows a higher complexity. The number and order of the different operations as well as the number and composition of the recycle streams lead to higher degrees of freedom. Hybrid separations are characterized by one or more recycle streams which connect the unit operations, e.g. the distillate stream leaving the column is fed into a pervaporation unit and one stream leaving this unit is recycled into the distillation column.

Unit operations are based on different phase equilibria (e.g. distillation is based on the vapor–liquid equilibrium, crystallization is based on the solid–liquid equilibrium) with different separation boundaries – azeotropes in the case of distillation and eutectic points in the case of crystallization. Combining different unit operations can therefore be used to circumvent these boundaries, as is the case for azeotropes in distillation. Furthermore, all unit operations can be applied within their optimal working range. The energy consumption and investment costs of hybrid separations should be less than those of the conventional sequential approach.

Combinations of distillation and pervaporation (membrane separation with a vapor permeate stream) have been applied to many separation problems. One example is the separation of ethanol–water mixtures. Ethanol and water form an azeotrope and therefore conventional distillation is not feasible. Using a combination including pervaporation, pure ethanol and water can be obtained as can be seen from Fig. 3.2-10. Water can be obtained from the bottom of the distillation column. The distillate stream consists of the azeotropic mixture which is fed into the pervaporation unit. The highly selective polyvinyl alcohol membrane is capable of separating water from ethanol. Pure ethanol can be obtained from the retentate, and the permeate stream, which consists of ethanol and water, is recycled into the distillation column.

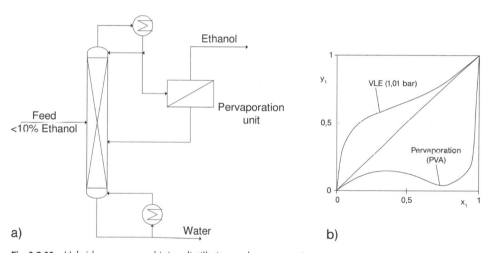

Fig. 3.2-10 Hybrid process combining distillation and pervaporation.

The combination of distillation with crystallization is often applied for the separation of mixtures of isomers. Separation of dichlorobenzene isomers has been described by Ruegg (Ruegg, 1989). Also diphenyl methane diisocyanate (MDI) (Stepanski and Fässler, 2002) and xylene isomers (Stepanski and Haller, 2000) have been separated by combining distillation and crystallization. Even though acetic acid and water do not form an azeotrope the separation factor is close to unity. Separating the two components by distillation therefore demands a large number of theoretical stages and a high reflux ratio, leading to high energy consumption. The components can also be separated by a combination of crystallization and distillation.

Short-cut methods as well as rigorous techniques have been used for the development and optimization of hybrid separation processes. Research continues on the development of generally applicable rules for the design and optimization of hybrid separations. A review of the current status has been published by Franke et al. (2004).

3.2.4
Column Internals

3.2.4.1
Internals for Conventional Distillation Processes

In principle, two different designs of column internals can be distinguished – tray columns and packed columns. In tray columns, vapor flowing from the bottom to the top is condensed on a plate, which in the simplest case consists of a perforated plate. Liquid flows down onto the stage below through a downcomer section. Fig. 3.2-11 shows a schematic drawing of a tray column (Stichlmair and Fair, 1998). The velocity of the vapor leads to a holdup of the liquid on each tray. If the vapor velocity increases, the liquid is blown off the tray and if the vapor load is too small, vapor does not flow through all of the holes of the perforated tray and liquid leaks through the tray. The liquid flowing through the downcomer section allows only up to a maximum volume flow, which can be calculated using Torricelli's equation (Stichlmair and Fair, 1998). The minimum liquid load is reached if the tray is not uniformly covered by liquid. The operation range of the column lies between the minimum and maximum vapor and liquid loads. Different trays are used in industrial practice. From the point of view of a green process it is important to minimize the pressure drop in the column in order to avoid superfluous energy consumption.

Column packings can be separated into structured and random packings. In contrast to tray columns, vapor and liquid phases are in direct contact along the complete packed section of the column. The main goal is to provide a large interfacial area for good contact between the two phases. Random packings are based on the Raschig ring, which was invented in 1907. The length of the ring equals its diameter. This leads to a very homogeneously dumped bed and to a lower pressure drop than that of tray columns. From a green point of view a high number of theoretical stages per meter and a low pressure drop over a wide range of liquid and vapor loads are important for sustainable distillation processes. Nowadays some 50 different types of random packings are available, with different cost, operating range and pressure drop. But a certain degree of inhomogeneity is unavoidable in random packings. Structured packings are formed of corrugated sheets, which were developed by Sulzer in the 1960s. Corrugated sheets here are assembled parallel to the vertical. The inclinations of the corrugation between the neighboring sheets al-

Vapor to condenser

Reflux from condenser

e

a

b

d

e

f

c

Vapor from reboiler

a: sieve trays
b: downcomer
c: outlet weir
d: inlet weir
e: man way
f: tray support

Liquid to reboiler

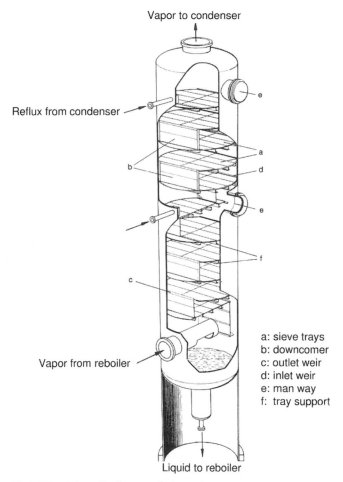

Fig. 3.2-11   Schematic diagram of a tray column.

ter in a specific manner. The packings provide a homogeneous bed structure and combine a low pressure drop with high separation efficiencies, and hence are suitable for developing sustainable distillation processes.

The operating range of structured packings differs considerably from that of tray columns. Both column types therefore have their specific application range. In packed columns, the vapor load can be reduced to very low values but a certain liquid load has to be maintained in order to ensure wetting of the packing elements.

Liquid and vapor flow cannot be regarded independently from each other. The pressure drop in the column rises with increasing vapor load. This effect increases with increasing liquid load since void fraction of the packing for the vapor flow becomes smaller.

The vapor load is usually expressed with help of the *F*-Factor:

$$F = u_g \sqrt{\rho_g} \tag{30}$$

If the vapor load (*F*-Factor) is sufficient to ensure a build-up of liquid in the packing the loading point of the packing is reached. This is the preferred operating point of the packings since vapor rises through the liquid phase. The flooding point gives the upper limit of the operation range of a packed column. At this point the pressure drop of the vapor flow through the bed increases so dramatically that liquid is no longer able to flow downwards. The countercurrent flow of the two phases breaks down.

Selecting an appropriate packing or tray is an ambitious task and should be considered with regard to energy and resource consumption and the flexibility and versatility of the resulting process. Details about designing columns can be found in the literature (Stichlmair and Fair, 1998).

The latest developments focus on higher theoretical stages per meter over a wide range of liquid and vapor load. This can be done by optimizing the packings with regard to pressure drop. Examples are Mellapak-Plus, which shows a very low pressure drop with excellent separation capabilities (Moser, 1999). Other developments combine packings with different specific surfaces and hydrodynamic properties in alternating layers within a distillation column. This leads to the residence time of the liquid being increased in some column sections. Packings with a high surface are operated at maximum liquid load and guarantee a good separation efficiency whereas the sections with a low surface are operated at low liquid level and mainly have the role of a demister (Kashani et al., 2004).

3.2.4.2
Internals for Reactive Distillation Processes

Choice of the correct equipment is crucial for developing a reactive distillation process. The most important factors influencing the equipment are relative volatilities and the reaction velocity. Some rules of thumb have been published in open literature (Gmehling and Brehm, 1996).

Whereas homogeneously catalyzed reactive distillation can be carried out in conventional tray columns (sometimes modified to ensure sufficient residence time of the reactants), a heterogeneous catalyst has to be fixed in the reactive section with the help of special internals. These internals have to combine good wetting characteristics to achieve a good contact between the liquid and vapor phases with a large amount of catalyst that is readily accessible by the liquid in order to avoid macro-kinetic influences.

The first type of internals is the so-called "tea bag" (Smith, 1984), which consists of wire-mesh bags that can be filled with the catalyst. These bags are stored on the trays of conventional distillation columns. The main drawback of this packing type is that transport limitations cannot be avoided, thus leading to smaller reaction rates. Furthermore the bags on the trays result in high additional pressure drops.

Bags filled with catalyst do not necessarily have to be stored on the trays. Bags have also been stored in the downcomer, which is a region of liquid holdup (Carland, 1994).

Structured packings with small bags in which the catalyst is stored (e.g. Katapak from Sulzer Chemtech or Katamax from Koch-Glitsch) combine the separation efficiency from conventional structured packings with good contact between the liquid phase and the catalyst. Katapak is made of corrugated wire-mesh sheets. Any heterogeneous catalyst with a size between about 0.5 and 2 mm can be immobilized between two of these sheets, forming a sandwich-like structure. This packing shows good radial distribution with only minor axial distribution of the liquid. This allows the column to operate with a uniform residence time.

Another possibility for storing the catalyst in the reactive section of the column is to make the internal itself catalytically active. This concept was introduced by Flato and Hoffmann (1992), who formed Raschig rings out of ionic ion-exchange resins. However, swelling of the polymer led to breakage of these rings. Modifications use glass Raschig rings on which an ion-exchange resin is precipitated as catalyst. The surface of conventional packings can also be made catalytically active. Brehm et al. (2002) showed that zeolites can be formed by hydrothermal synthesis on the surface of conventional packings. This type of packing has been used for the synthesis of ethyl-*tert*-butyl ether (Oudshoorn et al., 1999).

3.2.5
Summary

The main issue with regard to green chemistry is the development of distillation processes with a minimum demand in energy. Recent developments of different packing types that show a high number of theoretical stages per meter over a wide range of liquid and vapor load are important for sustainable chemical processes. Process intensification (hybrid processes) is a promising way to significantly decrease the consumption of resources, including energy and raw materials. Reactive distillation processes show distinct advantages over the conventional sequential approach for many reactions of commercial interest. Current research focuses on the scale-up of these processes and on an in-depth understanding of the direct interaction of reaction and distillation. Several examples for producing chemicals from a natural source on an industrial scale using reactive distillation have been reported. Esters from fatty acids can be produced by homogeneously (Jeromin et al., 1989; Schleper et al., 1990, Bock et al., 1997) or heterogeneously catalyzed (Steinigeweg and Gmehling, 2003a, von Scala et al., 2003) reactions. Although distillation and evaporation are rather mature technologies, there is still a lot of potential for optimization with regard to the sustainable development of chemical processes.

Symbols

a	Attractive parameter (Pa mol^2/m^6)
b	Co volume (m^3/mol)
$\dot{F}$	Feed flow (mol/s)
f	Fugacity
F	F-Factor (Pa$^{0.5}$)
h	Molar Enthalpy (J/mol)
H	Heat balance
h^E	Heat of mixing (J/mol)
Δh_V	Heat of vaporization (J/mol)
K	Partition coefficient
k	Interaction parameter
k	Capacity
$\dot{L}$	Liquid flow (mol/s)
M	Material balance
P	Pressure (Pa)
p_i	Partial pressure of component i(Pa)
Poy	Poynting factor
$\dot{Q}$	Heat flow (W)
R	General gas constant (8.31433 J/(mol K))
$\dot{S}$	Side stream (mol/s)
S	Selectivity
T	Absolute Temperature (K)
u_g	Vapor velocity (m/s)
$\dot{V}$	Vapor flow (mol/s)
v	Molar volume (mol/m^3)
x	Liquid phase mole fraction
y	Vapor phase mole fraction
z	Feed composition

Greek letters

φ	Fugacity coefficient
γ	Activity coefficient
α	Separation factor
γ^∞	Activity coefficient at infinite dilution
ρ	Vapor density (kg/m^3)

Indices

0	Standard
E	Excess property
i	Component i
j	Component j
L	Liquid phase
s	Saturation
V	Vapor Phase

References 153

References

Abrams, D. S., Prausnitz, J. M., *AIChE J.* **1975**, 21, 116.

Ahlers, J., Gmehling, J., *Fluid Phase Equilib.* **2001**, 191, 177.

Anastas, P.T, Warner, J. C., *Green Chemistry Theory and Practice*, Oxford University Press, New York, **1998**.

Beßling, B.; Löning, J.M.; Ohligschläger, A.; Schembecker, G.; Sundmacher, K., *Chem. Eng. Technol.* **1998**, 21, 393.

Bock, H.; Wozny, G.; Gutsche, B., *Chem. Eng. Process.* **1997**, 36, 101.

Brehm., A.; Zanter K. D., *Chem. Eng. Technol.* **2002**, 25, 917.

Carland, R. J., U.S. Patent 5,308,451, **1994**.

DDBST www.ddbst.de (**2004**)

Doherty, M.F.; Malone M.F., *Conceptual Design of Distillation Systems*, McGraw-Hill, New York, **2001**.

Flato, J.; Hoffmann, U., *Chem. Eng. Technol.* **1992**, 15, 193.

Franke, M., Gorak, A., Strube, J., *Chem. Ing. Tech.* **2004**, 76, 199.

Fredenslund, Aa., Gmehling, J., Rasmussen, P., *Vapor-Liquid Equilibria Using UNIFAC*, Elsevier, Amsterdam, **1977**.

Gmehling, J. et al., *Activity Coefficients at Infinite Dilution*, 4 parts, DECHEMA Chemistry Data Series, Frankfurt, from **1986**.

Gmehling, J. et al., *Vapor-Liquid Equilibrium Data Collection*, DECHEMA Chemistry Data Series, 29 volumes, Frankfurt, from **1977**.

Gmehling, J. Kleiber, M., Steinigeweg, S., 'Grundzüge der thermischen Verfahrenstechnik'. In: Dittmeyer, R., Keim, W., Kreysa, G., Oberholz, A. Winnacker Küchler, eds., *Chemische Technik*, Band 1, Methodische Grundlagen, Wiley-VCH, Weinheim, **2004**.

Gmehling, J., Brehm, A., *Grundoperationen*, Thieme-Verlag, Stuttgart, **1996**.

Gmehling, J., Christensen, C., Holderbaum, T., Rasmussen, P., Weidlich, U., *Heats of Mixing Data Collection*, DECHEMA Chemistry Data Series, 4 parts, DECHEMA, Frankfurt, from **1988**.

Gmehling, J., Kolbe, B., *Thermodynamik*, VCH-Verlag, Weinheim, **1992**.

Gmehling, J., Menke, J., Krafczyk, J., Fischer, K., *Azeotropic Data*, 3 Vols., Wiley-VCH, Weinheim, 2nd edition, **2004**.

Gmehling, J., Möllmann, C., *Ind. Eng. Chem. Res.* **1998**, 37, 3112.

Gmehling, J.; Wittig, R.; Lohmann, J.; Joh, R., *Ind. Eng. Chem. Res.* **2002**, 41, 1678.

Holderbaum, T., Gmehling, J. *Fluid Phase Equilib.* **1991**, 70, 251.

Huron, M. J., Vidal, J., *Fluid Phase Equilib.* **1979**, 3, 255.

Jeromin, L.; Peukert, E.; Gutsche, B.; Wollmann, G.; Schleper, B. European Patent 332971, **1989**.

Kashani, N., Siegert, M., Sirch T., *Chem. Ing. Tech.* **2004**, 76, 929.

Kheawhom, S., Hirao, M. J., *Chem. Eng. Jpn.* **2004**, 37, 243.

Kister, H.Z., *Chem. Eng. Progress* **2002**, 98, 52.

Kojima, K., Tochigi, K., *Production of Vapor-Liquid Equilibria by the ASOG Method*, Kodansha-Elsevier, Tokio, **1979**.

Krummen, M., Wasserscheid, P., Gmehling, J., *J. Chem. Eng. Data* **2002**, 47, 1411.

Matsuyama, H., Nishimura, H., *Chem. Eng. Jpn.* **1977**, 10, 181.

Moritz, P., Hasse, H., *Chem. Eng. Sci.* **1999**, 54, 1367.

Moser, F., *Sulzer Technical Review* 3/1999, Winterthur, **1999**.

Müller, D., Ronge, G., Schäfer, J. P., Leimkühler, H. J., Lecture at "International Conference on Distillation and Absorption" Baden-Baden, **2002**.

Noeres, C., Hoffmann, A., Gorak, A., *Chem. Eng. Sci.* **2002**, 57, 1545.

Oudshoorn, O. L., Janissen, M., van Kooten, W. E. J., Jansen, J. C., van Bekkum, H., van den Bleek, C. M., Calis, H. P. A., *Chem. Eng. Sci.* **1999**, 54, 1413.

Pöpken, T., Steinigeweg, S., Gmehling, J., *Ind. Eng. Chem. Res.* **2001**, 40, 1566.

Renon, H., Prausnitz, J. M., *AIChE J.* **1968**, 14, 135.

Ruegg, P.J., *Chem. Ind.* **1989**, 11, 83.

Ruiz, C. A., Basualdo, M. S., Scenna, N. J. *Trans. Inst. Chem. Engr.* **1995**, 73, 363.

Sattler, K., *Thermische Trennverfahren*, Wiley-VCH, Weinheim, **2001**.

Schleper, B., Gutsche, B., Wnuck, J., Jeromin, L., *Chem. Ing. Tech.* **1990**, 62, 226.

Schreinemakers, F. A. H., *Z. Phys. Chem.* **1901a**, 36, 257.

Schreinemakers, F. A. H., *Z. Phys. Chem.* **1901b**, 36, 413.

Smith, Jr., L. A., US Patent Nr. 4,443,559, **1984**.

Sørensen, J.M., Arlt ,W., Macedo, E.A., Rasmussen, P., *Liquid-Liquid Equilibrium Data Collection*, DECHEMA Chemistry Data Series, 4 parts, DECHEMA, Frankfurt, from **1979**.

Steinigeweg, S., PhD Thesis, University of Oldenburg, **2003**.

Steinigeweg, S.; Gmehling, J., *Ind. Eng. Chem. Res.* **2003a**, 42, 3612.

Steinigeweg, S., Gmehling, J., *Chem. Eng. Proc.* **2003b**, 43, 447.

Steinigeweg, S.; Gmehling, J., *Ind. Eng. Chem. Res.* **2002**, 41, 5483.

Stepanski, M., Fässler, P., *Sulzer Technical Review* 4/2002, Winterthur, **2002**.

Stepanski, M., Haller, U., *Sulzer Technical Review* 3/2000, Winterthur, **2000**.

Stichlmair, J., Fair, J. R., *Distillation: Principles and Practice*, Wiley-Verlag, New York, **1998**.

Sundmacher, K., Hoffmann, U., *Chem. Eng. Sci.* **1996**, 51, 2359.

Taylor, R., Krishna, R. *Chem. Eng. Sci.* **2000**, 55, 5183.

Von Scala, C., Moritz, P., Fässler, P., *Chimia* 57, 799, **2003**.

Weidlich, U., Gmehling, J., *Ind. Eng. Chem. Res.* **1987**, 26, 1372.

Wilson, G. M. *J. Am. Chem. Soc.* **1964**, 86, 127.

3.3
Green Enantiomeric Separations by Inclusion Complexation

Fumio Toda

3.3.1
Introduction

When one enantiomer of a racemic guest compound is included selectively in an inclusion complex with a chiral host compound, the racemic compound can be separated into its enantiomers. In some cases, enantiomeric separation of a racemic compound can be accomplished by inclusion complexation with a racemic or an achiral host compound. In the special case of complexation between a racemic host and a racemic guest, simultaneous and mutual enantiomeric separation of both the components becomes possible. By a combination of easy racemization of the guest and enantiomeric separation by inclusion complexation with a chiral host, a racemic compound can be transformed into one enantiomer which can be separated in 200% yield. A kinetic resolution by a combination of solid-state reaction and enantiomeric separation by the complexation method is also possible. The arrangement of an achiral host molecule in a chiral form in its complex with an achiral guest is also an interesting phenomenon. By using the chiral arrangement of achiral host molecules in the complex, a racemic guest can be separated into its enantiomers by complexation with the achiral host. A more interesting observation is that a racemic host–guest complex crystal is transformed into a conglomerate crystal by heating or contact with solvent vapor. Green enantiomeric separations by using these phenomena are described.

Host–guest inclusion complexations are usually carried out in organic solvents. As a green process, inclusion complexation can be performed in a water suspension medium or in the solid state. When the solid-state reaction in a water suspension medium is combined with an enantioselective inclusion complexation in the same water medium, a one-pot green preparative method for obtaining optically active compounds can be designed. In all these cases, enantiomers separated as inclusion complexes are recovered by distillation of the inclusion complex. When enantioselective inclusion complexation in the solid state is combined with the distillation technique, a unique green process for enantiomeric separation can result.

Green Separation Processes. Edited by C. A. M. Afonso and J. G. Crespo
Copyright © 2005 WILEY-VCH Verlag GmbH & Co. KGaA, Weinheim
ISBN 3-527-30985-3

Since the host compound remaining after the distillation can be used again and again, the process is economically and ecologically favorable.

Some enantiomeric separations based on the host–guest inclusion technique have already been described so far in reviews and books [1] and this chapter describes enantiomeric separations summarized from the viewpoint of achieving more green processes.

3.3.2
Enantiomeric Separations

3.3.2.1
Enantiomeric Separation of Hydrocarbons and Their Halogeno Derivatives

Preparation of enantiomerically active hydrocarbons is difficult and only a few examples of the preparation of chiral hydrocarbons have been reported. For example, chiral 3-phenylcyclohexene has been derived from tartaric acid through eight synthetic steps. Enantiomeric separation by host–guest complexation with a chiral host is more fruitful for the preparation of chiral hydrocarbons. For example, when a solution of *(R,R)*-(–)-*trans*-4,5-bis(hydroxydiphenylmethyl)-1,4-dioxaspiro[4.4]-nonane (**1b**) [2] (3 g, 6.1 mmol) and *rac*-3-methylcyclohexene (**2a**) (0.58 g, 6.1 mmol) in ether (15 ml) was kept at room temperature for 12 h, a 2:1 inclusion complex of **1b** and **2a** (2.5 g, 75%) was obtained as colorless prisms in the yield indicated. The crystals were purified by recrystallization from ether to give the inclusion complex (2.4 g, 71%), which upon heating *in vacuo* gave (–)-**2a** of 75% ee by distillation (0.19 g, 71%) [3]. By the same inclusion complexation, (+)-4-methyl- (**2b**) (33% ee, 55%), (–)-4-vinylcyclohexene (**2c**) (28% ee, 73%), (–)-bicyclo[4.3]-nonane-2,5-diene (**3**) (ee value not determined, 90%), and (–)-3-chloro (**4a**) (56% ee, 48%) and (+)-3,4-dichloro-1-butene (**4b**) (ee value not determined, 42%) were obtained in the optical and chemical yields indicated [3]. In these cases, some quantities of ether are used as solvent only once in the experiment. Since the separated enantiomers can be recovered by distillation without using any solvent, the process is economically and ecologically favorable. When a solution of **1a** (50 g, 0.1 mol) in *rac-trans*-1,2-dichlorocyclohexane (**5**) (50 g, 0.33 mol) was kept at room temperature for 12 h, a 2:1 inclusion complex of **1a** and (–)-**5** was obtained as colorless prisms (45.7 g, 80% yield based on **1a**). Heating of the complex *in vacuo* gave (–)-**5** of 43% ee by distillation (6 g, 77% yield based on **1a**). The same treatment of the (–)-**5** of 43% ee (6 g) with **1a** followed by distillation gave (–)-**5** of 72% ee (0.7 g, 90% based on **1a**). When the (–)-**5** of 72% ee was treated again with **1a** as described above, (–)-**5** of 90% ee was finally obtained as a colorless oil [4]. In this case, solvent is not necessary since **5** itself also functions as a solvent.

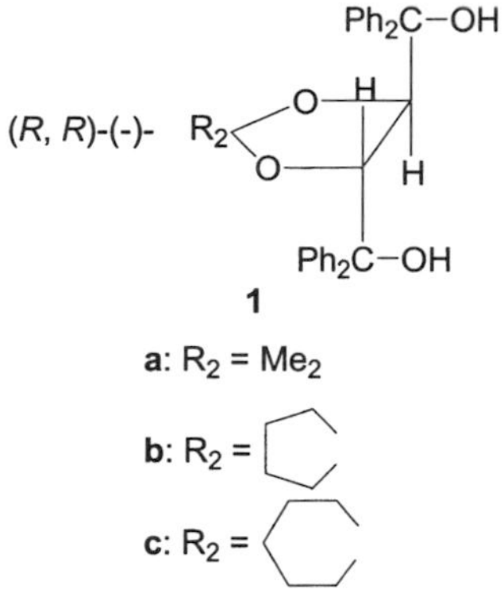

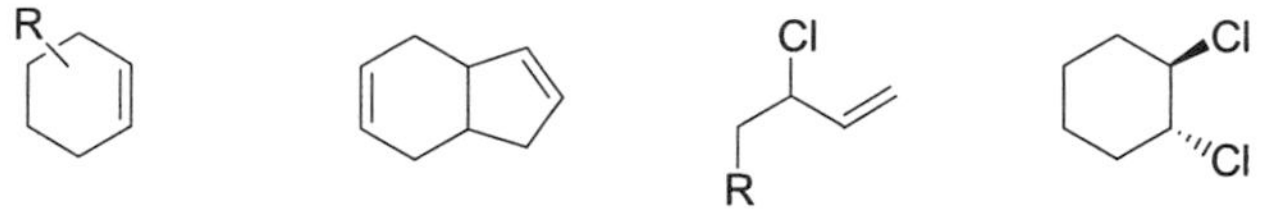

$$Ph_2C-OH$$

$(R, R)-(-)-$ R$_2$

1

a: R$_2$ = Me$_2$

b: R$_2$ =

c: R$_2$ =

2

a: R = 3-Me
b: R = 4-Me
c: R = 4-CH=CH$_2$

3

4

a: R = H
b: R = Cl

5

3.3.2.2
Amines, Amine N-Oxides, Oximes, and Amino Acid Esters

Inclusion complexation of *(S)-(–)-*1-(*o*-chlorophenyl)-1-phenylprop-2-yn-1-ol **(6)**[5] and *rac*-2-methylpiperazine **(7a)** in BuOH gave a 2:1 complex of **6** and *(S)-(+)-*enantiomer **(7c)**, which upon heating *in vacuo* after three recrystallizations from BuOH finally afforded optically pure **7c** in 19% yield [6]. Inclusion complexation of *(S,S)-*(–)-1,6-bis(*o*-chlorophenyl)-1,6-diphenylhexa- 2,4-diyne-1,6-diol **(8)** [7] and **7a** in Bu-OH gave a 1:1 complex of **8** and *(R)-(–)-*enantiomer **(7b)**, which after three recrystallizations gave pure crystals in 26% yield. Heating of the crystals *in vacuo* gave **7b** of 100% ee by distillation in 25% yield. The host compounds that remain after the distillation can be used again for the enantioseparation. Treatments of the filtrate remaining after the former and the latter separation experiments with **8** and **6**, respectively, gave optically pure **7b** and **7c**, respectively in yields of around 20% [5]. X-ray analysis of the 2:1 complex of **6** and **7c** showed that two **6** molecules bind to one **7c** molecule by the formation of two OH—N hydrogen bonds. The data also showed that the combination of **6** in the *(S)-*configuration and **7c** in the *(S)-*configuration is important. This agrees with the fact that **6** does not form a complex with **7b** in the *(R)-*configuration [6].

Enantiomers of amine *N*-oxides **(9a–e)** were efficiently separated by complexation with *(R)-(+)-*2,2'-dihydroxy-1,1'-binaphthyl **(10b)**. In this case, both enantiomers of **9** were obtained in an optically pure form [8]. For example, when a solution

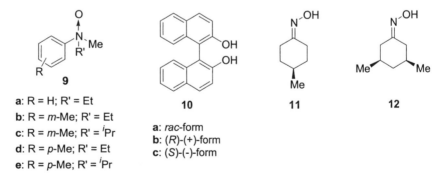

(S)-(-)- **6**

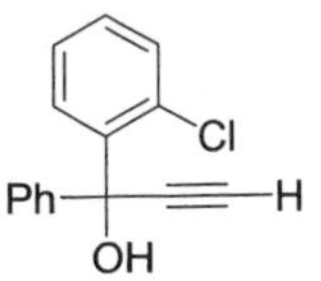

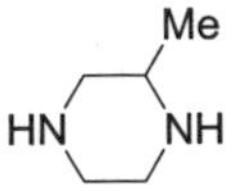

7
a: *rac*-form
b: (S)-(+)-form
c: (R)-(-)-form

(S, S)-(-)- **8**

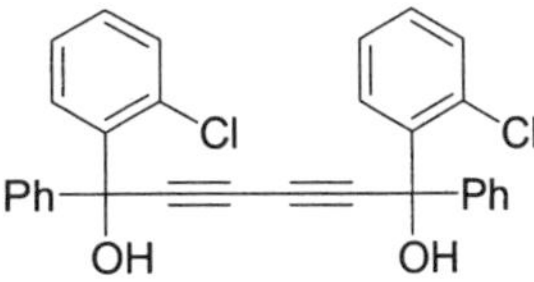

of **10b** (1 g) and two molar equivalents of *rac*-**9b** (1.2 g) in THF–hexane was kept at room temperature for 5 h, a 1:1 complex of **10b** and (+)-**9b** was obtained as color-less prisms. The crystals were recrystallized from THF–hexane to give pure crys-tals in 53% yield. The complex was separated into its components by column chro-matography on silica gel. Firstly, **10b** (0.5 g) was recovered from a fraction eluted by ethyl acetate–benzene. Secondly, (+)-**9b** of 100% ee (0.29 g, 48%) was obtained from a fraction eluted by MeOH in the yield indicated. Evaporation of the residual filtra-te after separation of the complex between **10b** and (+)-**9b**, gave crude (–)-**9b**. Treat-ment of the crude (–)-**9b** with **10c** in a similar manner to that described above, fol-lowed by column chromatography, finally yielded (–)-**9b** of 100% ee (0.24 g, 40%) in the yield indicated [8]. By a similar complexation method with **10b**, (+)-enantio-mers of **9a** and **9c-e** were separated to give (+)-**9a** (100% ee, 21%), (+)-**9c** (73% ee, 39%), (+)-**9d** (100% ee, 30%) and (+)-**9e** (100% ee, 68%) in the optical and chemical yields indicated [6]. Mechanism of the precise chiral recognition in the complex was studied by X-ray analysis of the complex **10b**-(+)-**9b** and **10c**-(–)-**9b**. [8].

9
a: R = H; R' = Et
b: R = *m*-Me; R' = Et
c: R = *m*-Me; R' = ^{i}Pr
d: R = *p*-Me; R' = Et
e: R = *p*-Me; R' = ^{i}Pr

10
a: *rac*-form
b: (R)-(+)-form
c: (S)-(-)-form

11

12

Enantiomeric separation of two oximes of 4-methyl-1-(hydroxyimino)cyclohexa-ne (**11**) and *cis*-3,5-dimethyl-1-(hydroxyimino)cyclohexane (**12**) was accomplished by complexation with **8** [9]. Complexation of **8** and *rac*-**11** gave a 1:1 complex of **8** and (+)-**11** as colorless needles. Treatment of the complex with alkylamine gave an alkylamine complex of **8** and (+)-**11**. Since the optical purity of the (+)-**11** was not determined directly, its *O*-benzoyl derivative was prepared and its optical purity was determined to be 79% ee by high-performance liquid chromatography (HPLC). Therefore, the optical purity of the (+)-**11** obtained by the enantiomeric separation procedure can be estimated to be higher than 79% ee. By the same treatment of *rac*-**12** with **8**, (+)-**12** of approximately 59% ee was obtained [9].

Since amino acids themselves are included by host compounds only with difficulty owing to their ionic character, enantiomeric separations are performed using their ester derivatives. For example, inclusion complexation of **1c** and *rac*-*N*-ethylethoxycarbonylaziridine (**13**) in benzene–hexane gave a 1:1 complex of **1c** and (–)-**13** as colorless needles in 59% yield, which upon distillation *in vacuo* gave (–)-**13** of 100% ee in 34% yield [10]. By a similar enantiomeric separation of *rac*-**14** by complexation with **1b**, (–)-**14** of 100% ee was obtained in 44% yield. By a similar complexation with **1c**, *rac*-**15** and *rac*-**16** were separated to give optically pure (–)-**15** (28%) and (–)-**16** (33%), respectively, in the yields indicated [10]. The *rac*-methyl esters of alanine (**17**) and phenylalanine (**18**) were easily separated into enantiomers by inclusion complexation with **1c** and **1b**, respectively, to give enantiomerically pure (+)-**17** and (+)-**18** in 12 and 40% yields, respectively [11].

EtN〈CO_2Et **13** ^{n}PrN〈CO_2Me **14** ^{n}PrN〈CO_2Me Me   **15** ^{i}PrN〈CO_2Me Me   **16**

$MeCH(NH_2)COOMe$ **17** $PhCH_2CH(NH_2)COOMe$ **18**

3.3.2.3
Alcohols and Cyanohydrins

Alcohol and cyanohydrin guests easily form inclusion complexes with brucine host (**19**) through hydrogen-bond formation between the OH group of the guest and the tertiary amino nitrogen atom of **19**. Inclusion complexation allows efficient enantiomeric separation of racemic guests. For example, inclusion complexation of **19** (1.8 g) and an equimolar amount of *rac*-*m*-chlorophenylphenylcarbinol (**20c**) in MeOH–hexane gave a 1:1 complex of **19** and (+)-**20c** (1.21 g). Two recrystallizations of the crude complex from MeOH–hexane gave pure complex (0.8 g), which upon chromatography on silica gel (AcOEt) gave (+)-**20c** of 99.2% ee (0.28 g, 56% yield) [12]. Evaporation of the solvent from the filtrate left after separation of the crude complex of **19** and (+)-**20c**, followed by chromatography on silica gel and distillation, gave (–)-**20c** of 72.4% ee in 98% yield. The optical purities and yields of optically active **20a–i** obtained by the same separation method applied to **20c** are summarized in Table 3.3-1 [12].

Table 3.3-1 Optical purity and yield of the
(+)-enantiomer of **20** obtained by complexation
with **19**.

	20	(+)-Enantiomer	
	X	ee (%)	yield (%)
a	*m*-Me	92.1	61.3
b	*p*-Me	92.6	2.6
c	*m*-Cl	99.2	56.0
d	*p*-Cl	97.0	30.4
e	*m*-Br	98.0	44.0
f	*p*-Br	100.0	20.0
g	*m*-OMe	93.0	32.0
h	*m*-NO$_2$	99.6	47.8
i	*p*-NO$_2$	85.5	72.0

rac-Alcohols substituted with one aryl and one sterically bulky group, such as **21** and **22**, were also separated into enantiomers efficiently by complexation with **19**. For example, when a solution of **19** and an equimolar amount of *rac*-**21** in MeOH was kept at room temperature for 12 h, a 1:1 complex of **19** and (–)-**21** was obtained, after three recrystallizations from MeOH, as colorless prisms, which upon heating *in vacuo* gave (–)-**21** of 100% ee in 19.8% yield by distillation. Similarly, *rac*-**22** gave finally (–)-**22** of 100% ee in 38% yield [12].

19

20   **21**   **22**

rac-Alcohols substituted with one aryl and one sterically less bulky group were separated into enantiomers by complexation with **1a** or **1b**. The (–)-enantiomers of **23–29** obtained by enantioselective complexation with **1a** or **1b** are summarized in Table 3.3-2 [13]. The most interesting application of enantiomeric separation of alcohols by complexation with **1b** or **1c** was accomplished for glycerol acetals, **30a–c**. Enantiomerically pure (+)-**30a** (30%), (+)-**30b** (80%) and (+)-**30c** (20%) were obtained in the yields indicated [13]. Enantiomeric separation of *rac*-1,3-butanediol (**31**) by complexation with **1c** was also successful. X-ray analysis of a 1:1 complex of **1c** and (S)-(+)-**31** showed that host and guest molecules are tightly associated by hydrogen bonding, which facilitates mutual chiral recognition [14].

PhCH(OH)CH₃ **23**   H₃C—⟨benzene⟩—CH(OH)C₂H₅ **24**   PhCH(OH)C₂H₅ **25**

26   **27**   PhCH(OH)C≡CH **28**

PhO—⟨benzene⟩—CH(OH)C≡CH **29**   Me—O / Me—O ⟩—OH **30**   OH ... OH **31**

Table 3.3-2 Enantiomer separation of alcohols **23–30** by complexation with host **1a** or **1b**.

Alcohols	Hosts			
	1a		1b	
	yield (%)	ee (%)	yield (%)	ee (%)
23	72.5	75.1		
24			79.1	63.6
25	39.7	91.4		
26			79.4	78.8
27			45.3	78.4
28	69.5	97.0		
29			65.0	92.1

In an enantiomeric separation of cyanohydrins, it was found that **19** is a suitable host for cyanohydrins substituted with one aryl group and one bulky alkyl group. In this case, not only did a simple enantiomeric separation of *rac*-cyanohydrin occur, but also its transformation into one enantiomer, so that one pure enantiomer was obtained in a yield of more than 100%. For example, when a solution of *rac*-1-cyano-2,2-dimethyl-1-phenylpropanol (**32a**) and an equimolar amount of **19** in MeOH was kept at room temperature for 12 h, a 1:1 brucine complex of (+)-**32a** was obtained in 134% yield. Decomposition of the complex with dilute HCl gave (+)-**32a** of 97% ee in 134% yield. From the filtrate, *rac*-**32a** was obtained [15]. The yield of (+)-**32a** of more than 100% shows a transformation of (–)-**32a** to (+)-**32a** through racemization during the complexation due to the base (**19**)-catalyzed equilibrium shown in Scheme 3.3-1. The yield of (+)-**32a** could be increased to 200% by leaving the MeOH solution to evaporate gradually during the complexation. For example, when a solution of *rac*-**32a** (1 g) and an equimolar amount of **19** (2.1 g) in MeOH (2 ml) was kept in an uncapped flask at room temperature for 1, 3, 6, 12 and 24 h, the amount of MeOH decreased to 1.9, 1.8, 1.6, 1.3 and 0.6 ml, respectively, and the (+)-**32a** was obtained in the optical and chemical yields indicated after complexation for 1 (55% ee, 40%), 3 (80% ee, 80%), 6 (95% ee, 110%), 12 (96% ee, 160%) and 24 h (97% ee, 200%) [15]. By the same procedure, enantiomerically pure (+)-**32b** (177%), (–)-**32c** (94%), (–)-**32d** (132%), (–)-**32e** (110%) and (+)-**32f** (84%) were obtained in the yields indicated [15].

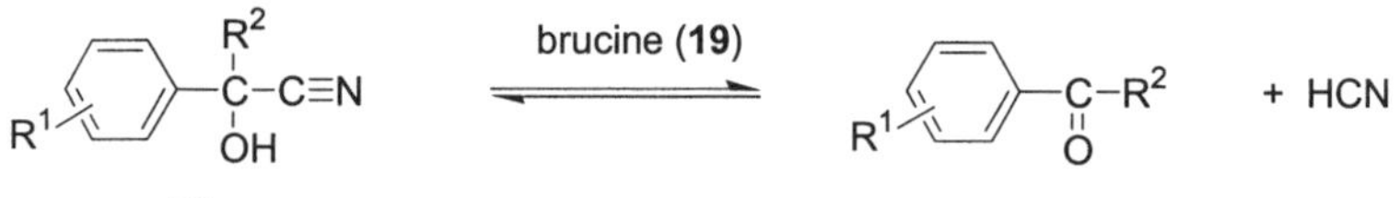

a: R^1 = H; R^2 = *t*Bu   **g**: R^1 = *m*-PhO; R^2 = H

b: R^1 = *p*-Cl; R^2 = *t*Bu   **h**: R^1 = H; R^2 = *n*Bu

c: R^1 = *p*-Me; R^2 = *t*Bu   **i**: R^1 = H; R^2 = *i*Pr

d: R^1 = *p*-OH; R^2 = *t*Bu   **j**: R^1 = H; R^2 = *n*Pr

e: R^1 = *m*-OH; R^2 = *t*Bu   **k**: R^1 = H; R^2 = Et

f: R^1 = H; R^2 = CCl$_3$ **l**: R^1 = H; R^2 = Me

                                 m: R^1 = H; R^2 = H

Scheme 3.3-1

Very interestingly, all cyanohydrins (**32h–m**) that are substituted with one phenyl group and one less bulky alkyl group or a hydrogen atom do not form complexes with **19**. However, the cyanohydrins that are substituted with two alkyl groups (**33a–c**) or with one alkyl group and one hydrogen atom (**33d–f**) did form complexes with **19**, and their enantiomers were separated [15].

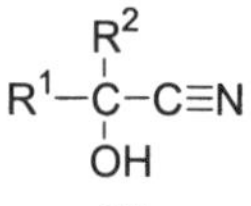

a: R^1 = *t*Bu; R^2 = Me

b: R^1 = *i*Pr; R^2 = Me

c: R^1 = Et; R^2 = Me

d: R^1 = *s*Bu; R^2 = H

e: R^1 = *i*Pr; R^2 = H

f: R^1 = ClCH$_2$; R^2 = H

g: R^1 = Me; R^2 = H

The chiral hosts **1a** and **1b** were found to be useful for the enantiomeric separation of cyanohydrins that cannot be separated with **19**. For example, **32g** and **32m** were separated by complexation with **1b** and **1a**, respectively, to give (–)-**32g** of 72.5% ee (70%) and (+)-**32m** of 100% ee (47.6%), respectively, in the yields indicated. The most simple chiral cyanohydrin derived from acetaldehyde (**33g**) was obtained in enantiomerically pure (+)-form in 52.6% yield by complexation of *rac*-**33g** with **1a** [13].

3.3.2.4
Epoxides and Oxaziridines

2,3-Epoxycyclohexanone of 20% ee was first prepared in 1980 by an enantioselective epoxidation of 2-cyclohexenone. However, enantiomerically pure 3-methyl- (34), 3,5-dimethyl- (35) and 3,5,5-trimethyl-2,3-epoxycyclohexanone (36) can easily be prepared in 18, 30 and 35% yields, respectively, by inclusion complexation of their racemic compounds with the chiral host 8 [16].

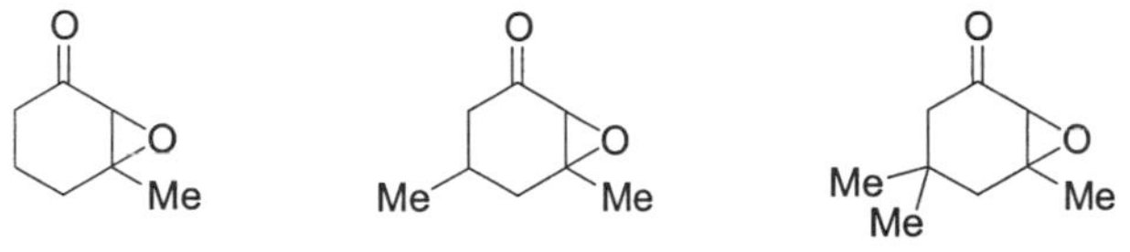

34　　　　　　　　**35**　　　　　　　　**36**

Preparation of enantiomerically active β-ionone epoxide by a solid-state kinetic resolution in the presence of the chiral host **1c** is also possible. When a mixture of **1c**, β-ionone (37) and *m*-chloroperbenzoic acid (MCPBA) is ground by mortar and pestle in the solid state, (+)-**38** of 88% ee was obtained [17]. The mechanism of the kinetic resolution in the solid state is as follows. Firstly, oxidation of **37** with MCPBA gives *rac*-β-ionone epoxide (38). Secondly, enantioselective inclusion of (+)-**38** with **1c** occurs. Thirdly, uncomplexed (−)-**38** is further oxidized to give the Baeyer–Villiger oxidation product (−)-**39** of 72% ee. This is the first example of an enantiomeric separation by enantioselective complexation in the solid state [17].

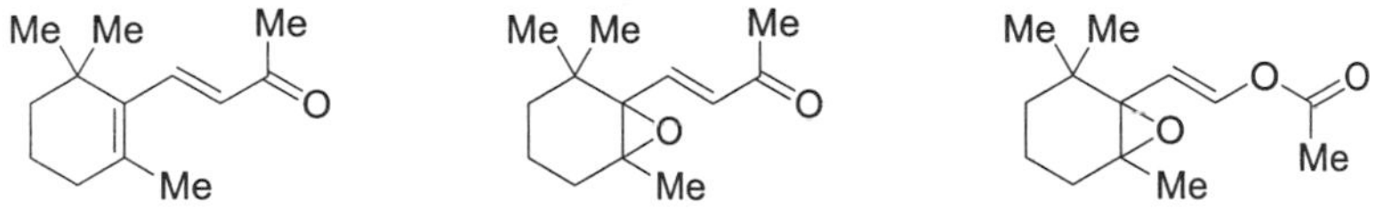

37　　　　　　　　**38**　　　　　　　　**39**

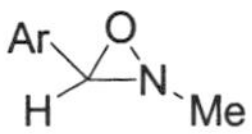

40

a: Ar = C_6H_5
b: Ar = $p\text{-}MeC_6H_4$
c: Ar = $p\text{-}ClC_6H_4$

d: Ar =

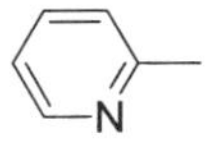

Oxaziridines (40a-d) were also separated into enantiomers efficiently by complexation with **1c**, and (+)-**40a** (99% ee, 69%), (+)-**40b** (98% ee, 68%), (+)-**40c** (90% ee, 56%) and (+)-**40d** (90% ee, 59%) were obtained in the optical and chemical yields indicated [18].

3.3.2.5
Ketones, Esters, Lactones and Lactams

In 1983, enantiomeric separations of *rac*-3-methylcyclopentanone (**41**), *rac*-3-me-thyl- cyclohexanone (**42**) and *rac*-5-methyl-γ-butyrolactone (**43**) with **8** were repor-ted as the first successful examples of enantiomeric separation by complexation with a chiral host compound [7]. By this method, enantiomerically pure (+)-**41** (13%) and (+)-**42** (6%) and (−)-**43** (5%) were obtained in the yields indicated. Alt-hough the enantiomerically pure 4-hydroxycylo-2-pentenone (**44a**) is a very impor-tant starting material for the synthesis of prostaglandins, it is not easy to obtain the enantiomerically pure **44a** efficiently. (*R*)-(+)-10,10'-Dihydroxy-9,9'-biphenanthryl (**45**) [19] was found to be a good host for enantiomeric separation of ester derivati-ves (**44b–d**) and the tetrahydropyranyl ether (**44e**) of **44a**. By complexation with **45**, enantiomerically pure (−)-**44b** (45%), (−)-**44c** (28%), (−)-**44d** (51%) and (−)-**44e** (51%) were obtained in the yields indicated [19]. Although **45** is not useful for the enan-tiomeric separation of *rac*-**44a**, hydrolysis of (−)-**44e** of 100% ee with dilute HCl ga-ve (−)-**44a** of 100% ee in almost quantitative yield [19].

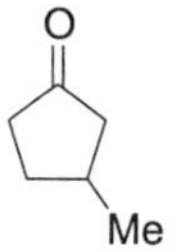

O
41

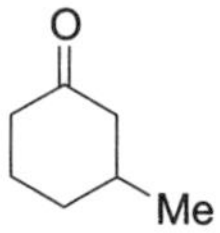

O
Me
42

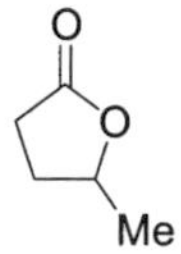

O
O
Me
43

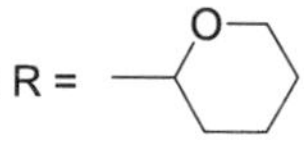

O
OR
44

a: R = H
b: R = COCH₃
c: R = COC₂H₅
d: R = COC₃H₇

e: R =

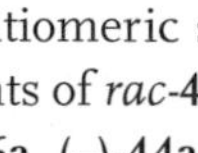

(*R*)-(+)-

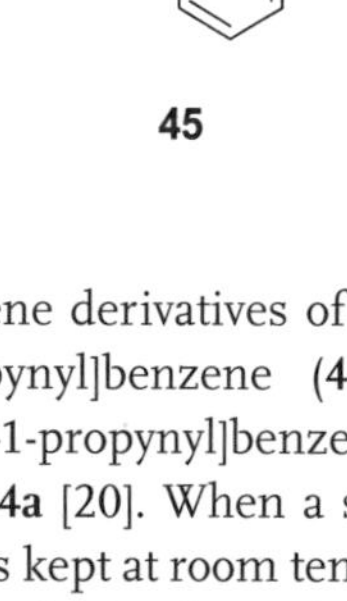

OH
OH

45

It was found that simple benzene derivatives of **6**, (*S,S*)-(−)-1,4-bis[3-(*o*-chloro-phenyl)-3-hydroxy-3-phenyl-1-propynyl]benzene (**46a**) and (*S,S*)-(−)-1,3-bis[3-(*o*-chlorophenyl)-3-hydroxy-3-phenyl-1-propynyl]benzene (**46b**) are useful for the di-rect enantiomeric separation of **44a** [20]. When a solution of **46a** and two molar equivalents of *rac*-**44a** in EtOH was kept at room temperature for 12 h, a 1:1:1 com-plex of **46a**, (−)-**44a** and EtOH was obtained as colorless prisms, which upon hea-ting *in vacuo* gave (−)-**44a** of 100% ee in 55% yield. Interestingly, when the comple-xation was carried out in toluene, a 1:2 complex of **46a** and (+)-**44a** was formed.

From the complex, (+)-**44a** of 100% ee was obtained by distillation. By the same complexation of *rac*-**47** with **46a**, (+)-**47** of 100% ee was obtained in 47% yield. Bicyclic ketones **48** and **49**, which are also important materials for prostaglandin synthesis, were separated into their enantiomers by complexation with **46b** to give enantiomerically pure (–)-**48** and (+)-**49** in 26 and 29% yields, respectively [21].

For enantiomeric separation of lactones (**50–53**) and ketones (**54–55**), **8** is a useful host. By complexation with **8**, enantiomerically pure (+)-**50** (10%), (+)-**51** (10%), (+)-**52** (13%), (+)-**53** (32%), (–)-**54** (20%) and (+)-**55b** (24%) were obtained in the yields indicated [21]. Interestingly, however, *trans*-isomers of **53** and **54** did not form complexes with **8**. This shows that the shape of the molecule may be important for the formation of the inclusion complex. For enantiomeric separation of **55–58**, the chiral host **1a** is very effective, and enantiomerically pure (–)-**55a** (41%), (–)-**55b** (62%), (–)-**55c** (43%), (+)-**56** (80%), (–)-**57** (58%) and (–)-**58** (70%) were obtained by complexation in the high yields indicated [3]. The mechanisms of these precise chiral recognitions between **1a** and bicyclic ketones have been clarified by X-ray study of two inclusion complexes of **1a** with (–)-**55b** and with (–)-**58** [22].

Enantiomeric separations of bicyclo[2.2.1]heptanones (**59a, 61–62**), bicyclo[2.2.2]octanones (**63–66**) and bicyclo[3.2.1]octanone (**67**) have also been well accomplished by complexation with appropriate chiral host compounds. Enantiomeric separations of **60, 61, 62** and **64** were carried out by complexation with **8** to give, finally, enantiomerically pure (+)-**60** (33%), (–)-**61** (16%), (+)-**62** (60%) and (–)-**64** (41%), respectively, in the yields indicated [23]. However, enantiomeric separations of **65** and **67** were accomplished efficiently by complexation with **1a** to give enantiomerically pure (–)-**65** and (–)-**67** in 56 and 48% yields, respectively. On the other hand, enantiomeric separation of **66** can be accomplished only by complexation with **45** to give, finally, (–)-**66** of 100% ee in 31% yield [23]. The mechanism of this efficient chiral recognition in the inclusion complex has been studied by X-ray analysis [23]. Nevertheless, none of **1a**, **8** and **45** is applicable to the enantiomeric separation of **59a** and **63**. Although *rac*-**59a** was separated into its enantiomers by complexation with brucine (**19**) to give finally (1*R*,4*R*)-(+)-**59a** of 27% ee in 40% yield [24], all attempts at enantiomeric separation of **63** failed [23]. However, *rac*-**59b** was separated to afford (1*R*,5*S*)-(–)-**59b** of 92% ee by complexation with **19**. The X-ray crystal structure of the complex of **19** and (1*R*,5*S*)-(–)-**59b** has been studied [24].

59

a: X = CH$_2$
b: X = NH

60   **61**   **62**

63   **64**   **65**   **66**   **67**

Inclusion complexations between chiral hosts and *rac*-guests for enantiomeric separations have been carried out in organic solvents. However, the complexation can also be carried out in a water suspension medium. This is a more "green" procedure. For example, when a suspension of finely powdered **1c** (1.5 g, 2 mmol) and *rac*-5-ethoxyfuran-2(5*H*)-one (**68a**) (0.5 g, 4 mmol) in water (10 ml) containing hexadecyltrimethylammonium bromide (0.005 g) as a surfactant was stirred at room temperature for 6 h, a 1:1 inclusion complex of **1c** and (+)-**68a** was formed as crystals, which upon heating *in vacuo* (220°C/20 mmHg) gave (+)-**68a** of 98% ee (0.06 g, 24%). When recrystallization was used for a complex of **1c** and *rac*-**68a**, (–)-**68a** of 100% ee was obtained in 10% yield [25]. By similar inclusion complexation of *rac*-

68b and *rac-***68c** with **1c** in a water suspension medium, (–)-**68b** (94% ee, 25%) and (+)-**68c** (90% ee, 46%), respectively, were obtained in the enantiomeric and chemical yields indicated [25].

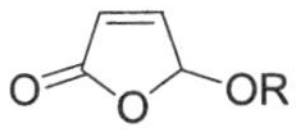

a: R = Me
b: R = Et
c: R = *n*Pr
d: R = *i*Pr
e: R = C$_6$H$_{11}$

68

Enantiomeric separations of bicyclic acid anhydride **69**, lactones **70** and **71** and carboximides **72** and **73** by complexation with **1a–c** in organic solvents were also successful (Table 3.3-3) [26]. These complexations can probably be carried out in a water suspension medium and hence be described as green processes. *rac*-Pantolactone (**74**) was separated to produce (*S*)-(–)-**74** of 99% ee in 30% yield by complexation with **1c** [27]. Enantiomerically impure monoterpenes were purified by inclusion complexation with a chiral host compound. For example, (1*S*,5*S*)-(–)-verbenone (**75a**) of 78% ee gave 99% ee enantiomer by complexation with **1a**. By similar treatment of **75b** of 91% ee with **1a** as above, (1*R*,5*R*)-(+)-**75b** of 98% ee was obtained [28].

Table 3.3-3   Enantiomer separation of **69–73** by complexation with **1a–c**.

Guest	Host		Product	
			optical purity (% ee)	yield (%)
69	1a	(+)-69	100	52
70a	1b	(+)-70a	43	59
70b	1b	(+)-70b	100	56
71a	1c	(–)-71a	67	63
71b	1c	(+)-71b	100	28
72a	1a	(+)-72a	33	90
72a	1b	(–)-72a	100	16
72b	1c	(+)-72b	100	50
72c	1c	(+)-72c	100	38
73a	1c	(+)-73a	100	30
73b	1a	(+)-73b	100	26
73b	1b	(+)-73b	100	63
73c	1b	(+)-73c	98	19
73d	1b	(+)-73d	98	40

Enantiomeric separations of glycidic esters (**76**), which are important synthons for various biologically active substances, were accomplished efficiently by complexation with the chiral hosts **1a–c**. For example, when a solution of **1b** and an equimolar amount of *rac*-ethyl 2,2-diethylglycidate (**76g**) in ether–hexane was kept

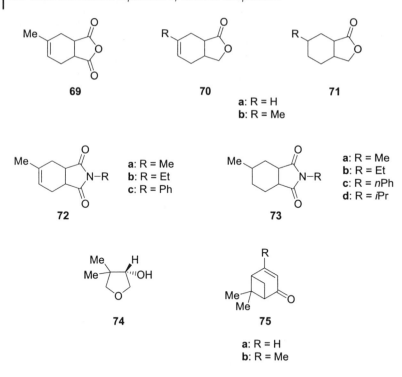

69

70

a: R = H
b: R = Me

71

72

a: R = Me
b: R = Et
c: R = Ph

73

a: R = Me
b: R = Et
c: R = *n*Ph
d: R = *i*Pr

74

75

a: R = H
b: R = Me

at room temperature for 12 h, a 2:1 complex of **1b** and (+)-**76g** was produced as co-lorless prisms, which upon heating *in vacuo* gave (+)-**76g** of 100% ee in 62% yield [29]. Enantiomeric separation experiments of *rac-***76a-f** and *rac-***76h-j** were carried out using the same procedure (Table 3.3-4). Although the efficiencies of enantio-meric separations of *rac-***76e–h** and *rac-***76j** were excellent, those of **76a**, **76d** and **76i** were only moderate, and those of **76b** and **76c** were poor.

The biphenanthrol host **45** was found to be useful for enantiomeric separations of simple esters. By inclusion complexation with **45**, *rac-***77–82** gave (+)-**77** (69.4% ee, 75%), (+)-**78** (100% ee, 86%), (–)-**79** (92.7% ee, 86%), (–)-**80** (95.3% ee, 57%), (–)-**81** (60% ee, 81%) and (+)-**82** (58% ee, 85%), respectively, in the optical and chemi-cal yields indicated [19]. In order to clarify the mechanism of the precise chiral re-cognition between **45** and the chiral ester, X-ray crystal structures of the 1:1 com-plexes of **45** with (*S*)-(–)-**77** and with (*S*)-(–)-**80** were studied [30].

By complexation with **45**, *rac*-β-lactams **83** and **84** were also separated efficiently into (–)-**83** of 100% ee (77%) and (–)-**84** of 100% ee (63%), respectively, in the good yields indicated [31].

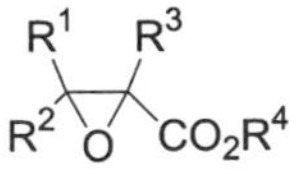

a: $R^1 = R^2 = R^3 = R^4 = Me$
b: $R^1 = R^2 = R^3 = Me; R^4 = Et$
c: $R^1 = R^2 = R^4 = Me; R^3 = Et$
d: $R^1 = R^2 = Me; R^3 = R^4 = Et$
e: $R^1 = R^2 = R^4 = Me; R^3 = H$
f: $R^1 = R^2 = Me; R^3 = H; R^4 = Et$
g: $R^1 = R^2 = R^4 = Et; R^3 = H$
h: $R^1R^2 = (CH_2)_5; R^3 = H; R^4 = Me$
i: $R^1R^2 = (CH_2)_5; R^3 = H; R^4 = Et$
j: $R^1 = R^3 = R^4 = Me; R^2 = H$

76

MeCHClCOOMe **77**

MeCH(OPh)COOMe **78**

MeCH(OH)CH₂COOEt **79**

ClCH₂CH(OH)CH₂COOMe **80**

MeOOCCH(OH)CH₂COOMe **81**

MeCH(NH₂)COOEt **82**

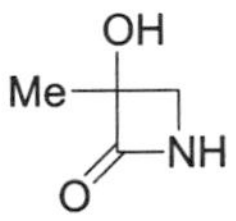

83

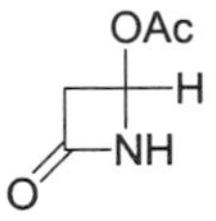

84

Table 3.3-4 Enantiomer separation of **76a–j** through inclusion complexation with **1b** or **1c**.

Host	Guest	Product	optical purity (% ee)	yield (%)
1b	76a	(+)-76a	100	9
1b	76b	(+)-76b	10	37
1c	76b	(+)-76b	18	38
1b	76c	(−)-76c	10	72
1b	76d	(+)-76d	66	69
1c	76d	(+)-76d	47	40
1b	76e	(+)-76e	100	63
1c	76f	(+)-76f	96	32
1b	76g	(+)-76g	100	62
1c	76g	(+)-76g	100	23
1b	76h	(−)-76h	100	31
1c	76h	(−)-76h	100	51
1c	76l	(−)-76i	54	63
1b	76j	(−)-76j	93	42

3.3.2.6
Sulfoxides, Sulfinates, Sulfoximines, Phosphinates and Phosphine Oxides

The binaphthol host **10b** was found to be very effective for enantiomeric separation of some sulfoxides. When a solution of **10b** and two molar equivalents of *rac*-methyl *m*-methylphenyl sulfoxide (**85c**) in benzene–hexane was kept at room temperature for 12 h, a 1:1 complex of **10b** and (+)-**85c** was obtained, after one recrystallization from benzene, as colorless prisms in 77% yield. Chromatography of the complex on silica gel gave (+)-**85c** of 100% ee in 77% yield [32]. By the same procedure, *rac*-**85d** was separated by **10b** to give (+)-**85d** of 100% ee in good yield. However, *rac*-**85a** was poorly separated with **10b**, giving approximately 5% ee enantiomer, while **85b** and **85e** did not form complexes with **10b**. In order to establish why the chirality of the *m*-substituted derivatives **85c** and **85d** is so precisely recognized by **10b**, the crystal structure of the complex of **10b** and (+)-**85c** was studied by X-ray analysis [33].

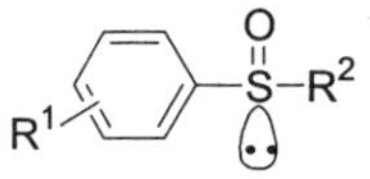

85

a: R^1 = H; R^2 = Me
b: R^1 = *o*-Me; R^2 = Me
c: R^1 = *m*-Me; R^2 = Me
d: R^1 = *m*-Me; R^2 = Et
e: R^1 = *p*-Me; R^2 = Me

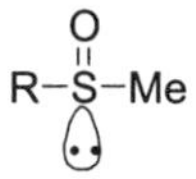

86

a: R = *n*Bu
b: R = *i*Bu
c: R = *sec*-Bu
d: R = *n*Pr
e: R = *i*Pr
f: R = Et

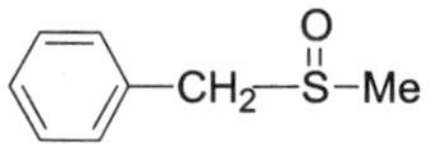

87

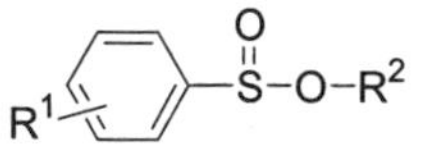

88

a: R^1 = H; R^2 = Me
b: R^1 = *p*-Me; R^2 = Me
c: R^1 = *p*-Me; R^2 = Et

Some dialkyl sulfoxides (**86**) were also separated into enantiomers by complexation with **10b**. *n*-Butyl methyl sulfoxide (**86a**) and methyl *n*-propyl sulfoxide (**86d**) were easily separated with **10a** to give enantiomerically pure (+)-**86a** and (–)-**86d**, respectively, in good yields. However, **86b** and **86f** were poorly separated with **10b**, and **86c** and **86e** did not form complexes with **10b** [32].

Although **10b** is not effective for separation of *rac*-**85a**, host **1c** is very effective for the separation of not only **85a** but also methyl sulfoxide (**87**) and alkyl phenylsulfinates (**88**). By complexation of these racemic compounds with **1c**, (*R*)-(+)-**85a**

(100% ee, 56%), (–)-**87** (100% ee, 15%), (+)-**88a** (69% ee, 18%), (+)-**88b** (77% ee, 20%) and (–)-**88c** (56% ee, 27%) were obtained in the optical and chemical yields indicated [34].

89

a: R^1 = H; R^2 = Me
b: R^1 = H, R^2 = Et
c: R^1 = o-Me; R^2 = Me
d: R^1 = m-Me; R^2 = Me
e: R^1 = m-Me; R^2 = Et
f: R^1= m-Me; R^2 = nPr
g: R^1 = p-Me; R^2 = Me

90

a: R^1 = Me; R^2 = iC_5H_{11}
b: R^1 = Et, R^2 = nC_6H_{13}

a: R = H
b: R = o-Me
c: R = m-Me
d: R = p-Me

91

a: R = H
b: R = m-Me

92

The chiral host **10b** was effective for enantiomeric separations of alkyl aryl sulfoximines (**89a–g**). By complexation of *rac*-**89b**, *rac*-**89d** and *rac*-**89e** with **10b**, (–)-**89b** (100% ee, 37%), (–)-**89d** (100% ee, 70%) and (+)-**89e** (100% ee, 50%), respectively, were obtained in the optical and chemical yields indicated [35]. However, separation of *rac*-**89a** with **10b** was not effective and (–)-**89a** of 35% ee was obtained in 45% yield after five recrystallizations of the complex of (–)-**89a** and **10b** from benzene. **89c, 89f** and **89g** did not form complexes with **10b**. These results show that the efficiency of the enantiomeric separation is highest when the alkyl group is methyl or ethyl and the aryl group is *m*-tolyl. Since this tendency is similar to that in the case of sulfoxide, the efficiency of the enantiomeric separation of **89** probably depends on the packing of **10b** and **89** molecules in the crystalline lattice of their inclusion complex, as has been reported for the complex of **10b** and (+)-**85c** [32]. Although **10b** did not form complexes with dialkyl sulfoximines (**90**), **8** formed complexes with some of them, and some were separated into enantiomers efficiently by the complexation. For example, by complexation of **8** with *rac*-**90a** and *rac*-**90b** in ether, (–)-**90a** (100% ee, 80%) and (–)-**90b** (100% ee, 88%), respectively, were finally obtained in the optical and chemical yields indicated [35].

Enantiomeric separations of phosphinates and phosphine oxides are not easy by the usual method. However, these were easily separated by complexation with **10b**. For example, by inclusion complexation of *rac*-phosphinates *(***91a–d***)* with **10b**, 100% enantiomerically pure (–)-**91a** (12%), (–)-**91b** (47%), (–)-**91c** (50%) and (–)-**91d**

(32%), respectively, were obtained in the yields indicated. [36]. By the same method, *rac*-phosphine oxides (**92a–b**) were also separated efficiently by complexation with **10b** to give enantiomerically pure (+)-**92a** and (+)-**92b** in 60 and 33% yields, respectively [36]. The X-ray crystal structure of the 1:1 complex of **10b** with (+)-**92a** has been analyzed [36].

3.3.3
Green One-Pot Preparative Process for Obtaining Optically Active Compounds by a Combination of Solid-state Reaction and Enantiomeric Separation in a Water Suspension Medium

A new green preparative method for obtaining optically active compounds can be designed using one-pot solid-state reactions and enantiomeric separation processes, carried out continuously in a water suspension medium. Some successful examples are described in this section.

When a mixture of acetophenone (**93a**) (1 g, 8.3 mmol), $NaBH_4$ (0.94 g, 24.9 mmol) and water (10 ml) was stirred at room temperature for 2 h, *rac*-1-phenylethanol (**94a**) was produced. To the water suspension of rac-**94a** was added powdered **1a** (3.87 g, 8.3 mmol), and the mixture was stirred for a further 3 h to give a 2:1 inclusion complex of **1a** with (–)-**94a**. The inclusion complex formed was filtered and dried. Heating of the complex *in vacuo* gave (–)-**94a** of 95% ee (0.42 g, 85%) [37]. From the filtrate remaining after separation of the inclusion crystals by filtration, (+)-**94a** of 77% ee (0.35 g, 70%) was obtained by extraction with ether. Optically active alcohols **94b–g** were prepared by the same procedure (Table 3.3-5). This is an excellent combination of a well-established solid-state reaction [1, 38] and host–guest inclusion complexation in the solid state [39].

a: Ar = C_6H_5, R = CH_3
b: Ar = C_6H_5, R = CH_2CH_3
c: Ar = C_6H_5, R = $CH_2CH_2CH_3$
d: Ar = C_6H_5, R = CF_3
e: Ar = 2-pyridyl, R = CH_3
f: Ar = 3-pyridyl, R = CH_3
g: Ar = 4-pyridyl, R = CH_3
h: Ar = 2-furyl, R = CH_3
i: Ar = 2-thiophenyl, R = CH_3

a: R^1 = H, R^2 = CH_3, R^3 = H
b: R^1 = R^2 = R^3 = H
c: R^1 = R^2 = CH_3, R^3 = H
d: R^1 = R^2 = H, R^3 = CH_3

R–S–Me $\xrightarrow[\text{H}_2\text{O}]{\text{H}_2\text{O}_2}$ rac- R–$\overset{\text{O}}{\underset{}{\overset{\uparrow}{\text{S}}}}$–Me

97   **98**

a: R = C$_6$H$_5$
b: R = 4-CH$_3$C$_6$H$_4$
c: R = C$_5$H$_{11}$
d: R = C$_6$H$_{13}$

Table 3.3-5 Result of one-pot preparation of optically active *sec*-alcohols **94a,b** and **94e–i** by a combination of ketone reduction and enantiomeric resolution in a water suspension medium.

Host	Ketone	host : guest	From complex			From filtrate		
			product	yield (%)	enantio-meric purity (% ee)	product	yield (%)	enantio-meric purity (% ee)
1a	93a	2 : 1	(–)-94a	85	95	(+)-94a	70	77
1a	93b	2 : 1	(–)-94b	96	62	(+)-94b	50	52
1a	93e	3 : 1	(–)-94e	26	76	(+)-94e	156	18
1b	93e	2 : 1	(–)-94e	44	99	(+)-94e	134	40
1c	93e	1 : 1	(–)-94e	92	88	(+)-94e	76	62
1a	93f	1 : 1	(+)-94f	88	>99	(–)-94f	86	73
1b	93f	1 : 1	(+)-94f	86	96	(–)-94f	82	66
1c	93f	1 : 1	rac-94f	78	0	rac-94f	86	0
1a	93g	1 : 1	(+)-94g	82	47	(–)-94g	78	26
1b	93g	1 : 1	(+)-94g	80	77	(–)-94g	82	36
1c	93g	1 : 1	rac-94g	90	0	rac-94g	80	0
1a	93h	2 : 1	(–)-94h	76	93	(+)-94h	96	50
1a	93l	2 : 1	(–)-94i	84	86	(+)-94i	61	43

Very interestingly, in some cases, the enantioselective inclusion complexation of the *rac*-product with a chiral host in aqueous medium is more efficient than recrystallization from an organic solvent. For example, inclusion complexation of *rac*-**94e** with **1a** or **1b** did not occur by their recrystallization from organic solvent, but enantioselective inclusion complexation between *rac*-**94e** and **1b** occurred efficiently in aqueous medium to give, finally, (–)-**94e** of 99% ee.

The one-pot method can be applied to the preparation of optically active epoxides and sulfoxides. For example, when a mixture of 30% aqueous H$_2$O$_2$ (2.61 g, 23 mmol), 8 M aqueous NaOH (30 ml), 3-methyl-2-cyclohexen-1-one (**95a**) (2.5 g, 23 mmol) and water (10 ml) was stirred at room temperature for 3 h, *rac*-**96a** was produced. **1b** (3.7 g, 7.6 mmol) was added to the water suspension medium of *rac*-**96a**, and the mixture was stirred for a further 48 h to give a 1:1 inclusion complex of **1b** with (–)-**96a**. Heating the filtered inclusion complex *in vacuo* gave (–)-**96a** of 97% ee (0.88 g, 61%). From the residual filtrate after separation of the inclusion complex, (+)-**96a** of 48% ee (1.43 g, 99%) was obtained by extraction with ether. By the same procedure, optically active **96b–d** were obtained (Table 3.3-6). In the case of **95b** and **95c**, (+)-**96b** and (+)-**96c** were obtained in an enantiopure state [37].

When a mixture of methyl phenyl sulfide (**97a**) (1 g, 8.1 mmol), 30% aqueous H$_2$O$_2$ (1.84 g, 16.2 mmol) and water (10 ml) was stirred at room temperature for

Table 3.3-6 Result of one-pot preparation of optically active epoxides **96a–d** by a combination of cyclohexenone epoxidation and enantiomeric resolution in a water suspension medium.

Host	Cyclo-hexanone	Inclusion complex	From complex			From filtrate		
		host : guest	product	yield (%)	enantio-meric purity (% ee)	product	yield (%)	enantio-meric purity (% ee)
1a	95a	1 : 1	(–)-96a	61	97	(+)-96a	99	48
1c	95a	1 : 1	(–)-26a	56	97	(+)-96a	79	34
1b	95b	1 : 1	(+)-96b	38	100	(–)-96b	78	51
8	95c	1 : 1	(+)-96c	57	100	(–)-96c	62	87
1b	95d	2 : 1	(–)-96d	63	63	(+)-96d	96	47

24 h, *rac*-methyl phenyl sulfoxide (**98a**) was produced. The chiral host **1c** (2 g, 4 mmol) was added to the water suspension medium of *rac*-**98a**, and the mixture was stirred for a further 15 h to give a 1:1 inclusion complex of **1c** with (+)-**98a**. Heating the filtered inclusion complex *in vacuo* gave (+)-**98a** of 57% ee (0.45 g, 82%). From the filtrate remaining after separation of the inclusion crystals, (–)-**98a** of 54% ee (0.4 g, 73%) was obtained by extraction with ether. By the same procedure, optically active **98b–d** were prepared (Table 3.3-7). In the case of (+)-**98b** and (–)-**98c**, the efficiency of the enantiomeric separation is very high [37].

Table 3.3-7 Result of one-pot preparation of optically active sulfoxides **98a–d** by a combination of sulfide oxidation and enantiomeric resolution in a water suspension medium.

Host	Sulfide	Inclusion complex	From complex			From filtrate		
		host : guest	product	yield (%)	enantio-meric purity (% ee)	product	yield (%)	enantio-meric purity (% ee)
1c	97a	1 : 1	(+)-98a	82	57	(–)-98a	73	54
1c	97b	1 : 1	(+)-96b	75	98	(–)-98b	89	78
1c	97c	1 : 1	(–)-96c	70	96	(+)-98c	80	55
1c	97d	3 : 2	(–)-96d	55	49	(+)-98d	100	31

Since the chiral host compounds remain after separation of the chiral products from their inclusion complexes by distillation, they can therefore be used repeatedly, so this one-pot method in water is ecologically advantageous and economical.

3.3.4
Enantiomeric Separation by Inclusion Complexation in Suspension Media and by Fractional Distillation

In this section, one-pot preparations of optically active compounds by a combination of solid-state reaction and enantioselective inclusion complexation in a water suspension medium are described. In order to establish the suspension procedure as a general enantiomeric separation method, enantiomeric separations of various compounds by complexation in hexane and water suspension media were studied. Furthermore, by combining enantioselective inclusion complexation with a chiral host in the solid state with distillation, a fascinating enantiomeric separation method by fractional distillation was established.

For example, when a suspension of powdered **1a** (1 g) and an equimolar amount of *rac*-1-phenylethanol (**94a**) (0.262 g) in hexane (10 ml) was stirred at room temperature for 6 h, a 2:1 inclusion crystal of **1a** and (–)-**94a** (1.12 g) was obtained. Heating of the crystal *in vacuo* gave (–)-**94a** of 95% ee (0.112 g, 85%) [40]. Various alcohols and epoxides were separated efficiently by a similar procedure (Table 3.3-8). In most cases, the chiral hosts **1a**, **1b**, **1c** and **8** are very effective, as shown in Table 3.3-8. The efficiencies of hexane and water as suspension media are not significantly different [40].

Table 3.3-8 Results of enantiomer separation using the method of inclusion crystallization by suspension.

Host	Guest	Medium	Product	yield (%)	(%) ee
1a	23	Hexane	(–)-23	85	95
1a	23	H$_2$O	(–)-23	85	98
1a	25	Hexane	(–)-25	75	100
1a	25	H$_2$O	(–)-25	75	98
1a	28	Hexane	(+)-28	89	92
1a	28	H$_2$O	(+)-28	76	100
1b	103a	Hexane	(+)-103a	80	80
1b	103b	Hexane	(+)-103b	93	78
1b	96a	Hexane	(–)-96a	82	100
1a	96a	H$_2$O	(–)-96a	73	100
1b	76b	Hexane	(+)-76b	78	75
8	96c	Hexane	(+)-96c	57	98
8	96c	H$_2$O	(+)-96c	85	97
1b	76g	Hexane	(+)-76g	75	100
1b	76g	H$_2$O	(+)-76g	89	100
1c	76d	Hexane	(+)-76d	78	100
1c	76d	H$_2$O	(+)-76d	80	100
1b	76e	Hexane	(+)-76e	59	70
1b	76e	H$_2$O	(+)-76e	52	86
1c	100	Hexane	(+)-100	76	75
1c	100	H$_2$O	(+)-100	74	47

When enantiomeric separation by the suspension method is combined with distillation, the two enantiomers can be separated easily by fractional distillation. For example, after a suspension of **1a** (1 g) and *rac*-**94a** (0.26 g) in hexane (1 ml) had been kept at room temperature for 1 h, the mixture was distilled *in vacuo* to give initially (+)-**94a** of 59% ee (0.26 g, 125%) at lower boiling temperature. The residue was then distilled at elevated temperature and (–)-**94a** of 97% ee (0.09 g, 69%) was obtained at higher boiling temperature. When the distillation is repeated for the enantiomerically impure sample, the enantiomerically pure product can be obtained.

Finally, it was found that no liquid as a suspension medium is necessary for enantiomeric separation by fractional distillation. For example, heating a mixture of **1c** and two molar equivalents of *rac*-1-(*p*-tolyl)ethylamine (**99c**) in the Kugelrohr apparatus at 70°C, 2 mmHg gave (+)-**99c** of 98% ee in 102% yield by distillation, and further heating of the residue at 150°C, 2 mmHg gave (+)-**99c** of 100% ee in 98% yield. By mixing **1c** and *rac*-**99c** in the solid state, inclusion complexation between **1c** and (+)-**99c** occurs. On heating, uncomplexed (–)-**99c** distills out first at a relatively low temperature. At elevated temperature, the inclusion complex of **1c** and (+)-**99c** decomposed and (+)-**99c** distills out. Similarly, **99a** and **99b** were also easily separated into the corresponding enantiomers by distillation in the presence of a chiral host (Table 3.3-9). By the same procedure, enantiomeric separations of epoxides (**88, 96a–c, 76f–g, 100**), alcohols (**94a–b, 28, 102**), hydroxycarboxylic acid esters (**103**), amino alcohols (**104–105**) and amine (**106**) were accomplished efficiently (Table 3.3-9) [41]. For enantioseparation of **102** and **105**, a new chiral host **101** was designed. Table 3.3-9 lists only the enantiomers that form inclusion complexes with the chiral host used and are then distilled at a relatively high temperature.

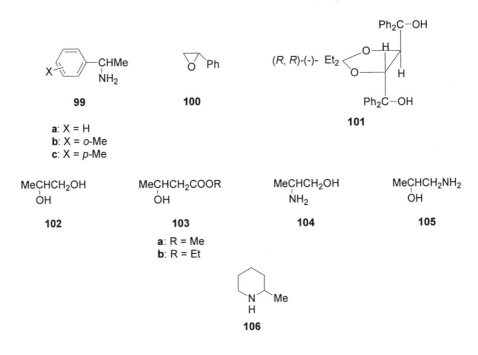

Table 3.3-9   Enantiomer separation by fractional distillation in the
presence of a chiral host.

Host	*rac*-Guest	Distillation times repeated	Chiral product		
				yield (%)	ee (%)
1b	96b	2	(+)-96b	60	94
1b	96a	2	(−)-96a	60	93
1c	96a	2	(−)-96a	55	98
8	96c	3	(+)-96c	31	95
1c	76d	1	(+)-76d	58	92
1b	76g	1	(+)-76g	74	92
1c	76g	1	(+)-76g	78	90
1c	100	4	(−)-100	10	90
1a	23	2	(−)-23	40	96
1a	25	2	(−)-25	47	92
1a	28	3	(+)-28	17	96
101	102	2	(−)-102	39	90
1b	103a	3	(+)-103a	30	94
1b	103b	3	(+)-103b	39	92
1b	104	2	(+)-104	33	100
1c	104	2	(+)-104	37	100
101	105	1	(−)-105	95	25
1a	99a	1	(−)-99a	3	62
1c	99b	3	(−)-99b	18	95
1c	99c	1	(−)-99c	98	100
1a	106	1	(+)-106	98	42

3.3.5
Enantiomeric Separation Without Using a Chiral Source

Enantiomeric separation of some compounds can be accomplished by complexation with achiral compounds. Some examples are described below.

3.3.5.1
Enantiomeric Separation of *rac*-7-Bromo-1,4,8-triphenyl-2,3-benzo[3.3.0]octa-2,4,7-trien-6-one

The title compound (**107**) [42] has been found to form inclusion complexes with a
wide variety of solvent molecules [43]. In the complexation, racemates or conglo-
merates of **107** were formed, depending on the choice of solvent. In the latter case,
the inclusion crystals consisting of one enantiomer of **107** were formed preferentially and the enantiomeric separation of **107** could be performed. For example, recrystallization of *rac*-**107** from the solvents shown in Table 3.3-10 gave a 1:1 complex of the *rac*-**107** with the solvent as yellow crystals. On the other hand, recrystallization of the *rac*-**107** from the solvents shown in Table 3.3-11 gave a 1:1 complex

of optically active **107** with the solvent as yellow crystals. By seeding with one complex crystal of the enantiomerically active **107** during the recrystallization of *rac*-**107**, a large quantity of the chiral complex crystal could be obtained [43]. For example, when one piece of a crystal of the (+)-**107**-THF complex was added to a solution of *rac*-**107** (10 g) in THF (50 ml) and the mixture was kept at room temperature for 12 h, the THF complex of enantiomerically pure (+)-**107** (0.81 g, 14%) was obtained. Distillation of THF from the complex *in vacuo* gave (+)-**107** of 100% ee. To the filtrate left after filtration of the (+)-**107**-THF complex, one piece of the crystal of (−)-**107**-THF complex was added and the solution was kept at room temperature for 12 h to give the THF complex of enantiomerically pure (−)-**107** (1.1 g, 19%). Distillation of THF from the complex *in vacuo* gave (−)-**107** of 100% ee.

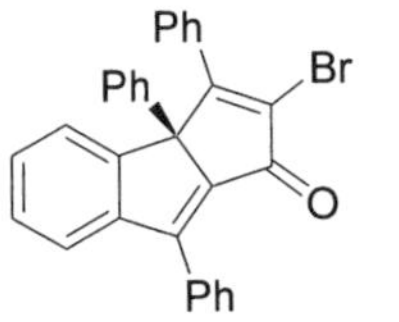

107

Table 3.3-10 Formation of racemic 1:1 inclusion crystals of **107** with solvent (guest) molecule.

Solvent (guest)	mp (°C)	Solvent (guest)	mp (°C)
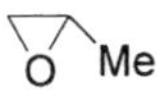 O Me	112–115	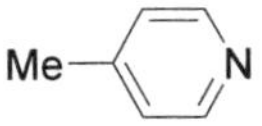 Me N	100–102
 O	nd	PhNH$_2$	59–61
 O	nd	 N N	145–149
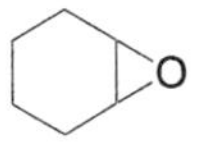 O	109–112	 Me$_2$N NMe$_2$	177–179
CHCl$_3$	101–105	 MeO OMe	159–165
CHBr$_3$	nd	PhCl	88–90
CH$_2$Cl$_2$	127–131	 Cl Cl	132–138

nd – not distinct.

Table 3.3-11 Formation of conglomeratic 1:1 inclusion crystals of
107 with solvent (guest) molecule.

Solvent (guest)	mp (°C)	Solvent (guest)	mp (°C)
 (tetrahydrofuran)	112–114	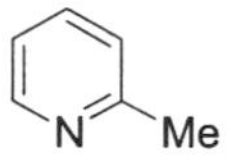 (2-methylpyridine)	135–139
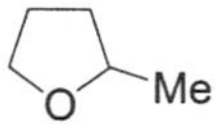 (2-methyltetrahydrofuran)	nd	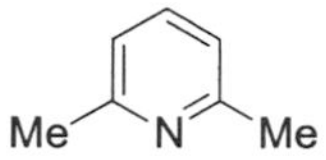 (2,6-dimethylpyridine)	98–102
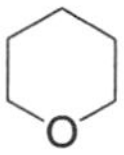 (tetrahydropyran)	120–125	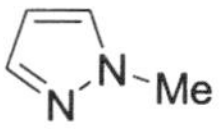 (1-methylpyrazole)	119–123
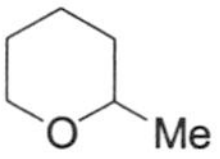 (2-methyltetrahydropyran)	nd	MeI	nd
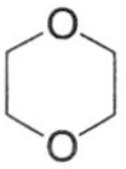 (1,4-dioxane)	122–128	EtBr	nd
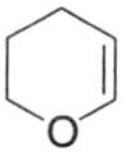 (3,4-dihydro-2H-pyran)	nd	EtI	nd
		CCl₄	127–130
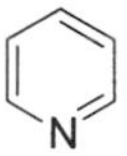 (pyridine)	nd	CBr₄	129–134

nd – not distinct.

It is interesting to note that the solvents that form complexes with *rac*-**107** do not form complexes with optically active **107** and vice versa. The big difference in the roles of the various solvents shown in Tables 3.3-10 and 3.3-11 is also interesting. X-ray crystal-structure analysis of a 1:1 complex of (S)-(–)-**107** and 1,2-dichloroethane showed that the latter is accommodated in the cavity of the inclusion complex in the form of a nearly eclipsed chiral rotamer, but no significant interaction between the host and guest molecules is present [44].

3.3.5.2
Enantiomeric Separation by Complexation with Achiral 2,3,6,7,10,11-Hexahydroxytriphenylene

The title achiral molecule (**108**) was found to be arranged in a chiral form in its inclusion complex with a guest compound. By using this phenomenon, some *rac*-guests were separated into their enantiomers by complexation with achiral **108** [45].

Firstly, it was found that **108** forms inclusion complexes with various solvent molecules. In the inclusion complexes with PrⁱOH, cyclopentanone, 2-cyclopentenone and 2-cyclohexenone, **108** molecules were found to be ordered in a chiral form from measurement of CD spectra in the solid state of the complexes formed. X-ray analysis of a 1:3 complex of **108** and 2-cyclopentenone showed that **108** molecules form a chiral helical column through hydrogen-bond formation, and 2-cyclopentenone molecules bind to the OH groups of **108** in the chiral column [45]. Using the complexation of the helices of **108** with the guest allowed enantiomeric separation of *rac*-guest. For example, recrystallization of **108** from 2-methylcyclopentanone (**109**) gave their 1:3 inclusion complex. Heating one piece of the complex crystal that shows a (+)-Cotton effect in the region of 300 nm *in vacuo* gave (+)-**109** of 34% ee by distillation. Heating the other piece of crystal that shows a (–)-Cotton effect in the region of 300 nm *in vacuo* gave (–)-**108** of 37% ee [45].

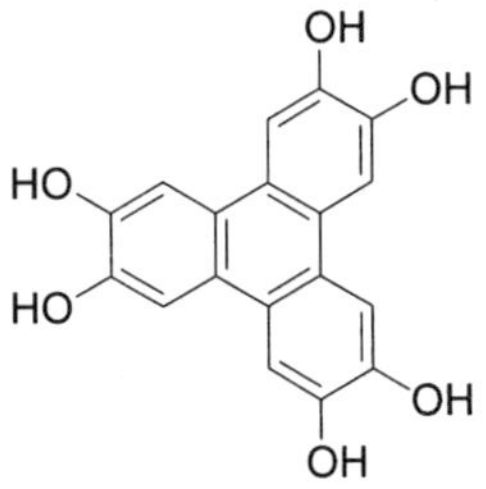

108 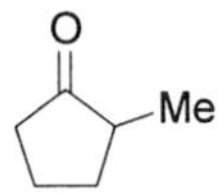   **109**

3.3.5.3
Enantiomeric Separation of 2,2'-Dihydroxy-1,1'-binaphthyl by Complexation with Racemic or Achiral Ammonium Salts

Efficient enantiomeric separation of the title racemic compound (**10a**) by complexation with (+)-*N*-benzylcinchonidinium chloride [46], chiral *N,N,N*-trimethyl-*N*-(2-hydroxy-1-alkylethyl) ammonium bromide [47] and chiral *N*-(3-chloro-2-hydroxypropyl)-*N,N,N*-trimethylammonium chloride [48] has been reported. However, it was disclosed that chirality of these ammonium salts is not necessary for the enantiomeric separation of **10a**, and **10a** can easily be separated into enantiomers (**10b** and **10c**) by complexation with racemic or achiral ammonium salts.

$Me_3N^+CH_2CH(OH)CH_2Cl\ Cl^-$ $Me_3N^+CH_2CH_2OH\ Cl^-$ $Me_4N^+\ Cl^-$
110   **111**   **112**

a: conglomerate
b: (S)-(–)-form
c: (R)(+)-form

When complexation of **10a** and the conglomerate salt, *N,N,N*-trimethyl-*N*-(2-hydroxy- ethyl)ammonium chloride (**110a**) in EtOH was carried out by seeding 1:1 complexes of **10c** with **110b** and of **10b** with **110c**, efficient simultaneous and mutual enantiomeric separation occurred to give **10b, 10c, 110b** and **110c** simultaneously in good optical and chemical yields. For example, powdered complex of **10c** and **110b** (3 mg) was added to a solution of **10a** (7.15 g, 25 ml) and **110a** (3.30 g, 17.5 mmol) in EtOH (90 ml) and the solution was kept at room temperature for 12 h, to give a 1:1 complex of **10c** and **110b** (0.77 g, 19%). The complex was dissolved in a mixture of ether and water. From the ether phase **10c** of 96% ee was obtained, and from the aqueous phase **110b** of 60% ee was obtained. To the EtOH solution left after separation of the complex of **10c** and **110b** by filtration, was added powdered complex of **10b** and **110c** (3 mg) and the solution was kept at room temperature for 14 h to give a 1:1 complex of **10b** and **110c** (0.8 g, 19%). The complex was dissolved in a mixture of ether and water. From the ether solution **10b** of 97.5% ee was obtained, and from the aqueous solution **110c** of 66% ee was obtained (Scheme 3.3-2) [49]. By repeating the seeding experiments a total of eight times, as shown in Sche-

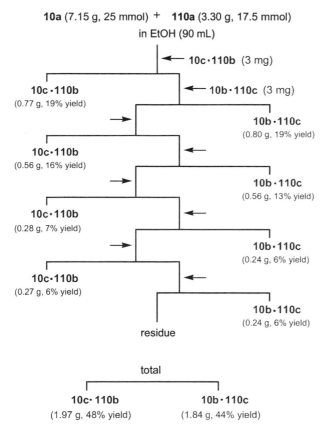

Scheme 3.3-2 Simultaneous and mutual enantiomeric separation of **10a** and **110a** by preferential crystallization in the presence of chiral seed crystals.

me 3.3-2, **10c** of 96% ee (34%), **10b** of 59% ee (48%), **110b** of 96.5% ee (31%) and **110c** of 62.5% ee (44%) were obtained in the yields indicated. In the simultaneous and mutual enantioseparation of **10a** and **110a**, the efficiency for the separation of **10a** was quite good, although that of **110a** was only moderate. Nevertheless, the efficient enantioseparation of **10a** by using a small amount of the seed crystal of **10c-110b** and **10b-110c** complexes is valuable [49].

The achiral salt, N-(2-hydroxyethyl)-N,N,N-trimethylammonium chloride (**111**) was found to form a mixture of two 1:1 inclusion complexes **10b-111** and **10c-111** as conglomerate crystals, but not the racemic complex **10a-111**. The result strongly suggests that enantiomeric separation of **10a** can easily be done by complexation with **111**. **10b-111** and **10c-111** crystals can easily be separated mechanically. By repeating the seeding experiments as shown in Scheme 3.3-3, **10b** of 99% ee and **10c** of 99.5% ee were finally obtained in 66 and 64% yields, respectively [49].

Furthermore, the much simpler achiral ammonium salt tetramethylammonium chloride (**112**) was found to be useful for the enantiomeric separation of **10a**, although **112** forms both racemic and conglomerate complexes with **10a** and **10b** (or **10c**), respectively. Firstly, the racemic and conglomerate complexes of **10** and **112**

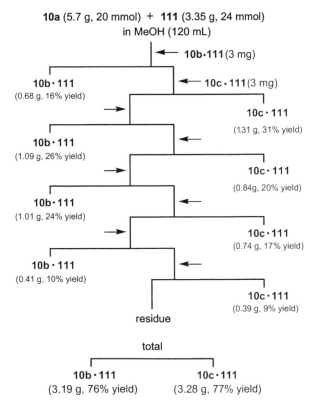

Scheme 3.3-3 Enantiomeric separation of **10a** by inclusion complexation with **111** in the presence of chiral seed crystals.

were prepared. **10a** (1.43 g, 5 mmol) and **112** (0.55 g, 5 mmol) were recrystallized from MeOH (7 ml), and a racemic 1:1:1 complex, **10a-112**-MeOH (**113**) crystallized out as colorless needles. However, when **10a** and **112** in the same amounts as above were recrystallized from MeOH (15 ml), a 1:1 mixture of the conglomerate complexes **10b-112** and **10c-112** (**114**) was formed as colorless prisms. **10b-112** and **10c-112** crystals were separated mechanically and their spectral data compared with those of corresponding authentic samples of **10b-112** or **10c-112** prepared by complexation of **10b** or **10c** with **112** in MeOH. The X-ray crystal structure of **10c-112** was also determined. Secondly, an enantiomeric separation experiment by seeding with chiral complex crystals was carried out as shown in Scheme 3.3-4. By repeating the seeding processes eight times, **10b** of 99% ee and **10c** of 99% ee were finally obtained in 50 and 53% yields, respectively [49].

Study of the enantiomeric separation of **10a** by complexation with **112** revealed two interesting modes of transformation of racemic complex **113** into conglomerate complex **114** in the solid state. One is a thermal transformation. Thermogravimetric and differential thermal analysis measurements of **113** showed the evaporation of MeOH at 120°C, phase transfer from MeOH free-**113** to **114** at 183°C and

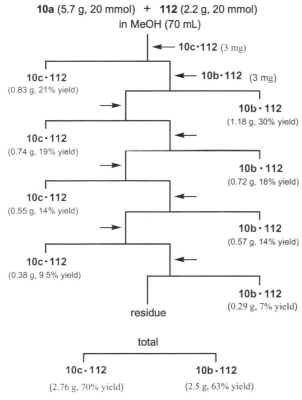

Scheme 3.3-4 Enantiomeric separation of **10a** by inclusion complexation with **112** in the presence of chiral seed crystals.

finally melting of **114** at 130°C [49]. The transformation of **113** into **114** also occurred simply by contact of **113** with MeOH vapor in the solid state for 30 min at room temperature. Contact of MeOH-free **113** with MeOH vapor for 30 min also gave **114**. Very interestingly, contact of a mixture of powdered **10a** and **112** with MeOH vapor for 30 min gave **114**, although simple mixing of powdered **10a** with an equimolar amount of **112** using a mortar and pestle for 30 min did not give any complex. The transformation from **113** into **114** also occurred by contact with EtOH vapor; however, CHCl₃, Et₂O, toluene or hexane vapor did not cause any change. When transformation from a racemic crystal into a conglomerate crystal in the solid state is applicable to an enantiomeric separation, a new green enantioseparation method can be designed.

3.3.6
Conclusions and Perspectives

There is no doubt that enantiomeric separation of racemic compounds will become increasingly important. The enantiomeric separation method should, however, be simple, green and economical. In this chapter, some examples of separations by inclusion complexation with chiral or achiral host compounds are described. In future, chiral host compounds for enantiomeric separation should be improved. More green and cheap chiral sources should be developed for more efficient and safe enantioseparations. For example, naturally occurring chiral sources such as sugars, amino acids, terpenes, alkaloids and cellulose could become useful and important.

References

1 (a) F. Toda, *Top. Curr. Chem.* **1987**, *140*, 43–69; (b) F. Toda, *Acc. Chem. Res.* **1995**, *28*, 480–486; (c) Z. Urbanczyk-Lipkowska, F. Toda, in *Separations and Reactions in Organic Supramolecular Chemistry*, F. Toda, R. Bishop, eds., J. Wiley & Sons, Chichester, **2004**, pp. 1–31.

2 H. Miyamoto, M. Sakamoto, K. Yoshioka, R. Takaoka, F. Toda, *Tetrahedron Asymm.* **2000**, *11*, 3045–3048.

3 F. Toda, K. Tanaka, *Tetrahedron Lett.* **1988**, *29*, 551–554.

4 M. Kato, K. Tanaka, F. Toda. *Supramol. Chem.* **2001**, *13*, 175–180.

5 (a) F. Toda, K. Tanaka, *Tetrahedron Lett.* **1981**, *22*, 4669–4672; (b) F. Toda, K. Tanaka, H. Ueda, T. Oshima, *J. Chem. Soc., Chem. Commun.* **1983**, 743–744; (c) F. Toda, K. Tanaka, H. Ueda, T. Oshima, *Israel. J. Chem.* **1985**, *25*, 338–345.

6 F. Toda, K. Tanaka, M. Kido, *Chem. Lett.* **1988**, 513–516.

7 F. Toda, K. Tanaka, T. Omata, K. Nakamura, T. Oshima, *J. Am. Chem. Soc.* **1983**, *105*, 5151.

8 F. Toda, K. Mori, Z. Stein, I. Goldberg, *Tetrahedron Lett.* **1989**, *30*, 1841–1844.

9 F. Toda, H. Akai, *J. Org. Chem.* **1990**, *55*, 4973–4974.

10 K. Mori, F. Toda, *Tetrahedron Asymm.* **1990**, *1*, 281–284.

11 F. Toda, A. Sano, L. R. Nassimbeni, M. L. Niven, *J. Chem. Soc., Perkin Trans. 2* **1991**, 1971–1975.

12 F. Toda, K. Tanaka, K. Koshiro, *Tetrahedron Asymm.* **1991**, *2*, 873–874.

13 F. Toda, S. Matsuda, K. Tanaka, *Tetrahedron Asymm.* **1991**, *2*, 983–986.

14 K. Nishikawa, A. Matsumoto, H. Tsukada, M. Shiro, F. Toda, *Acta Crystallogr. Sec. C*, **1997**, 351–353.

15 F. Toda, K. Tanaka, *Chem. Lett.* **1983**, 661–664.

16 K. Tanaka, F. Toda, *J. Chem. Soc., Chem. Commun.* **1983**, 1513–1514.

17 F. Toda, K. Mori, Y. Matsuura, H. Akai, *J. Chem. Soc., Chem. Commun.* **1990**, 1591–1593.

18 F. Toda, M. Ochi, *Enantiomer*, **1996**, *1*, 85–88.

19 F. Toda, K. Tanaka, *Tetrahedron Lett.* **1988**, *29*, 1807–1810.

20 K. Tanaka, O. Kakinoki, F. Toda, *J. Chem. Soc., Perkin Trans. 1* **1992**, 307.

21 (a) K. Tanaka, O. Kakinoki, F. Toda, *Tetrahedron Asymm.* **1992**, *3*, 517–520; (b) F. Toda, K. Tanaka, *Chem. Lett.* **1985**, 885–888.

22 L. R. Nassimbeni, M. L. Niven, K. Tanaka, F. Toda, *J. Crystallogr. Spectrosc. Res.* **1991**, *21*, 451–457.

23 F. Toda, K. Tanaka, D. Marks, I. Goldberg, *J. Org. Chem.* **1991**, *56*, 7332–7335.

24 K. Tanaka, M. Kato, F. Toda, *Heterocycles*, **2001**, *54*, 405–410.

25 F. Toda, K. Tanaka, C. W. Leung, A. Meetsma, B. L. Feringa, *J. Chem. Soc., Chem. Commun.* **1994**, 2371–2372.

26 F. Toda, H. Miyamoto, H. Ohta, *J. Chem. Soc., Perkin Trans. 1.* **1994**, 1601–1604

27 F. Toda, A. Sato, K. Tanaka, T. C. W. Mak, *Chem. Lett.* **1989**, 873–876.

28 F. Toda, K. Tanaka, M. Watanabe, T. Abe, N. Harada, *Tetrahedron Asymm.* **1995**, *6*, 1495–1498.

29 F. Toda, H. Takumi, K. Tanaka, *Tetrahedron Asymm.* **1995**, *6*, 1059–1062.

30 G-H. Lee, Y. Wang, K. Tanaka, F. Toda, *Chem. Lett.* **1988**, 781–784.

31 F. Toda, K. Tanaka, M. Yagi, Z. Stein, I. Goldberg, *J. Chem. Soc., Perkin Trans. 1* **1990**, 1215–1216.

32 F. Toda, K. Tanaka, S. Nagamatsu, *Tetrahedron Lett.* **1984**, *25*, 4929.

33 F. Toda, K. Tanaka, T. C. W. Mak, *Chem. Lett.* **1984**, 2085–2088.

34 F. Toda, K. Tanaka, T. Okuda, *J. Chem. Soc., Chem. Commun.* **1995**, 639–640.

35 K. Mori, F. Toda, *Chem. Lett.* **1988**, 1997–2000.

36 F. Toda, K. Mori, Z. Stein, I. Goldberg, *J. Org. Chem.* **1988**, *53*, 308–312.

37 H. Miyamoto, S. Yasaka, R. Takaoka, K. Tanaka, F. Toda, *Enantiomer.* **2001**, *6*, 51–55.

38 (a) F. Toda, *Top. Curr. Chem.* **1988**, *149*, 211–238; (b) K. Tanaka, F. Toda, *Chem. Rev.* **2000**, *100*, 1025–1074; F. Toda ,ed., *"Organic Solid-State Reactions*, Kluwer, Dordrecht, 2002.

39 F. Toda, K. Tanaka, A. Sekikawa, *J. Chem. Soc., Chem. Commun.* **1987**, 279–280.

40 F. Toda, Y. Tohi, *J. Chem. Soc., Chem. Commun.* **1993**, 1238–1240.

41 F. Toda, H. Takumi, *Angew. Chem., Int. Ed. Engl.* **1996**, *33*, 728–729.

42 F. Toda, M. Sasaoka, Y. Todo, K. Iida, T. Hino, Y. Nishiyama, H. Ueda, T. Oshima, *Bull. Chem. Soc. Jpn.* **1983**, *56*, 3314–3318.

43 F. Toda, K. Tanaka, *Tetrahedron Asymm.* **1990**, *1*, 359–362.

44 F. Toda, K. Tanaka, R. Kuroda, *Chem. Commun.* **1997**, 1227–1228.

45 F. Toda, K. Tanaka, T. Matsumoto, T. Nakai, I. Miyahara, K. Hirotsu, *J.Phys. Org. Chem.* **2000**, *13*, 39–45.

46 K. Tanaka, T. Okada, F. Toda, *Angew. Chem., Int. Ed. Eng.* **1993**, *32*, 1147–1148.

47 F. Toda, K. Tanaka, *Chem. Commun.* **1997**, 1087–1088.

48 F. Toda, K. Yoshizawa, S. Hyoda, S. Toyota, S. Chatziefthimiou, I. M. Mavridis, *Org. Biomol. Chem.* **2004**, *2*, 449–451.

49 (a) Y. Yoshizawa, S. Yoyota, F. Toda, *Chem. Comm.* **2004**, 1844–1845; *Tetrahedron* **2004**, *60*, 7767–7774.

3.4
Chromatography: a Non-analytical View

Alirio E. Rodrigues and Mirjana Minceva

3.4.1
Introduction

This chapter deals with chromatographic processes, with emphasis on perfusion chromatography and simulated moving beds (SMBs).

Permeable particles containing large pores are used in separation and reaction engineering as adsorbents and catalysts. Perfusion chromatography, developed in 1990 [1] for the separation of proteins, is based on the concept of "augmented diffusivity by convection" [2], which combines the contributions of mass transport by convection and diffusion in adsorbent pores. An example of flow-through particles is given in Fig. 3.4-1 where wide pores are of the order of 7000 Å and polymeric microspheres contain small diffusive pores.

The importance of intraparticle convection in the area of catalytic reaction engineering was addressed a long time ago by Wheeler [3], who claims that convection will only be important for large pores (10000 Å) or high-pressure (100 atm) gas-

microspheres

pore

1 μm **Fig. 3.4-1** An example of a perfusive particle.

Green Separation Processes. Edited by C. A. M. Afonso and J. G. Crespo
Copyright © 2005 WILEY-VCH Verlag GmbH & Co. KGaA, Weinheim
ISBN 3-527-30985-3

phase reactions. This chapter provides order-of-magnitude analysis and the model equation for intraparticle forced convection, diffusion and reaction in isothermal pellets at the steady state. The new model parameter is the intraparticle mass

Peclet number $\lambda = {v_0 \ell}/{D_e}$

which relates the intraparticle convective velocity v_0 and diffusion D_e in a slab particle of half-thickness ℓ. Wheeler did not solve the model equation and so he missed the point that catalyst effectiveness is enhanced by intraparticle forced convection when reaction and diffusion rates are of similar orders of magnitude, as recognized in 1977 by Nir and Pismen [4].

Interestingly enough, the concept of "augmented diffusivity by convection" was quantified in relation to the measurement of effective diffusivity in catalyst-permeable particles when the chromatographic method was being used in attempts to understand why analysis of experimental data by a conventional model (with only mass transport by diffusion inside pores) was leading to values that changed with the flow-rate through the bed [2]. However, the practical application of the concept of "augmented diffusivity by convection" was in the separation engineering area, namely in perfusion chromatography for protein separation.

The Simulated Moving Bed (SMB) technique is a powerful technique for preparative scale chromatography, known since 1961 [5]. This technology, known as the Sorbex process, based on the UOP patent by Broughton and Gerhold, was originally developed in the areas of petroleum refining and petrochemicals [6]. Recently, SMB technology has found new applications in the areas of biotechnology, pharmaceuticals and fine chemistry; the first industrial unit was installed at UCB Pharma (Belgium) in 1999 by Novasep [7]. SMB is now a key technology for chiral separations. It is interesting to compare "old" and "new" applications of SMB technology. In the Parex process (Fig. 3.4-2) for *p*-xylene recovery from a mixture of xylene isomers the column diameter $D = 10$ m (maximum), bed height $H = 1$ m and particles are 600 μm in diameter; in chiral separations $D = 1$ m (maximum) and $H = 0.1$ m (similar D/H ratio) with particles of 20-μm diameter. The adsorbent capacity is 200 kg m^{-3} in the Parex process compared with 10 kg m^{-3} in chiral separations; the productivity is 120 kg m^{-3} h^{-1} in the Parex and 1–10 kg m^{-3} h^{-1} in chiral separations.

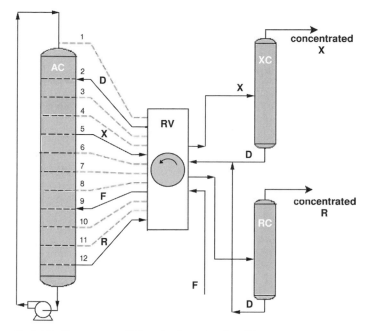

Fig. 3.4-2 Parex process (AC – adsorbent chamber, RV – rotary valve, XC – extract column, RC – raffinate column, D – desorbent, X – extract, F – feed, R – raffinate).

3.4.2
Perfusion Chromatography

The behavior of a chromatographic column depends on equilibrium factors (nature of the adsorption isotherm) and kinetic factors: dispersion (axial, radial) and mass/heat transfer resistances.

The simplest equilibrium model of a chromatographic column assumes isothermal operation, plug fluid flow, infinitely fast mass transfer between fluid and solid phases (instantaneous equilibrium at the interface), negligible pressure drop and trace system. The model equations are the mass balance in a bed volume element and the equilibrium law at fluid/solid interface:

$$u_0 \frac{\partial c_i}{\partial z} + \varepsilon \frac{\partial c_i}{\partial t} + (1-\varepsilon) \frac{\partial \langle q_i \rangle}{\partial t} = 0 \tag{1}$$

$$\langle q_i \rangle = q_i^* = f(c_i) \tag{2}$$

where $\langle q_i \rangle$ is the average concentration in the adsorbent and $q_i^* = f(c_i)$ is the adsorbed concentration at the particle surface in equilibrium with the fluid concentration c_i. Using the cyclic relation between partial derivatives we get the De Vault equation (Eq. 3) [8]:

$$u_{c_i} = \left(\frac{\partial z}{\partial t}\right)_{c_i} = \frac{u_i}{1 + \dfrac{1-\varepsilon}{\varepsilon} f'(c_i)} \qquad (3)$$

This important result shows that adsorption in fixed beds is a phenomenon of propagation of concentration waves and that the nature of the equilibrium isotherm is the main factor influencing the shape of the breakthrough curve (Fig. 3.4-3). The physical concepts to be retained are: dispersive waves are formed when isotherms are unfavorable; each concentration propagates with a velocity given by the De Vault equation. Compressive waves are formed for favorable isotherms and the physical limit is a shock which propagates with a velocity

$$u_{sh} = \frac{u_i}{1 + \dfrac{1-\varepsilon}{\varepsilon} \dfrac{\Delta q_i}{\Delta c_i}} ,$$

where the slope of the chord linking the feed state and the bed initial state appears instead of the local slope of the equilibrium isotherm.

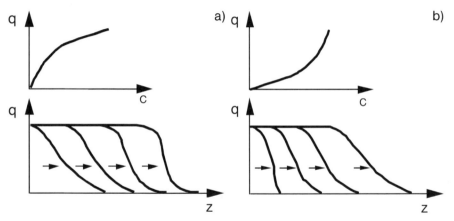

Fig. 3.4-3 (a) Unfavorable isotherms and dispersive fronts; (b) favorable isotherm and compressive front.

Kinetic factors will lead to dispersion of the fronts being much more important for favorable isotherms. Intraparticle mass-transfer resistance can be eliminated or decreased by using pellicular packings, reducing particle size or increasing particle permeability as shown in Fig. 3.4-4.

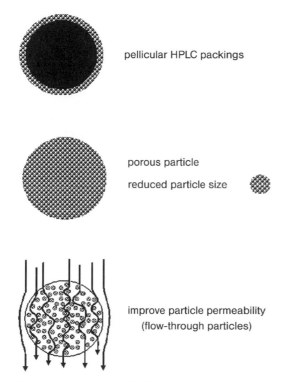

pellicular HPLC packings

porous particle

reduced particle size

improve particle permeability
(flow-through particles)

Fig. 3.4-4 Strategies to eliminate or decrease intraparticle
mass-transfer resistance.

3.4.2.1
The Concept of "Augmented Diffusivity by Convection"

The use of large-pore, permeable particles has been increasing in relation to protein separation by high-pressure liquid chromatography (HPLC). The key concept
behind the improved performance of flow-through packings is that of *augmented
diffusivity by convection* inside transport pores as shown by Rodrigues et al. early in
1982 [2]. In fact, the augmented diffusivity $\tilde{D}_e$ is related to the effective diffusivity
D_e by:

$$\tilde{D}_e = D_e \frac{1}{f(\lambda)} \tag{4}$$

Analysis of tracer experiments in chromatographic columns with a conventional
model that combines convection and diffusion inside pores is based on the mass
balance in a volume element for a slab adsorbent particle:

$$\tilde{D}_e \frac{\partial^2 c}{\partial x^2} = \varepsilon_p \frac{\partial c}{\partial t} \tag{5}$$

The particle transfer function relating the average concentration inside particle and the particle surface concentration is:

$$\tilde{g}_p(s) = \frac{<\bar{c}>}{\bar{c}_s} = \frac{\tanh\sqrt{\tilde{\tau}_d s}}{\sqrt{\tilde{\tau}_d s}} \tag{6}$$

with an apparent diffusion time constant $\tilde{\tau}_d = \varepsilon_p \ell^2 / \tilde{D}_e$.

On the other hand the analysis by a detailed diffusion/convection model leads to a mass-balance equation in a slab adsorbent particle which includes terms relating to diffusive flux, convective flux and accumulation:

$$D_e \frac{\partial^2 c}{\partial x^2} - v_0 \frac{\partial c}{\partial x} = \varepsilon_p \frac{\partial c}{\partial t} \tag{7}$$

The particle transfer function

$$g_p(s) = \frac{\left(e^{2r_2}-1\right)\left(e^{2r_1}-1\right)}{\left(e^{2r_2}-e^{2r_1}\right)} \frac{\sqrt{\left(\frac{\lambda}{2}\right)^2 + \tau_d s}}{\tau_d s} \tag{8}$$

with $r_{1,2} = \frac{\lambda}{2} \pm \sqrt{\left(\frac{\lambda}{2}\right)^2 + \tau_d s}$, $\tau_d = \varepsilon_p \ell^2 / D_e$ and $\lambda = \frac{v_0 \ell}{D_e}$ (intraparticle Peclet number).

Model equivalence (Fig. 3.4-5) leads to Eq. 4 where the enhancement factor for pore diffusivity due to convection is $1/f(\lambda)$ shown in Fig. 3.4-6, with

$$f(\lambda) = \frac{3}{\lambda}\left[\frac{1}{\tanh\lambda} - \frac{1}{\lambda}\right] \tag{9}$$

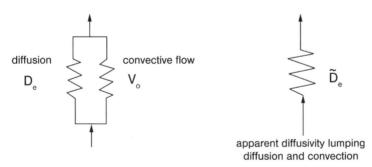

diffusion D_e convective flow V_o

$\tilde{D}_e$

apparent diffusivity lumping diffusion and convection

Fig. 3.4-5 Detailed convection/diffusion model (a) and lumped model (b) for adsorbent particle.

The key parameter is the intraparticle Peclet number defined as the ratio between the time constant for pore diffusion and the time constant for intraparticle convection. There are two limiting situations:

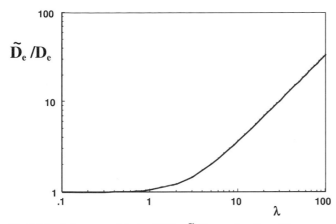

Fig. 3.4-6 Enhancement factor $1/f(\lambda) = \tilde{D}_e/D_e$ as a function of the intraparticle Peclet number λ.

(i) The diffusion-controlled case – at low superficial velocities, the intraparticle convective velocity v_0 is also small ; therefore $1/f(\lambda) = 1$ and so $\tilde{D}_e = D_e$;

(ii) The convection-controlled case – at high superficial velocities and so high v_0 and high λ, we get $f(\lambda) = 3/\lambda$; the augmented diffusivity is then $\tilde{D}_e = v_0\ell/3$ and depends on the particle permeability. Equation 4 is valid for spherical particles provided that the half thickness of the slab is replaced by $R_p/3$ where R_p is the radius of the sphere [9].

3.4.2.2
The Efficiency of a Chromatographic Column Measured by its HETP

The efficiency of chromatographic columns is usually measured under conditions of linear adsorption equilibria from pulse tests. The Height Equivalent to a Theoretical Plate (HETP) is calculated from the first moment μ_1 and the variance σ^2 of a chromatographic peak obtained in a column of length L by:

$$HETP = \sigma^2 L \Big/ \mu_1^2 \qquad (10)$$

For conventional packings and linear isotherms Van Deemter et al. [10] developed the well-known equation for HETP including contributions from eddy dispersion (*A*-term), molecular diffusion (*B*-term) and intraparticle kinetics (*C*-term).

$$HEPT = A + B/u + Cu \qquad (11)$$

or

$$HETP = A + \frac{B}{u} + \frac{2}{3}\frac{\varepsilon_p(1-\varepsilon_b)b^2}{[\varepsilon_b + \varepsilon_p(1-\varepsilon_b)b]^2}\tau_d u \qquad (12)$$

where ε_p is the particle porosity, ε is the interparticle porosity and $b = 1+ (1-\varepsilon_p)m/\varepsilon_p$. The slope of the equilibrium isotherm is m. For spherical particles of radius R_p the Van Deemter equation is obtained by replacing the factor 2/3 in Eq. 12 by 2/15 where the time constant for diffusion in a sphere is now $\tau_d = \varepsilon_p Rp^2/D_e$. By equating HETP equations for slab and spheres we get the equivalence between those geometries in the diffusion-controlled region, i.e., $\ell = R_p/\sqrt{5}$.

For large-pore particles an extension of the Van Deemter equation was derived by Rodrigues [11, 12] since, $\tilde{\tau}_d = \tau_d f(\lambda)$ and therefore we get:

$$HETP = A + \frac{B}{u} + Cf(\lambda)u \tag{13}$$

At low velocities $f(\lambda) \approx 1$ and both equations lead to similar results. However, at high superficial velocities, $f(\lambda) \approx 3/\lambda$ and so the last term in Rodrigues' equation becomes a *constant* since the intraparticle convective velocity v_0 is proportional to the superficial velocity u. The HETP reaches a plateau that does not depend on the value of the solute diffusivity but only on the particle permeability and pressure gradient (convection-controlled limit).

Two important features result from the use of large-pore, permeable packings: the column performance improves since HETP is reduced when compared with conventional packings (the C term in the Van Deemter equation is reduced); and the speed of separation can be increased (by increasing the superficial velocity) without losing column efficiency.

The Van Deemter equation for conventional supports (dashed line) and Rodrigues' equation for large-pore supports (full line) are shown in Fig. 3.4-7 for a typical HPLC process.

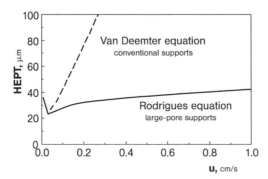

Fig. 3.4-7 *HETP* versus *u* (Van Deemter equation for conventional packings and Rodrigues equation for large-pore packings).

However, to be able to use the extended Van Deemter equation the intraparticle convective velocity v_0 must be estimated for calculation of the parameter λ. The simplest way is to write the equality of pressure drop across the particle and along the bed, i.e. $\Delta p/d_p = \Delta P/L$. By using Darcy's law for the flow in the column and in the pores [13] we get:

$$v_0 = \frac{B_p}{B_b} u \tag{14}$$

where B_p and B_b are the particle and bed permeabilities, respectively.

3.4.3
Simulated Moving Bed (SMB) Processes

3.4.3.1
The Concept of SMB

The principle of SMB operation can be easily understood by analogy with the equivalent True Moving Bed (TMB) process. The TMB unit (Fig. 3.4-8) is divided into four sections: section 1, between the eluent and extract ports; section 2, between the extract and feed ports; section 3, between the feed and raffinate points; and section 4, between the raffinate and the eluent inlet. In the ideal TMB operation, liquid and solid flow in opposite directions, and are continuously recycled: the liquid flowing out of section 4 is recycled to section 1, while the solid coming out of section 1 is recycled to section 4. The feed is continuously injected between sections 2 and 3, and two product lines can be continuously collected: the extract, rich in the more-retained species A, and the raffinate, rich in the less-retained species B, which move upwards with the liquid phase. Pure eluent or desorbent is injected at the beginning of section 1, with the liquid recycled from the end of section 4.

In a TMB, the solid flow-rate is constant all over the unit but the liquid flow-rates differ from section to section. For a binary feed (A+B), species A, the more-retained component, will be recovered in the extract and species B, the less-retained

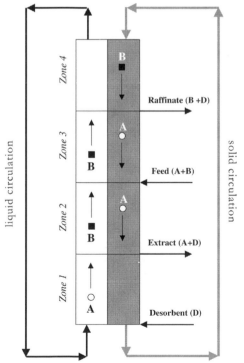

Fig. 3.4-8 The True Moving Bed (TMB).

species, will be recovered in the raffinate. In sections 2 and 3 the two components must move in opposite directions. The less-retained component B must be desorbed and carried with the liquid phase, while the more-retained species A must be adsorbed and carried with the solid phase. Section 2 is the zone of desorption of the less-retained species B, while section 3 is the zone of adsorption of the more-retained component A. The role of section 4 is to clean the eluent, which is then recycled to section 1 where the adsorbent is regenerated.

The TMB steady-state internal concentration profiles are presented in Fig. 3.4-9.

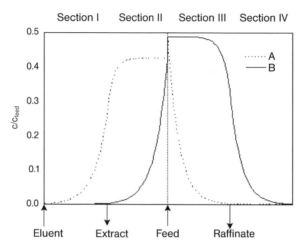

Fig. 3.4-9 Concentration profiles in a TMB in the steady state (A – more-retained species; B – less-retained species).

The ideal TMB has several limitations associated with the movement of the solid phase, namely because of adsorbent attrition, equipment erosion and difficulty in obtaining uniform solid and liquid flow. Simulated moving bed technology was developed to overcome these difficulties and to retain the process advantages of continuous and countercurrent flow. In the SMB process the adsorbent is fixed and the positions of the inlet and outlet streams are shifted periodically in the direction of the liquid-phase flow, to simulate the movement of the solid. In the Sorbex SMB technology developed by UOP (Fig. 3.4-2), a rotary valve is used to periodically change the position of the eluent, extract, feed and raffinate lines along the adsorbent bed. Alternative techniques to perform the port switching uses a set of individual on–off valves connecting the inlet and outlet streams to each node between columns; this is the case for NovaSep equipment shown in Fig. 3.4-10.

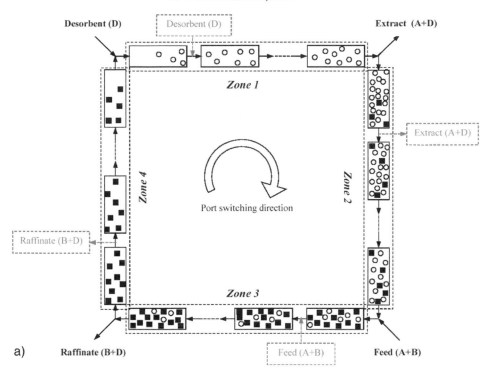

○ A more adsorbed component
■ B less adsorbed component

Fig. 3.4-10   (a) SMB technology for chiral separations; b) SMB unit *Licosep 12-26* (Novasep) at the authors' laboratory.

3.4.3.2
Modeling of SMB

There are two strategies of modeling SMB: as an equivalent TMB (the solid movement is taken into account and equivalence relations are used to relate the results to a real SMB (Table 3.4-1) or as a real SMB; each bed is analyzed individually and the periodic change in boundary conditions is taken into account [14]. Owing to the switching of inlet and outlet ports, each column plays a different role during the whole cycle, depending on its location. The model of the SMB unit is constituted of k (k = number of columns) column models, connected with each other by simple material balances on the connecting nodes. With each switching of the inlet and outlet ports each column should be updated in terms of flow-rate and inlet concentration. The flow-rate in each column, according to its location (section), can be calculated by mass balance around the inlet and outlet nodes. The inlet concentration of each column is equal to the outlet concentration of the previous column, except for the feed and desorbent nodes.

Table 3.4-1 Equivalence relations between an SMB and a TMB.

	SMB	TMB
Solid-phase velocity	0	$u_s = L_c/t^*$
Liquid-phase velocity	v^{SMB}	$v^{TMB} = v^{SMB} - u_s$

The main difference between the TMB and SMB approaches is related to the stationary regime. The time dependence of the boundary conditions in the SMB leads to a cyclic steady state instead of a real steady state as occurs in the TMB model. The cyclic steady state is reached after a certain number of cycles, but the system states are still varying over time because of the periodic movement of the inlet and outlet ports along the columns (Fig. 3.4-11).

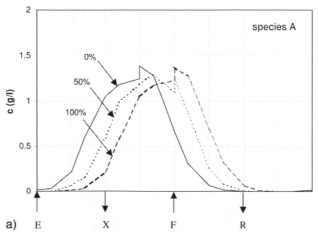

Fig. 3.4-11 Cyclic steady-state concentration profiles at the beginning, half-way and end of the switching time period. (a) Species A; (b) species B; (c) extract concentration history; (d) raffinate concentration history.

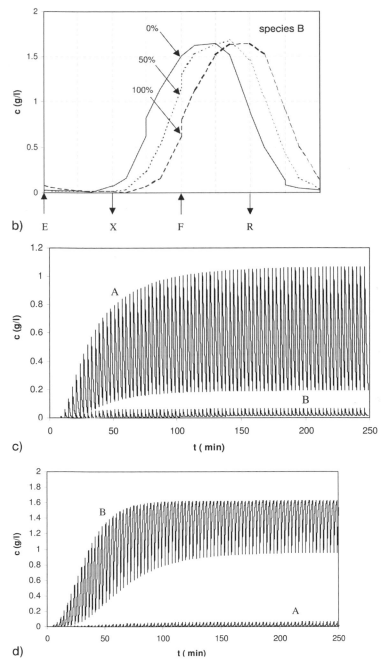

Fig. 3.4-11 Continued

The degree of complexity of the model depends on the process description (staged or distributed system), the mass-transfer resistance (equilibrium stage or mass-transfer resistance within the fluid and/or solid phase) and adsorption equilibria (linear, Langmuir, bi-Langmuir or modified Langmuir). The SMB and TMB model equations when axial dispersion plug flow for the liquid phase, plug flow for the solid phase, homogeneous linear driving force (LDF) for internal mass transfer [15] and multicomponent adsorption isotherm are assumed are presented in Table 3.4-2.

Table 3.4-2 Transient SMB and TMB model equations.

Simulated Moving Bed model equations

Mass balance over a volume element of the bed k:

$$\frac{\partial C_{ik}}{\partial \theta} = \gamma_k^* \left\{ \frac{1}{Pe_k} \frac{\partial^2 C_{ik}}{\partial x^2} - \frac{\partial C_{ik}}{\partial x} \right\} - \frac{(1-\varepsilon)}{\varepsilon} \alpha_k (q_{ik}^* - q_{ik})$$

Mass balance in the particle:

$$\frac{\partial q_{ik}}{\partial \theta} = \alpha_k (q_{ik}^* - q_{ik})$$

Initial conditions:

$\theta = 0$: $C_{ik} = q_{ik} = 0$

Boundary conditions for column k:

$$x = 0: \quad C_{ik} - \frac{1}{Pe_k} \frac{\partial C_{ik}}{\partial x} = C_{ik,0}$$

where $C_{ik,0}$ is the inlet concentration of species i in column k.

$x = 1$:

For a column inside a section and for extract and raffinate nodes: $C_{ik} = C_{ik+1,0}$

For the eluent node: $C_{ik} = \frac{v_1^*}{v_4^*} C_{ik+1,0}$ For the feed node: $C_{ik} = \frac{v_3^*}{v_2^*} C_{ik+1,0} - \frac{v_F}{v_2^*} C_i^F$

Global balances:

Eluent node: $v_1^* = v_4^* + v_E$ Extract node: $v_2^* = v_1^* - v_X$

Feed node: $v_3^* = v_2^* + v_F$ Raffinate node: $v_4^* = v_3^* - v_R$

Multicomponent adsorption equilibrium isotherm:

$$q_{Ak}^* = f_A(C_{Ak}, C_{Bk})$$
$$q_{Bk}^* = f_B(C_{Ak}, C_{Bk})$$

True Moving Bed model equations

Mass balance over a volume element of the section j:

$$\frac{\partial C_{ij}}{\partial \theta} = \gamma_j \left\{ \frac{1}{Pe_j} \frac{\partial^2 C_{ij}}{\partial x^2} - \frac{\partial C_{ij}}{\partial x} \right\} - \frac{(1-\varepsilon)}{\varepsilon} \alpha_j (q_{ij}^* - q_{ij})$$

Mass balance in the particle:

$$\frac{\partial q_{ij}}{\partial \theta} = \frac{\partial q_{ij}}{\partial x} + \alpha_j (q_{ij}^* - q_{ij})$$

Initial conditions:

$\theta = 0$: $C_{ij} = q_{ij} = 0$

Table 3.4-2 Continued

Boundary conditions for section j:

$x = 0$: $C_{ij} - \dfrac{1}{Pe_j}\dfrac{\partial C_{ij}}{\partial x} = C_{ij,0}$

where $C_{ij,0}$ is the inlet concentration of species i in section j.

$x = 1$:

For the eluent node: $C_{i4} = \dfrac{v_1}{v_4}C_{i1,0}$ For the extract node: $C_{i1} = C_{i2,0}$

For the feed node: $C_{i2} = \dfrac{v_3}{v_2}C_{i3,0} - \dfrac{v_F}{v_2}C_i^F$ For the raffinate node: $C_{i3} = C_{i4,0}$

and $q_{i4} = q_{i1,0}$, $q_{i1} = q_{i2,0}$, $q_{i2} = q_{i3,0}$, $q_{i3} = q_{i4,0}$

Global balances:

Eluent node: $v_1 = v_4 + v_E$ Extract node: $v_2 = v_1 - v_X$

Feed node: $v_3 = v_2 + v_F$ Raffinate node: $v_4 = v_3 - v_R$

Multicomponent adsorption equilibrium isotherm:

$q_{Aj}^* = f_A(C_{Aj}, C_{Bj})$
$q_{Bj}^* = f_B(C_{Aj}, C_{Bj})$

3.4.3.3
Design of SMB

The choice of operating conditions (flow-rates and switching time) of an SMB chromatographic process is not a simple task. Some constraints have to be met to recover the more-adsorbed species (A) in the extract and the less-adsorbed species (B) in the raffinate. These constraints are expressed in terms of net fluxes of components in each section considering an equivalent TMB. In section 1 species A must move upward to the extract port, in sections 2 and 3 species A must move downward to the extract port and species B must move to the raffinate port and in section 4 the net flux of species B has to be downwards (Fig. 3.4-8).

$$\frac{Q_I c_{AI}}{Q_S q_{AI}} > 1 \tag{15a}$$

$$\frac{Q_{II} c_{BII}}{Q_S q_{BII}} > 1 \qquad \frac{Q_{II} c_{AII}}{Q_S q_{AII}} < 1 \tag{15b}$$

$$\frac{Q_{III} c_{BIII}}{Q_S q_{BIII}} > 1 \qquad \frac{Q_{III} c_{AIII}}{Q_S q_{AIII}} < 1 \tag{15c}$$

$$\frac{Q_{IV} c_{BIV}}{Q_S q_{BIV}} < 1 \tag{15d}$$

The simplest case may be formulated for systems with linear adsorption isotherms using the equilibrium theory. The region of complete separation (100% extract and raffinate purity) predicted by the equilibrium theory is a triangle shown in Fig. 3.4-12. The section constraints are explicit inequality relations in terms of liquid to solid flow-rate ratios in the four TMB sections (m_1, m_2, m_3, m_4) [16, 17]:

$$K_B < m_2, m_3 < K_A \text{ and } m_1 > K_A \text{ and } m_4 < K_B \tag{16}$$

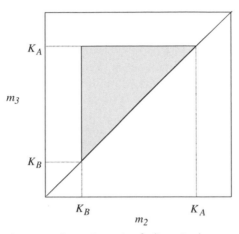

Fig. 3.4-12 Separation region for linear isotherms.

For nonlinear isotherms the "triangle" will be distorted. Higher mass-transfer resistances lead to a decrease of the separation region [18]. For systems where mass-transfer resistance inside particles is important, the "triangle theory" will provide initial estimates for the operating conditions. In some cases 100% product purity is not required; also, the approach based on an equilibrium model does not allow explicit prediction of the product purities, which are generally the main constraints for a feasible operating point. These are the reasons why the concept of separation volume was introduced and applied to sugars separations, xylenes separations and chiral separations [19–21]. The "separation volume" methodology uses a realistic mathematical model and explores the influence of the flow rates in zones 1 and 4 (desorbent flow rate). The "separation volume" methodology offers two possibilities: if the flow rate in zone 1 (m_1)) is fixed, the design leads to an ($m_2 \times m_3 \times m_4$) volume for a given separation requirement; if the flow rate in zone 4 (m_4) is fixed, the design will result in an ($m_2 \times m_3 \times m_1$) volume for a given separation requirement (see Fig. 3.4-13).

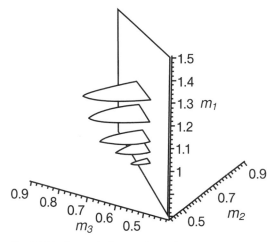

Fig. 3.4-13 Concept of separation volume.

3.4.3.4
Future Directions in SMB: Multicomponent Separations and SMBR

Simulated moving bed technology offers many advantages over preparative chromatography (cleaner, smaller, safer and faster processes) [22]. The main disadvantage is its limitation to the separation of binary mixtures or of one component from a multicomponent mixture. Progress in the area of binary separations by SMB technology involves asynchronous shifting of the inlet/outlet ports as in the Varicol process [23–25] as well as manipulation of feed concentration [26] and feed flow-rate [27].

The pseudo-simulated moving bed process – the JO process of Japan Organo Co. [28, 29] – has been successfully applied to the separation of a ternary mixture. The process cycle is divided into two steps (Fig. 3.4-14). In step 1, feed and eluent streams are introduced into the system, equivalent to a series of preparative chromatographic columns, and the intermediate component is produced. In step 2, similar to an SMB, there is no feed and the less-adsorbed species is collected in the raffinate while the more-retained species is collected in the extract [30].

The combination of a chemical or biochemical reaction with an SMB chromatographic separator has been the subject of considerable attention since the mid-1990s. This integrated reaction–separation technology has been named Simulated Moving Bed Reactor (SMBR) technology. The first application of an SMBR in a zeolite-catalyzed alkylation reaction was patented in 1977 by Zabrinsky and Anderson [31]. Integrating the reaction and separation steps in one single unit has the obvious economic advantage of reducing the cost of unit operations for downstream purification steps. In the case of a reversible reaction, where conversion is limited by the chemical equilibrium, removal of products as they are formed allows conversions well beyond equilibrium values to be achieved. For reactions in series or in

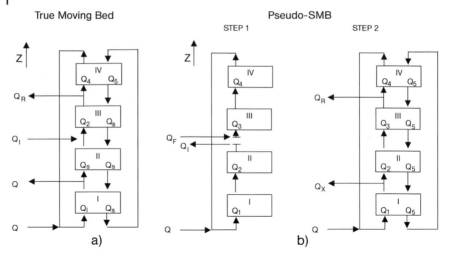

Fig. 3.4-14 Comparison between the two different techniques:
(a) TMB; (b) pseudo-SMB (JO process).

parallel, the selective separation of desired intermediate species may be possible. When a reaction product has an inhibiting or poisoning effect, its removal from the reaction medium also promotes enhanced yield.

The SMBR where reaction of type A → B + C (sucrose inversion) was studied in our laboratory [32]. The sucrose is introduced in the middle of the unit with the feed stream. The reaction is catalyzed by the enzyme invertase introduced in the unit with the eluent stream. The sucrose reacts near to the feed port; fructose and glucose are formed and separated in the extract and raffinate, respectively. Typical concentration profiles are shown in Fig. 3.4-15.

A more interesting example of SMBR is its application for the synthesis of diethylacetal from ethanol and acetaldehyde, with separation of the products acetal

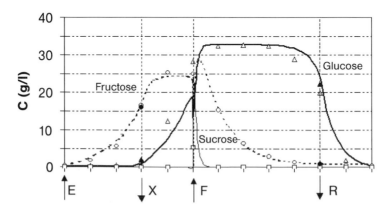

Fig. 3.4-15 SMBR cyclic steady-state internal concentration profiles.

and water. This is a reversible reaction catalyzed by acid polymeric resin, which is at the same time an adsorbent, allowing product separation [33] that would be difficult by other technologies such as catalytic distillation.

References

1 N. Afeyan, S. Fulton, N. Gordon, I. Mazsaroff, L. Varady, F. Regnier, *Biotechnol,* **1990**, *8*, 203–206

2 A.E. Rodrigues, B. Ahn, A. Zoulalian, *AIChE J.*, **1982**, *28*, 541–546

3 A. Wheeler, *Adv. Catal*, **1951**, *3*, 250–337

4 A. Nir, L. Pismen, *Chem. Eng. Sci.*, **1977**, *32*, 35–41

5 D.B. Broughton, C.G. Gerhold, U.S. Patent No. 2,985,589, **1961**

6 D.B. Broughton, *Chem. Engng Prog.*, **1977**, *73*, 49–51.

7 R.M. Nicoud, *Handbook of Bioseparations*, S. Ahuja (Ed.), Academic Press, San Diego, USA, **2000**, 475–509.

8 D. De Vault, *J. Am. Chem. Soc.*, **1943**, *65*, 532.

9 G. Carta, A.E. Rodrigues, *Chem. Eng. Sci.*, **1993**, *48*, 3927–3936

10 J. Van Deemter, F. Zuiderweg, A. Klinkenberg, *Chem. Eng. Sci.*, **1956**, *5*, 271–289.

11 A.E. Rodrigues, *LC-GC*, **1993**, *6*, 20–29.

12 A.E. Rodrigues, L. Zuping, J. Loureiro, *Chem. Eng. Sci.*, **1991**, *46*, 2765–2773

13 H. Komiyama, H. Inoue, *J.Chem. Eng. Jpn.*, **1974**, *7*, 281–286

14 L.S. Pais, J.M. Loureiro, A.E. Rodrigues, *AIChE J.*, **1998**, *44*, 561–569.

15 E. Glueckauf,, *Trans. Faraday Soc.*, **1955**, *51*, 1540–1551.

16 D.M. Ruthven, C.B. Ching, *Chem. Eng. Sci.*, **1989**, *44*, 1011–1038.

17 G. Storti, M. Mazzotti, M. Morbidelli, S. Carrà, *AIChE J.*, **1993**, *39*, 471–492.

18 L.S. Pais, J.M. Loureiro, A.E. Rodrigues, *Chem. Eng. Sci.*, **1997**, *52*, 245–257.

19 D.C.S. Azevedo, A.E. Rodrigues, *AIChE J.*, **1999**, *45*, 956–966.

20 M. Minceva, A.E. Rodrigues, *Ind. Eng. Chem. Res.*, **2002**, *41*, 3454–3461.

21 A.E. Rodrigues, L.S. Pais, *Sep. Sci. Technol.*, **2004**, *39*, 241–266.

22 R.M. Nicoud, *Recent Advances in Industrial Chromatographic Processes*, R.-M. Nicoud, ed., NOVASEP, Nancy, France, **1997**, 4–5.

23 P. Adam, R.M. Nicoud, M. Bailly, O. Ludemann-Hombourger, U.S. Patent No. 6,136,198, **2000**.

24 O. Ludemann-Hombourger, R.M. Nicoud, M. Bailly, *Sep. Sci. Technol.*, **2000**, *35*, 1829–1862.

25 L.S. Pais, A.E. Rodrigues, *J. Chromatogr. A*, **2003**, *1006*, 33–44.

26 H. Schramm, M. Kaspereit, A. Kienle, A. Seidel-Morgenstern, *Chem. Eng. Technol.*, **2002**, *25*, 1151–1155

27 H. Schramm, A. Kienle, M. Kaspereit, A. Seidel-Morgenstern, *Chem. Eng. Sci.*, **2003**, *58*, 5217–5227.

28 M. Ando, M. Tanimura, M. Tamura, U.S. Patent No. 4,970,002, **1990**.

29 T. Masuda, T. Sonobe, F. Matsuda, M. Horie, U.S. Patent No. 5,198,120, **1993**.

30 V.G. Mata, A.E. Rodrigues, *J. Chromatogr. A*, **2001**, *939*, 23–40.

31 R.F. Zabransky, R.F. Anderson, U.S. Patent No. 4,049,739 , **1977**.

32 D.C.S. Azevedo, A.E. Rodrigues, *Chem. Eng. J.*, **2001**, *82*, 95–107.

33 V.T.M. Silva, A.E. Rodrigues, "Processo Industrial de Produção de Acetais num Reactor Adsorptivo de Leito Móvel Simulado", patent pending (**2004**), PT 103123.

3.5
Fluid Extraction

3.5.1
Supercritical Fluids

Anna Banet Osuna, Ana Šerbanović, and Manuel Nunes da Ponte

3.5.1.1
Introduction

Supercritical fluids (SCFs) are gases at pressure and temperatures (slightly) above those of the vapor-liquid critical point. As the critical pressures of known substances are (much) higher than atmospheric pressure, a supercritical fluid is always a high-pressure gas. The unique property of an SCF is that its density is very sensitive to small changes in pressure and temperature. Density is directly related to many other physical (and chemical) properties of a fluid. The most important in supercritical fluid applications is the solvent power, that is, the ability to dissolve other substances.

With small relative changes in pressure (or temperature), a supercritical fluid can be brought from gas-like densities, where it can hardly dissolve anything, to liquid-like densities, where the molecules of the fluid can cluster around the molecules of solutes and dissolve them into the gas phase.

A simple example may be given. Carbon dioxide is often used in separations around 40°C. At 315 K (42.8°C), its density doubles, from 0.3 kg dm^{-3} to 0.6 kg dm^{-3}, when pressure is brought from 8.36 to 10.15 MPa, that is, a 21% increase. Notice that a similar increase of density in a low-pressure gas, commonly regarded as a highly compressible state, can only be achieved with a 100% increase in pressure.

So, with small variations of pressure or temperature, a supercritical fluid can be moved from a density where it dissolves reasonably high amounts of a solute to another one where it precipitates (almost) everything it had dissolved before. These characteristics make the supercritical-fluid phase attractive for solubility-based separation processes.

Green Separation Processes. Edited by C. A. M. Afonso and J. G. Crespo
Copyright © 2005 WILEY-VCH Verlag GmbH & Co. KGaA, Weinheim
ISBN 3-527-30985-3

3.5.1.2
Supercritical Fluids and Clean Separations

Gases at high pressures have been used in the chemical, oil and polymer industries for a long time. The production of low-density polyethylene is a good example of a reaction in a supercritical fluid. Ethylene (critical temperature 9°C) is used as both reactant and supercritical solvent.

The connection between supercritical fluids and clean separations is, however, more recent. It appears during the 1970s, when it was shown that supercritical carbon dioxide could become an important extraction solvent for food-related applications. Since then, the field has expanded enormously. A wide range of fundamental studies have been published and many patents submitted, with potential applications in separation processes, chemistry, and materials science. Many books and reviews have been published on different aspects of supercritical fluids.

The scope of this section will be strictly limited to describing the recent advances in "separation" and "chemical reaction plus separation" using carbon dioxide as the main solvent, which, in the authors' view, will set the trends in this area for the next few years.

3.5.1.3
Extraction with Carbon Dioxide

The advantages of carbon dioxide as solvent have been well publicized: it is, in fact, classified as GRAS – generally regarded as safe, it has low toxicity (threshold limit value – TLV = 5000 ppm), it is supercritical just above ambient temperature (critical temperature 31°C) and it is cheap. Also, like other supercritical fluids, it has advantageous gas-like transport properties, such as low viscosity and high diffusivity.

Carbon dioxide's solvent properties are somewhat intriguing. It is a polar substance, because it possesses a permanent quadrupole – an asymmetry of charge distribution where both oxygen atoms are more electronegative than the carbon one. However, it mainly solubilizes low molecular weight, non-polar compounds. It is a poor solvent for most other molecules, with some notable exceptions such as perfluorinated ones. This makes extractions and fractionations using carbon dioxide very selective processes of the few substances that it can dissolve easily, because they can be taken out from very complex matrices and mixtures.

Some general qualitative rules for solubility in carbon dioxide have been known for quite some time:

- Solubility is generally low, except close to the critical line of mixtures.
- Solubility decreases as the molar mass of the solute increases.
- Aliphatic hydrocarbons are more soluble than aromatics with similar molar masses.
- Double bonds increase solubility (alkenes are more soluble than alkanes).
- Branched hydrocarbons are more soluble than linear ones.
- Functional polar groups are bad for solubility (-COOH is worse than –OH).

- Fluorinated compounds are highly soluble – fluorinated tails increase solubility.
- Polymers with low intermolecular cohesion (such as silicones) are highly soluble.

Supercritical carbon dioxide extraction of natural products from solid plant matrices is currently an established application, with over one hundred industrial facilities of various sizes operating throughout the world. Several books have been published describing this process in detail – see, for example Brunner [1]. The main "green" credential of these processes is the replacement of volatile organic solvents. Their implementation resulted, however, from definite technological advantages.

A recently developed industrial-scale application is a textbook example of the main advantages of supercritical carbon dioxide in extractions from natural products. The Diamond process, jointly developed by the Centre d'Énergie Atomique and the French company Sabaté, [2] extracts the contaminant trichloroanisol (TCA) from cork stoppers for wine bottles. This contaminant is produced by fungi and is responsible for the infamous cork taint taste of wines, which has brought serious financial losses to the wine industry and damaged the image of cork as the ideal material for bottle stoppers.

The selectivity of such a weak solvent as carbon dioxide is used in this process to great advantage. At moderate pressures, not very far above critical, the solubilities may be finely tuned. The volatile TCA is soluble in CO_2 and extractable [3], unlike the natural constituents of cork. TCA can therefore be removed from it without affecting its fundamental properties as a unique material. Moreover, the high diffusivity of CO_2 allows it to penetrate the elaborate cell structure of cork to remove TCA, an effect that cannot be obtained with liquid solvents. Finally, the bactericidal and fungicidal properties of carbon dioxide reduce the cork microbial load, and prevent further contamination. All these form a unique combination, resulting in a technological advantage that probably cannot be matched by other processes.

Although most extraction applications are focused on natural products for human consumption, where the GRAS status of carbon dioxide represents a clear benefit, uses in cleaning of man-made materials, such as cleaning of mechanical precision parts or dry-cleaning of clothes, have also gone commercial. Dry-cleaning requires the solubilization of many substances that are essentially insoluble in carbon dioxide. It was made possible by the discovery that fluorinated surfactants are highly soluble in CO_2 and can act as solubility enhancers for many types of molecule [4].

A recently devised new application is the cleaning of silicone wafers in the microprocessor industry. This subject was thoroughly reviewed by Beckman [5]. It also involves chemical etching and photolithography, and, although technical hurdles are still unresolved, its potential "green" character is very high. In fact, its implementation would avoid the huge quantities of contaminated cleaning waters that are presently produced. One of the technical advantages of the use of carbon dioxide is the drastic reduction of interfacial tensions, which allows the cleaning of extremely thin crevices.

3.5.1.4
Fractionation of Liquid Mixtures

Fractionation of liquid mixtures with supercritical carbon dioxide in counter-current columns can be operated continuously, because liquids can be easily pumped into and out of a column. This represents a big advantage over extraction from solid materials, as it allows real process intensification – large quantities of feed can be processed with only a small volume under high pressure at any given time. Fractionation, mostly of natural products or extracts, has been extensively studied at the laboratory and pilot-plant scale. The design principles of this type of column have been established, and scale-up procedures devised [1,6]. They can be operated with reflux, as in distillation, and fractionation can therefore become an extremely selective process. Difficult separations can be effectively carried out.

The fractionation of extracts previously obtained by water–ethanol extraction was described in detail by Brunner and Budich [7]. This development suggests that fractionation with carbon dioxide can be advantageously used as a secondary separation process to obtain highly pure, high-value-added compounds from natural products.

The main contribution to the cost of the operation is the recycling of the solvent. Owing to the usually low solubilities, large solvent to liquid-feed mass ratios must be used, of the order of 50 to 100, or even higher. The solutes in the streams leaving the column are, in most cases, separated by pressure reduction and precipitation. Large amounts of solvent must therefore be recompressed into the column. Methods of separating the solutes from the solvent stream that would avoid decompression, such as the use of membranes, would contribute significantly to improving the economics of this type of fractionation.

In any case, it may be said that this is a well-studied process, waiting for a first application to go commercial. Once this happens, it can safely be predicted that many others will follow the same route.

3.5.1.5
Supercritical, Near-critical and "Expanded" Solvents in Chemical Reactions

As already mentioned, supercritical solvents have been used as chemical reaction solvents for a long time. However, their study in the context of clean processes or Green Chemistry is relatively recent. It was during the 1990s that research on carbon dioxide as a reaction medium developed. The main goal was the reduction or complete elimination of volatile organic solvents used in laboratory and industrial processes.

One of the main advantages sought in chemical processes in supercritical fluids is the possibility of integrating reaction and separation, thereby combining into one single process a succession of different steps. This is especially desirable in homogeneous catalysis, where the use of the tunable solvent power of a supercritical fluid may help to easily recover the expensive transition metals contained in the catalytically active species. On the other hand, in heterogeneous catalysis, the formation of side-products on the catalyst surface, which lead to its deactivation, can be decreased in supercritical fluids. Desorption of primary products from the catalyst surface is enhanced, which prevents secondary reactions, extends catalyst lifetime and may promote product selectivity.

Many studies were conducted to investigate the advantages of performing reactions in one supercritical single phase, and the benefit of the enhanced mass transfer properties. An advantage of CO_2 is that it is inert to oxidation and free-radical chemistry, which makes it a desirable solvent for oxidation and free-radical polymerization reactions. The discovery that addition of fluorinated tails to otherwise conventional homogeneous catalysts could greatly increase their solubility in supercritical carbon dioxide led to a flurry of new publications. In 1999, a review of the field, edited by Noyori [8], appeared in *Chemical Reviews*. This comprehensive edition contains twelve chapters, bearing testimony to the progress that had been attained in little more than a single decade of research.

In that collection, the review chapter by Darr and Poliakoff [9] systematically called attention to the importance of phase equilibrium in some of the so-called "supercritical" systems. In fact, they pointed out that the simplest chemical reaction will include at least three components (starting material, product and solvent) and most reactions of chemical interest will involve even more. When different solutes are dissolved in a "supercritical" solvent, the whole reaction mixture may be a homogeneous, one-phase system, or a heterogeneous, multi-phase one, because the actual critical parameters of the mixture, which define the separation between a one-phase and a two-phase system, are usually considerably different from those of the pure solvent. Moreover, they change continuously along the reaction path, as reactants give way to products, and compositions change. The example they gave (later published by Ke et al. [10]) for the catalytic hydroformylation of propene in $scCO_2$ (C_3H_6, CO and H_2 as starting materials, $CH_3CH_2CH_2CHO$ and $(CH_3)_2CHCHO$ as products) is striking. Their measurements on model mixtures representative of the intermediate compositions from reactants to products, give critical temperatures that rise more than 20 K and critical pressures that increase up to around 25% conversion and then decrease.

When more complex molecules are involved, the changes can be more pronounced. The phase behavior itself can even become more complicated, as the formation of a third phase, involving liquid–liquid–vapor equilibrium, is very common at pressures and temperatures not very far from the critical values of the solvent.

This focus on phase behavior brought about the understanding that some of the reactions that had been successfully accomplished in "supercritical" carbon dioxide were actually taking place in the liquid phase (or at the interface) of a heterogeneous liquid–vapor system. Similarly, reactions carried out in liquid, near-critical propane had shown that this solvent displayed the same advantages as those of truly supercritical ones. The fact is that the densities of liquid mixtures close to the mixture's critical line are sufficiently lower than those of a "classical" liquid to exhibit the same type of property values (lower viscosity, higher diffusivity, higher solubility of gases) that confer enhanced mass-transfer capabilities to supercritical solvents.

Although the relatively poor solvent power of carbon dioxide may be used to advantage for selective separations, as explained above, it is a key technical issue that is limiting its widespread use. It can, in fact, spoil the economics of otherwise highly attractive processes, namely integrated reaction–separation. The recent trend is to tackle this problem of solubility by using biphasic mixtures of CO_2 and liquid compounds, where the liquid phase retains some of the advantages of supercritical fluids, but with enhanced solvent power.

Carbon dioxide is highly soluble in many organic liquids, even those that hardly dissolve at all in the gas phase. The high quantity of CO_2 in the liquid makes it an "expanded fluid", which avoids the need to generate a single phase of CO_2, reactants and products (and catalyst), in order to create a situation in which transport limitations are, at least, greatly minimized. Simply ensuring that a significant amount of CO_2 is present in the liquid phase may be sufficient to gain kinetic control over the reaction. Here, CO_2 functions as a diluent, which reduces viscosity and may even increase the solubility in the lower phase in the case of gaseous reactants. Another advantage of this type of approach is that reactions can be carried out at lower (sometimes much lower) pressures than in truly supercritical conditions.

Beckman's comprehensive review paper [5] describes chemical reactions in carbon dioxide-based solvents, and evaluates their "green" character. The title of the paper – "Supercritical and near-critical CO_2 in green chemical synthesis and processing" – is revealing of the new trends in the use of carbon dioxide.

Eckert's group has contributed much to the development of this concept – a recent example is given by Xie et al [11]. Another important contribution was given by Wei et al [12]. They performed oxidations with molecular O_2, using as solvent acetonitrile expanded by carbon dioxide, at pressures from 50 to 100 bar. Homogeneous cobalt and iron catalysts performed well in this environment, but would need pressures of hundreds of bar to dissolve in some significant amount in supercritical CO_2. On the other hand, oxygen solubility increased in the expanded solvent up to two orders of magnitude, when compared with neat acetonitrile.

3.5.1.6
Phase Equilibrium and Reaction-rate Control

These findings indicate that the control of phase behavior is one of the keys to controlling chemical reactions in near-critical fluids. Phase equilibrium plays an important role in dense fluid reaction systems, because the number of phases in equilibrium and their composition can be changed easily by pressure and temperature, and also as a consequence of the conversion of materials by chemical reactions.

The role of phase equilibria on homogeneously catalyzed reactions in dense fluids was discussed by Buchmüller et al [13] in the palladium-catalyzed coupling reaction of butadiene and CO_2 to form δ-lactone, as the main product.

At a selected temperature of 60°C and a pressure of 60 bar, a two-phase mixture of 1,3-butadiene and CO_2 exists, consisting of a liquid phase of 58 mol% CO_2 and a vapor phase of 16 mol% butadiene. As mentioned in the publication, the catalytic reaction is supposed to take place preferentially in the liquid phase. During reaction, the mixture is depleted in butadiene and CO_2. At the end of the reaction, the high conversion of 1,3-butadiene means that a liquid product phase (nearly free of butadiene) and a CO_2-rich vapor phase exist. The δ-lactone is found to be soluble in CO_2 as well as the other by-products The residual butadiene, as referred in the paper, is expected to act as an entrainer and will, therefore, lead to an increase in solubility of the products in the vapor phase. After reaction, the liquid product, consisting of δ-lactone and by-products, is removed from the autoclave using the residual CO_2.

The effect of high-pressure CO_2 on the melting point of a solid compound – the gas-induced melting-point lowering – provides another method whereby phase equilibrium control may lead to enhanced reaction rates, coupled with easy separation of product from solvent.

The melting temperature of the solid can be significantly lowered (up to 30–40 K) in the presence of a gas, in this case CO_2, especially at pressures approaching the critical pressure. Once again, the pressures used are much lower than if supercritical CO_2 was used to dissolve the solid, the concentration of reagent in the reaction phase is much higher, which can lead to higher rates, and homogeneous catalysts do not have to be designed to be CO_2-soluble. Jessop et al. [14] presented a good example with the hydrogenation of 2-vinylnaphthalene (mp 65–66°C) catalyzed by $RhCl(PPh_3)_3$ at 36°C.

3.5.1.7
Hydrogenations in CO_2

One of the main advantages of using supercritical fluids as solvents is their ability to dissolve large quantities of gases. Carbon dioxide and hydrogen, for example, mix in all proportions at temperatures above the critical temperature of CO_2 (304.1 K).

Hydrogenations in supercritical fluids are close to industrial-scale application. In the process developed by Härröd and colleagues [15], the hydrogenation of fatty acid methyl esters is carried out in propane, because the solubilities of the esters in supercritical CO_2 are too low. In propane, the reaction can be carried out with only one supercritical phase contacting the solid catalyst, and with enough hydrogen to ensure an extremely fast and selective reaction.

However, the hydrogenations described by Hitzler et al. [16] in carbon dioxide are carried out, at least in some cases, in a clearly biphasic, gas–liquid system. In any case, they manage to perform fast hydrogenations using a highly effective flow process, either at the laboratory level or at the large-scale pilot plant built at the Thomas Swan site. There has been considerable scientific argument on whether supercritical hydrogenation reactions proceed faster and more efficiently in a single phase or in multiple phases. It basically depends very much on the reaction system involved.

Chouchi et al. [17] carried out the hydrogenation of α-pinene in CO_2-containing mixtures. These mixtures can easily be turned either biphasic (gas–liquid) or single-phase (supercritical), by small changes in pressure, at moderate pressures – around 100 bar. Surprisingly, with a Pd catalyst, the hydrogenation of α-pinene was much faster in a two-phase system than when only a single-phase was in contact with the catalyst. This was probably the first study where a comparison was made using the same reactants and catalyst, with only small changes in the process parameters, but with totally different phase behavior. Milewska et al. [18] extended this study to other catalysts, and obtained results that were highly dependent on the catalysts used. Thus it seems that differential adsorption of hydrogen and pinene at the catalyst surface is the cause of the different rates observed in biphasic and monophasic mixtures. As for many other liquids, CO_2 is highly soluble in pinene; at pressures close to the critical pressure, the liquid phase in a biphasic mixture can contain 80 mol% or more of CO_2. Hence, the liquid reactant is an "expanded liquid", which can dissolve large quantities of hydrogen, thus significantly increasing its concentration in the vicinity of the solid catalyst. If the adsorption of pinene is related to the rate-limiting step, a biphasic system will lead to higher rates, because pinene will be more concentrated in the liquid around the catalyst than when diluted in a single supercritical phase. The opposite will happen for catalysts at which hydrogen adsorption controls the rate.

3.5.1.8
Ionic Liquids and Supercritical Carbon Dioxide

Ionic liquids (room-temperature molten salts, ILs) are reviewed in Section 3.5.3 of this book. They have an essentially negligible vapor pressure and the ability to dissolve a wide range of organic and inorganic compounds, which makes them an attractive alternative reaction medium to volatile organic compounds.

Blanchard et al. [19] were the first to suggest that the combination of supercritical CO_2 with an ionic liquid would generate very interesting biphasic mixtures in which to carry out reactions. Owing to their ionic nature and negligible vapor pressure, ILs exhibit no appreciable solubility in $scCO_2$. At the same time, CO_2 is remarkably soluble in the IL phase, and its solubility increases dramatically with increasing pressure up to about 100 bar. It can then be used to extract numerous organic substances from the liquid without contamination of solvent in the final product. Even at high operating pressures, a CO_2–IL system remains biphasic.

These biphasic systems show many advantageous properties that render them attractive for synthetic chemistry. Since 2000, studies of chemical reactions carried out in CO_2 + ILs have been published at an accelerating rate.

Although more research needs to be conducted on IL toxicity and environmental impact, both solvents are considered benign, and the need to use volatile organic compounds to separate reaction products may be eliminated. It is, in principle, possible to run a homogeneously catalyzed reaction in continuous mode. Ionic liquid and immobilized catalyst remain in the reactor at all times, which makes their recycling possible. That is very important, because they tend to be rather expensive, and unless they can be reused their application in large-scale industrial processes might not be feasible. At the same time, the insolubility of ILs in the $scCO_2$ phase prevents the product being contaminated with the solvent. Reaction and product separation from both catalyst and solvents can be achieved in a single operation, which minimizes the number of process steps needed.

Using ILs as solvents has one limitation – their viscosity, which is significantly higher than that of conventional solvents. By dissolving carbon dioxide, the resulting "expanded liquid" presents a distinctively lower viscosity, and also, as in the cases described above, enhanced mass transfer and higher solubility of permanent gases in the liquid phase.

The use of $scCO_2$ as an extraction medium therefore facilitates both product recovery and catalyst recycling, while maintaining a Green Chemistry concept. One way of using it is depicted in Fig. 3.5-1.

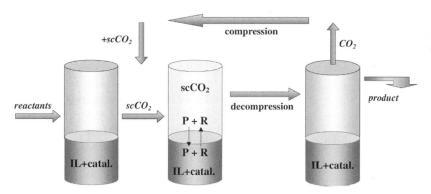

Fig. 3.5-1 Organometallic catalyst recycling and product recovery using an $scCO_2$-IL biphasic system.

In this case, CO_2 is introduced into the system after the reaction is finished, and it acts solely as an extraction medium. A more ambitious design would also use the high-pressure gas as a reactant carrier.

Webb et al. [20] used this approach to study alkene hydroformylations. They found a lower rate, yet higher selectivity towards the desired products, when compared to the reaction in IL alone. This increase in selectivity can be attributed to the partitioning of the product to the gas phase, which reduces its contact with the catalyst, and prevents further reaction. In order to have higher reaction productivity it is necessary to obtain a higher substrate concentration in the IL phase. This can be achieved by decreasing the CO_2 partial pressure, which decreases its solvent power, and leads to reactant partitioning more into the IL phase. Webb et al. also found that, under certain conditions, the system could operate continuously for several weeks without detectable catalyst degradation.

Dissolution of $scCO_2$ in the IL phase before the reaction starts was found to have both positive and negative effects on reaction selectivity and rate, depending on the nature of the reactants and the reaction itself.

Jessop et al. [21] studied a number of different solvents (ILs, ILs with cosolvents, scCO$_2$, CO$_2$-expanded ILs and conventional solvents), in order to evaluate the asymmetric hydrogenation of α,β-unsaturated carboxylic acids. They found that, depending on the nature of the reactant, the presence of CO$_2$ in the reaction system can cause either an increase or a decrease in the reaction selectivity. Some compounds exhibit higher enantioselectivity with a high H$_2$ concentration (*type I substrates*), while others exhibit higher enantioselectivity with a low H$_2$ concentration (*type II substrates*). Since scCO$_2$-expanded ILs have lower viscosity and increased transfer rate of H$_2$ into the liquid phase, it can be advantageous for running the hydrogenation reactions of type I substrates. Type II substrates, on the other hand, exhibit higher selectivity when the reaction is performed in ILs without the presence of CO$_2$.

Hou et al. [22] compared reaction selectivity and conversion for the Wacker oxidation of 1-hexene in four different reaction systems (without solvent, or using scCO$_2$, IL or CO$_2$–IL as solvent). The conversions in all reaction systems were similar. The selectivity in the CO$_2$- based expanded liquid was significantly higher, and was found to increase with pressure. The higher selectivity of this system can be explained by partial dissolution of a reactant in the CO$_2$ gas phase, which leaves less reactant in contact with catalyst in the liquid phase, decreasing isomerization. An enhanced mass transfer in the CO$_2$-expanded liquid may also lead to reduced isomerization. Recycling experiments were performed for supercritical CO$_2$ and CO$_2$ + IL. The catalyst was stable in both systems, but more stable in the latter. The conclusion that can be drawn is that not only is selectivity enhanced, but also catalyst stability.

These studies prove that the "CO$_2$-expanded liquid" concept is also applicable to ILs and can lead to versatile, integrated reaction–separation systems.

In some cases, unusual liquid–liquid phase equilibria in IL systems can be used to design reaction controlled by "phase switches".

Najdanovic-Visak et al. [23] took advantage of the unusual phase behavior of the (water + ethanol + the IL [bmim][PF$_6$]) system to perform the epoxidation of isophorone, catalyzed by an aqueous solution of hydrogen peroxide, in advantageous monophasic conditions. The epoxidation reaction rate was much faster than in the previously reported biphasic catalysis. Owing to a large cosolvent effect of ethanol on water + [bmim][PF$_6$], they managed to obtain single-phase reaction conditions, which introduce higher reaction rates. Carbon dioxide was then used, following the scheme of Fig. 3.5-1, to extract the product without loss of ionic liquid, which can be reused in the next reaction cycle. The solvent previously used to extract the product from the IL phase, ethyl acetate, is partially miscible with [bmim][PF$_6$], which may lead to product contamination. Thus the reaction carried out in the (water + ethanol + IL + scCO$_2$) system can benefit from single-phase reaction conditions, which introduce increased reaction rates, without losing the advantages of biphasic system for catalyst recycling and product separation.

The fine tuning of CO$_2$-expanded ILs should become an interesting tool in developing new, green solvents for synthetic chemistry, where clean separations are integrated into the reaction design. At higher pressures, CO$_2$ is denser, and hence has better solvent power. However, it is on lowering the pressure that the selectivity can be used to advantage. Scurto et al. [24] have recently reported that relatively low pressures of carbon dioxide, of the order of 50 bar, can induce separation of water from several ionic liquids. This sort of induced phase split will certainly lead to many ingenious integrated reaction separation schemes in the near future.

3.5.1.9
A Note on Supercritical Water

Supercritical water is a recent addition to the short list of intensively studied supercritical fluids. Solubilities in sc-water have some notorious differences from what might be expected. Namely, sc-water dissolves hydrocarbons and gases such as oxygen, but does not dissolve ionic salts such as sodium chloride. In fact, paradoxical as it may seem, supercritical water behaves like a non-aqueous solvent. Water has a high critical temperature ($T = 647.1$ K $= 374°C$) and pressure ($p = 22$ MPa). At room temperature, liquid water contains a highly developed network of hydrogen bonds between its molecules. As temperature increases, this network is progressively destroyed by the thermal kinetic energy, and water becomes a more "normal" polar solvent.

Owing to its high critical-point parameters, working with supercritical water requires high-pressure/high-temperature technology. Sc-water can become a highly corrosive medium, especially when chloride-containing solutes are dissolved, and unusual and expensive materials must be used. On the other hand, precipitation of salts may lead to complicated process strategies to avoid clogging of tubing. These problems have for some time held back the full industrial-scale implementation of potentially very interesting applications of sc-water.

High-temperature, high-pressure liquid water can, however, be used in a similar fashion as CO_2-expanded liquids, although with very different properties. As the temperature increases, the hydrogen-bond network is destroyed and water's properties change. Near-critical water can therefore be regarded as just another green expanded solvent, with tuneable properties waiting to be explored.

Acknowledgments

Anna Banet Osuna and Ana Šerbanović thank Fundação para a Ciência e Tecnologia (Lisbon, Portugal) and the POCTI programme for grants.

References

1 G. Brunner, *Gas Extraction*, Springer, New York, **1994**.

2 G. Lumiat, C. Perre, J.-M. Aracil, Patent WO 01/23155 A1 and PCT/FR00/02653, **1999**.

3 D. Chouchi, C. Maricato, M. Nunes da Ponte, A. Pires, V. San Romão, *Proceedings of the Fourth International Symposium on Supercritical Fluids*, Sendai, Japan, **1997**, vol. A, p. 379.

4 J. M. deSimone, Z. Guan, C. S. Elsbernd, *Science* **1992**, *257*, 945.

5 E.J. Beckman, *J. Supercrit. Fluids* **2004**, *28*, 121–191.

6 G. Brunner, Proceedings of the Fourth International Symposium on Supercritical Fluids, Sendai, Japan, **1997**, vol. C, p. 745.

7 G. Brunner, M. Budich, in *Supercritical Fluids as Solvents and Reaction Media*, G. Brunner, ed., Elsevier, Amsterdam, **2004**, pp. 489–522.

8 R. Noyori, *Chem. Rev.* **1999**, *99*, 353.

9 J.A. Darr, M. Poliakoff, *Chem. Rev.* **1999**, *99*, 495.

10 J. Ke, B.X. Han, M.W. George, H.K. Yan, M. Poliakoff, *J. Am. Chem. Soc.* **2001**, *123*, 3661.

11 X.F. Xie, C.L. Liotta, C.A. Eckert, *Ind. Eng. Chem. Res.* **2004**, *43*, 2605.

12 M. Wei, G. T. Musie, D.H. Busch, B. Subramanian, *J. Am. Chem. Soc.* **2002**, *124*, 2513.

13 K. Buchmüller, N. Dahmen, E. Dinjus, D. Neumann, B. Powietzka, S. Pitter, J. Schön, *Green Chem.* **2003**, *5*, 218.

14 P. Jessop, D.C. Wynne, S. DeHaai, D. Nakawatase, *Chem. Commun.*, **2000**, 693.

15 M. Härröd, S. van den Hark, M.-B. Macher, P. Møller, in *High Pressure Process Technology: Fundamentals and Applications*, A. Bertucco and G. Vetter, eds., Elsevier, Amsterdam, **2001**, pp. 496–508; www.harrod-research.se/

16 M.G. Hitzler, F.R. Smail, S.K. Ross, M. Poliakoff, *Org. Proc. Res. Dev.* **1998**, *2*, 137.

17 D. Chouchi, D. Gourgouillon, M. Courel, J. Vital, M. Nunes da Ponte, *Ind. Eng. Chem. Res.*, **2001**, *40*, 2551.

18 A. Milewska, D. Gourgouillon, D. Chouchi, I. Fonseca, M. Nunes da Ponte, *Proceedings of the 8th Meeting on Supercritical Fluids*, Bordeaux, **2002**, p. 129.

19 L.A. Blanchard, D. Hancu, E.J. Beckman, J.F. Brennecke, *Nature* **1999**, *399*, 28.

20 P.B. Webb, M.F. Sellin, T.E. Kunene, S. Williamson, A.M.Z. Slawin, D.J. Cole-Hamilton, *J. Am. Chem. Soc.* **2003**, *125*, 15577.

21 P.G. Jessop, R.R. Stanley, R.A. Brown, C.A. Eckert, C.L. Liotta, T.T. Ngo, P. Pollet, *Green Chem.*, **2003**, *5*, 123.

22 Z.S. Hou, B.X. Han, L. Gao, T. Jiang, Z.M. Liu, Y.H. Chang, X.G. Zhang, J. He, *New J. Chem.*, **2002**, *26*, 1246.

23 V. Najdanovic-Visak, A. Serbanovic, J.M.S.S. Esperança, H.J.R. Guedes, L.P.N. Rebelo, M. Nunes da Ponte, *Chem Phys Chem.*, **2003**, *4*, 520.

24 A. M. Scurto, S.N.V.K. Aki, J.F. Brennecke, *Chem. Commun.* **2003**, 572.

3.5.2
Fluorinated Solvents

Hiroshi Matsubara and Ilhyong Ryu

3.5.2.1
Introduction

The repertoire of fluorinated solvents available for reaction media has surpassed what was imagined in the mid-1990s. This is the result of the evolution of "fluorous chemistry" in organic synthesis, to which organic chemists are directing tremendous attention [1]. The wording "fluorous" expresses philicity to perfluoroalkanes, analogously to the corresponding wording "aqueous" expressing philicity to water [2]. Fluorous reactions that are based on fluorous reagents naturally require a variety of fluorous solvents (fluorinated organic solvents), but even before this new wave, some trifluoromethyl-substituted compounds, such as trifluoroacetic acid (CF_3CO_2H, TFA) and hexafluoroisopropanol ($CF_3CHOHCF_3$, HFIP), were familiar to organic chemists as polar solvents available for organic reactions. Because of its strongly acidic nature, TFA is used in many acid-catalyzed rearrangement reactions as a reagent-type reaction medium [3]. TFA can cleave many nitrogen- and oxygen-protecting groups including *N*-Boc, benzyl ether, *t*-butyl ether and triphenylmethyl ether. This is the reason why TFA is frequently used in peptide synthesis as a reaction medium that also effects the removal of protecting groups. HFIP is a highly acidic alcohol ($pK_a = 9.3$), which is miscible with water and many common organic solvents except hydrocarbons. HFIP can dissolve polymers such as polyesters, polyamides, poly(vinyl alcohols) and poly(acrylonitriles); gel permeation chromatography (GPC) for polymers that exhibit a low solubility in organic solvents, can often be performed with HFIP as an eluent [4]. HFIP has several unique properties, which include high polarity, high ionization power, high ability to make hydrogen bonds, and low nucleophilicity. Using these properties of HFIP, various reactions, in particular avoiding strongly basic or acidic conditions, are carried out in HFIP [5].

In contrast to these CF_3-substituted solvents, the use of perfluorinated hydrocarbons, represented by perfluorohexanes (FC-72™), in synthesis as solvent had been quite limited until recently. Perfluorocarbons are generally immiscible with water and organic solvents at room temperature except for some low molecular weight solvents such as pentane and ether. Thus, when they are mixed with water or organic solvents, the lighter organic or water phase forms the upper layer, while the denser fluorous phase forms the lower layer. However, there are many cases in which a mixed solvent system composed of perfluorinated solvents and other organic solvents becomes homogeneous on heating. This thermomorphic behavior in a fluorous/organic mixed system was applied elegantly by Horváth and Rábai in their landmark work in fluorous chemistry in 1994, which achieved facile separation of catalysts and products in Rh-catalyzed hydroformylation reaction of alkenes

(Scheme 3.5-1) [6]. They used a binary solvent system comprised of toluene and perfluoromethylcyclohexane (PFMC: $CF_3C_6F_{11}$). After the reaction, upon cooling these two solvents are separated out again, enabling facile separation of Rh-catalyst having a fluorous phosphine ligand in fluorous layer and organic products in organic layer. Nowadays, the thermomorphic behaviors are among the most attractive aspects of fluorous technologies and gives a basis to fluorous-tagged method which employs fluorous reagents and organic substrates [7].

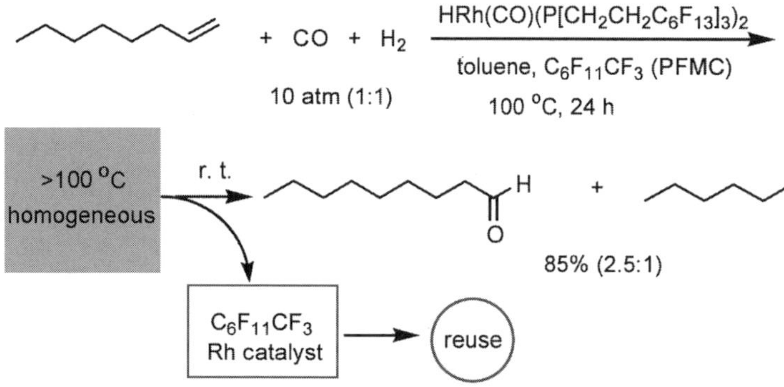

Scheme 3.5-1

The rapid progress in fluorous chemistry shed light on the use of fluorous-organic amphiphilic solvents, such as benzotrifluoride (BTF) [8] and a fluorous ether F-626 [9], which can mediate efficiently the reaction of organic substrates with fluorous reagents. What is interesting here is that these fluorous solvents also have a good potential as solvents for ordinary (non-fluorous) organic reactions.

In this short section, we focus mainly on four types of fluorinated solvents, traditional and new, which are available for ordinary organic synthesis. These are benzotrifluoride (BTF) **1**, fluorous ether F-626 **2**, fluorous dimethylformamide (F-DMF) **3**, and perfluorohexanes such as FC-72 **4**, whose physical properties are summarized in Table 3.5-1. The challenges to explore the *green* potentials of fluorous media have just begun but, no doubt, the concept is growing to constitute another important aspect of fluorous chemistry.

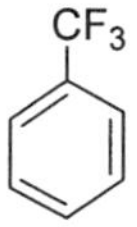

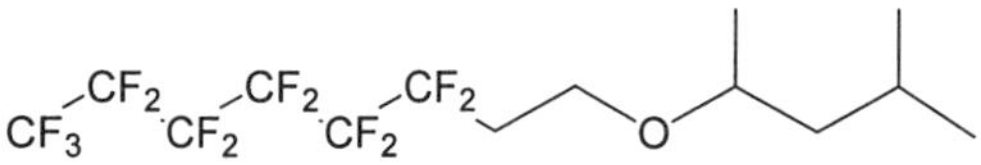

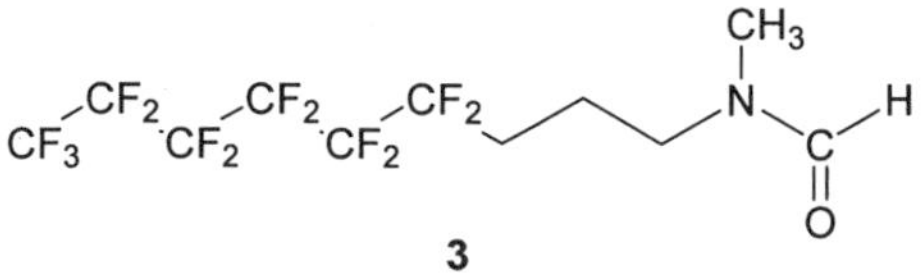

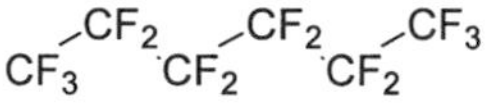

Table 3.5-1 Physical data for fluorinated solvents.

Solvent	bp (°C)	mp (°C)	n_D^{20}	d
BTF	102[a]	−29[b]	1.4150[c]	1.190[d]
F-626	214[e]	−110[#e]	1.3418[e]	1.354[e]
F-DMF	110 / 0.75mmHg[f]	−38[#f]	1.3593[f]	1.563[f]
FC-72	58−60[g]	−82[h]	1.2518i	1.670[h]

Transition temperature to glassy state.
[a] S. Baldwin, *J. Am. Chem. Soc.* **1967**, *89*, 1886. [b] A. Ogawa, D. P. Curran, *J. Org. Chem.*, **1997**, *62*, 450. [c] Y. A. Fialkov, L. I. Moklyachuk, M. M. Kremlev, L. M. Yagupol'skii, *J. Org. Chem. USSR (Engl. Transl.)* **1980**, *16*, 1269. [d] A. P. Rudenko, V. S. Sperkach, A. N. Timoshenko, L. M. Yagupol'skii, *Russ. J. Phys. Chem. (Engl. Transl.)* **1981**, *55*, 591. [e] H. Matsubara, S. Yasuda, H. Sugiyama, I. Ryu, Y. Fujii, K. Kita, *Tetrahedron*, **2002**, *58*, 4071. [f] H. Matsubara, R. Maeda, I. Ryu, Submitted to JP Patent, **2004**, 65754. [g] R. E. Banks, J. E. Burgess, R. N. Haszeldine, *J. Chem. Soc.*, **1965**, 2720. [h] R. D. Dunlap, C. J. Murphy, R. G. Bedford, *J. Am. Chem. Soc.*, **1958**, *80*, 83. [i] C. Brice, *J. Am. Chem. Soc.*, **1953**, *75*, 2921.

3.5.2.2
Benzotrifluoride (BTF)

Benzotrifluoride **1** is a colorless, low viscosity liquid. Its polarity is intermediate between those of dichloromethane and ethyl acetate, and it is miscible with common organic solvents and able to dissolve many organic compounds. Benzotrifluoride is used in a wide variety of reactions including radical reactions, oxidations and reductions, phase-transfer reactions, transition metal catalyzed processes and Lewis acid reactions. Ogawa and Curran reported that dichloromethane (CH_2Cl_2) is replaceable by BTF in many instances [8].

Thus, standard acylation, tosylation and silylation proceed in BTF in yields comparable to those obtained in CH_2Cl_2. Swern oxidation and Dess–Martin oxidation also occur in BTF to give the corresponding carbonyl compounds in high yield (Scheme 3.5-2). Benzotrifluoride can be used for oxidation reactions involving H_2O_2. As for Lewis acid-assisted reactions, mild Lewis acids such as $TiCl_4$ or $ZnCl_2$ can be used in BTF solution. Thus, Sakurai and Mukaiyama reactions as well as Friedel–Crafts acylations and Diels–Alder reactions using BTF are found to give good yields of the corresponding products (Scheme 3.5-3). Czifrák and Somsák reported that BTF can serve as an alternative reaction medium, in place of carbon tetrachloride, in the bromination of carbohydrates (Scheme 3.5-4) [10].

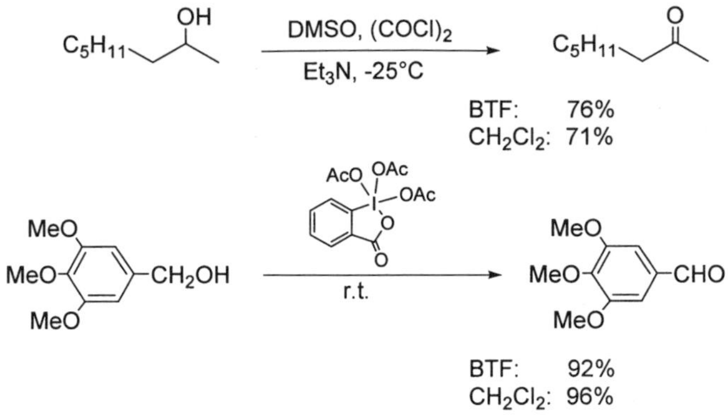

Oxidation of 2-heptanol to 2-heptanone via Swern oxidation:

C$_5$H$_{11}$—CH(OH)—CH$_3$ → (DMSO, (COCl)$_2$; Et$_3$N, -25°C) → C$_5$H$_{11}$—CH$_2$—CO—CH$_3$

BTF: 76%
CH$_2$Cl$_2$: 71%

Oxidation of trimethoxybenzyl alcohol to aldehyde (Dess–Martin periodinane, r.t.):

BTF: 92%
CH$_2$Cl$_2$: 96%

Scheme 3.5-2

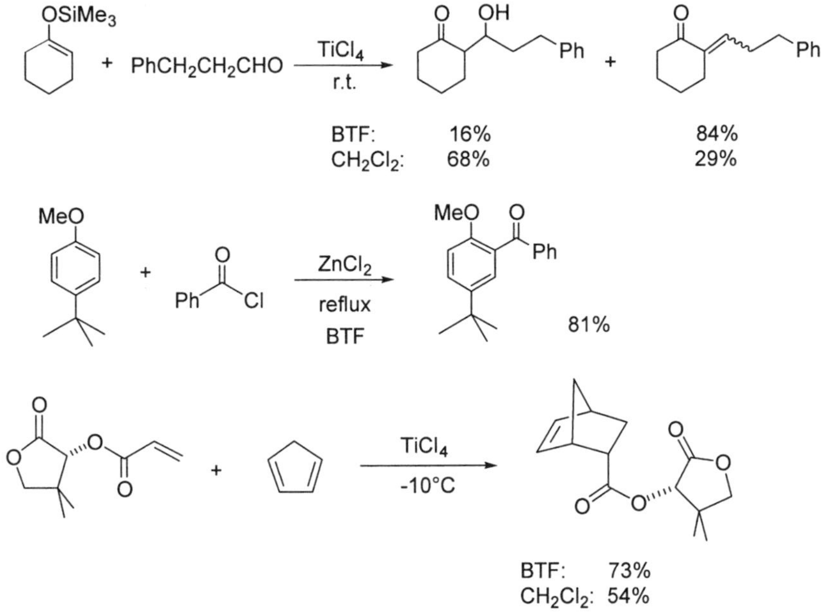

1-(trimethylsilyloxy)cyclohexene + PhCH$_2$CH$_2$CHO —(TiCl$_4$, r.t.)→ aldol product + enone

BTF: 16%   84%
CH$_2$Cl$_2$: 68%   29%

4-*tert*-butylanisole + PhCOCl —(ZnCl$_2$, reflux, BTF)→ ketone product   81%

Lactone acrylate + cyclopentadiene —(TiCl$_4$, -10°C)→ Diels–Alder product

BTF: 73%
CH$_2$Cl$_2$: 54%

Scheme 3.5-3

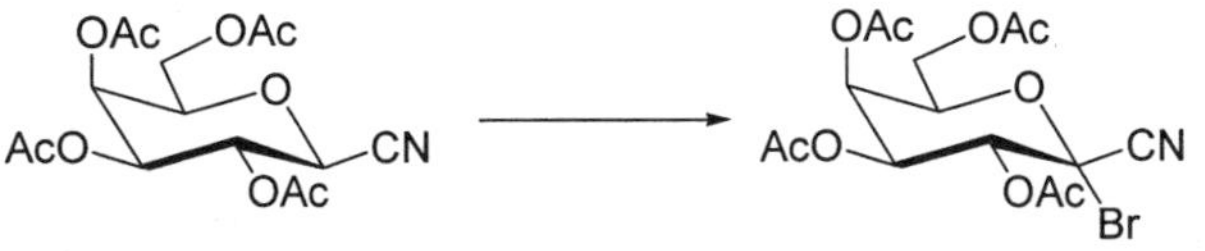

Conditions	Reaction time (h)	Yield (%)
Br$_2$, CCl$_4$, $h\nu$, reflux	0.5	88
KBrO$_3$-Na$_2$S$_2$O$_3$, BTF-H$_2$O, r.t.	27.0	88
Br$_2$, BTF, $h\nu$, K$_2$CO$_3$, reflux	1.0	69

Scheme 3.5-4

3.5.2.3
Fluorous Ether F-626

F-626 **2**: $1H,1H,2H,2H$-perfluorooctyl-1,3-dimethylbutyl ether, is found to be a useful fluorous/organic amphiphilic solvent; F-626 is easily removable from the reaction mixture by fluorous/organic biphasic workup. F-626 is a colorless, clear, slightly viscous liquid, miscible with common organic solvents but hardly soluble in water. The approximate partition coefficients of BTF, F-626 and F-DMF associated with biphasic treatment are shown in Table 3.5-2; these show that the majority of F-626 is distributed in the FC-72 phase except in the case when chloroform and ethyl acetate are used, while BTF is preferentially distributed in the organic solvents. Using F-626 as a solvent, $LiAlH_4$ reduction, catalytic hydrogenation and fluorous reductive radical reactions were successful. Furthermore, classical high-temperature reactions (up to 200°C), such as Vilsmeier formylation, Wolff–Kishner reduction and the Diels–Alder reaction, were also examined in F-626. The yields of the products in F-626 were almost comparable with those obtained in common organic solvents, thus suggesting that F-626 has the potential to be an easily removed and reusable high boiling solvent (Scheme 3.5-5).

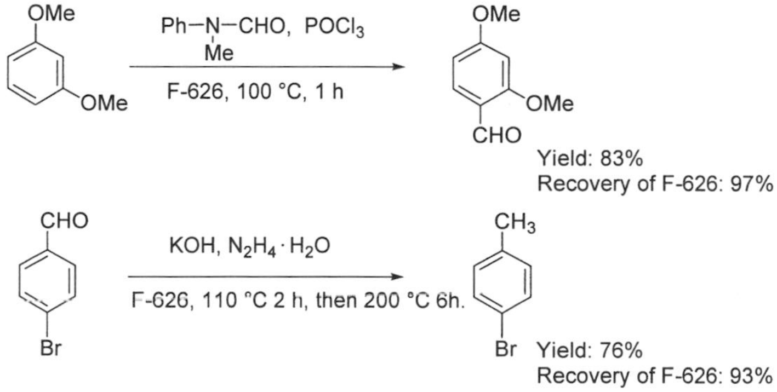

Scheme 3.5-5

Table 3.5-2 Partition coefficients of BTF, F-626 and F-DMF.

	Organic solvent / FC-72		
	BTF[a]	F-626[a]	F-DMF[b]
CH_3CN	1/0.13	1/7.3	1/0.08
MeOH	1/0.21	1.3.8	1/0.05
C_6H_6	1/0.18	1/1.6	1/1.13
Cyclohexane	_#	_#	1/8.30
Acetone	1/0.08	1/1.1	1/0.10
AcOEt	1/0.13	1/0.85	1/0.20
$CHCl_3$	1/0.16	1/0.85	1/0.13

Not yet determined.
[a] H. Matsubara, S. Yasuda, H. Sugiyama, I. Ryu, Y. Fujii, K. Kita, *Tetrahedron*, **2002**, *58*, 4071.
[b] H. Matsubara, R. Maeda, I. Ryu, Submitted to JP Patent, **2004**, 65754.

Recently, the present authors have achieved a facile recycling method for both catalyst and reaction medium using F-626 in a Mizoroki–Heck arylation reaction of acrylic acids [11]. The procedure employed a fluorous carbene complex, prepared in situ from a fluorous imidazolium salt, palladium acetate as the catalyst and F-626 as a single reaction medium. When acrylic acid was used as a substrate, separation of the product from the reaction mixture was performed simply by filtration with a small amount of FC-72. The FC-72 solution containing the fluorous Pd-catalyst and F-626 was evaporated and the residue containing the catalyst and F-626 (96% recovery) can be recycled for the next run (Scheme 3.5-6). They tried to reuse the catalyst, and observed no loss of catalytic activity in five re-use cycles.

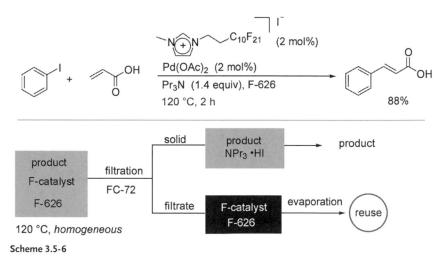

Scheme 3.5-6

3.5.2.4
F-DMF

Dimethylformamide represents a useful aprotic polar solvent frequently used for various organic processes including transition metal catalyzed reactions. Since DMF has a relatively high bp (152°C) and high solubility in water, an organic/aqueous biphasic treatment is normally used to separate this solvent from product. This aqueous treatment, although easy to carry out, makes the recovery of DMF from aqueous solution containing DMF difficult. This led us to develop a fluorous version of DMF (F-DMF) **3**, in which a $C_6F_{13}C_3H_6$ fluorous chain replaces one of the two methyl groups in the original DMF [12]. This solvent has a higher bp (110°C/0.75 mmHg) and is miscible with many organic solvent but hardly soluble in water. Although the approximate partition coefficients of F-DMF measured using polar solvent and FC-72, suggest that, although F-DMF is less fluorous than F-626 (Table 3.5-2), this solvent can be removed by fluorous/organic biphasic workup using non-polar solvents such as cyclohexane or benzene as the organic phase and FC-72 as the fluorous phase. Scheme 3.5-7 outlines the result of the Sonogashira coupling reaction, which proceeded successfully in F-DMF. Using fluorous carbene palladium complex as the catalyst and copper(I) iodide as a co-catalyst, iodobenzene reacted with phenylacetylene to give diphenylacetylene in 90% yield. After cooling, triphasic workup using cyclohexane, FC-72 and water was carried out under nitrogen. The resulting FC-72 solution containing the palladium catalyst and F-DMF (recovery >99%), was reused by adding copper iodide and diisopropylamine for the next run to give 81% yield of the product. Mizoroki–Heck reactions were also effected with F-DMF, affording similar results to those obtained using F-626 as solvent.

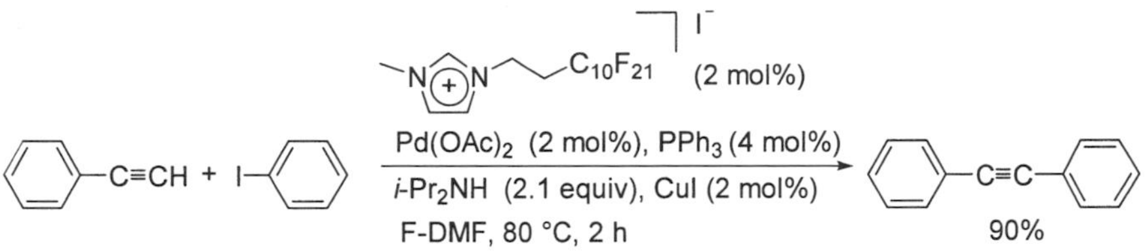

Scheme 3.5-7

3.5.2.5
FC-72 (Perfluorohexanes)

Non-fluorous, ordinary organic reactions that are carried out using perfluoroalkanes as reaction media are still scarce. However, the thermomorphic nature of perfluorocarbons with hydrocarbons allows for some synthetic reactions using perfluoroalkanes as a single reaction medium. For example, PFMC (perfluoromethylcyclohexane) was used as a single reaction medium for Rh-catalyzed hydrogenation and hydroformylation of 1-octene [13]. Thus, 1-octene dissolves at 80°C in PFMC to make a homogeneous reaction. Phase separation results when the reaction mixture is cooled. In the case of hydrogenation, such a phase separation requires cooling to 0°C, whereas in the case of hydroformylation the products nonanal and 2-methyloctanal separate from PFMC at room temperature.

Carbon tetrachloride is a harmful solvent, which has frequently been used for the bromination of alkenes by molecular bromine. Savage used FC-72 in place of carbon tetrachloride to dilute bromine as a reaction medium [14]. Ryu and coworkers used FC-72 for bromination of alkenes, although with a different concept [15]. The concept of the "phase-vanishing (PV) method" is illustrated in Scheme 3.5-8, taking as an example the bromination of alkenes. When a fluorous solvent such as perfluorohexanes ($d = 1.669$ g cm^{-3}) is mixed gently with a denser reagent such as bromine ($d = 3.12$ g cm^{-3}), two phases result with the reagent bromine on the bottom. When an organic solvent containing a substrate is added, the fluorous solvent then locates in the middle, screening the two otherwise miscible phases containing the substrate and the reagent from each other. In this triphasic reaction, the fluorous "phase screen" behaves as a liquid membrane that prevents mixing but permits passive transport of reagents from the bottom layer to the top layer and thereby regulates the reaction. As the reagent is consumed, the bottom phase vanishes. They reported that this PV method based on spontaneous reaction control by gravity and diffusion is indeed useful for typical synthetic reactions such as the dealkylation of aromatic ethers by boron tribromide and Friedel–Crafts acylation using tin tetrachloride. The PV method can be carried out successfully in a parallel synthesis by Friedel–Crafts reaction of thiophene [16].

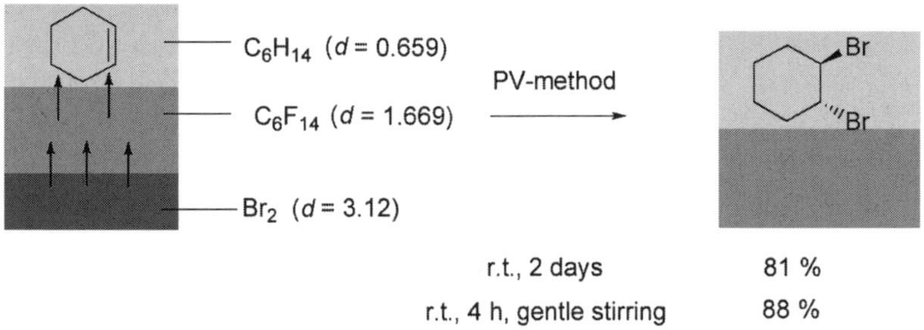

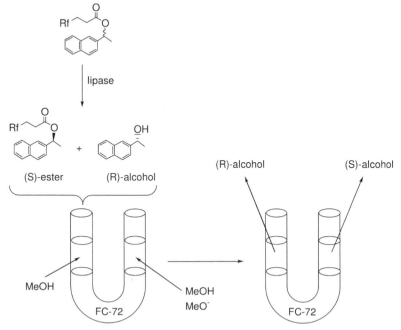

C_6H_{14} (d = 0.659)

C_6F_{14} (d = 1.669)

Br_2 (d = 3.12)

PV-method

r.t., 2 days 81 %

r.t., 4 h, gentle stirring 88 %

Scheme 3.5-8

Curran and coworkers reported on organic–fluorous–organic triphasic systems using a glass U-tube, which is useful for the separation of fluorous and non-fluorous compounds [17]. An example is shown in Fig. 3.5-2. A racemic ester with a fluorous ponytail was treated with a lipase to resolve the racemic ester into (*S*)-ester

lipase

(S)-ester   (R)-alcohol

(R)-alcohol   (S)-alcohol

MeOH

MeOH
MeO⁻

FC-72

FC-72

Fig. 3.5-2 Resolution of a racemic ester using lipase hydrolysis followed by a three-phase separation.

and (*R*)-alcohol. After removing the enzyme by filtration, the resulting mixture was injected into the source phase. Only (*S*)-ester with a fluorous tail could be transferred to the receiving phase. The reacting (*S*)-ester was hydrolyzed in the basic conditions of the phase (MeOH/MeO⁻) to give (*S*)-alcohol. The fluorous liquid membrane system can be applied to phase-vanishing type gas/liquid reactions. Iskra et al. reported that chlorination of olefins with Cl_2 gas could be performed in a U-tube with a fluorous solvent (FC-77: perfluoro-*n*-octanes) as the transport phase, affording the corresponding dichlorides in good yields (Fig. 3.5-3) [18].

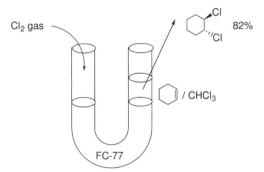

Fig. 3.5-3 Chlorination of olefins using the fluorous solvent FC-77 as the transport phase.

Using a glass U-tube as a reaction tool, Nakamura and coworkers expanded phase-vanishing methods to include reagents lighter than fluorous solvents, such as thionyl chloride and phosphorus trichloride [19].

3.5.2.6
Conclusion

The use of fluorinated solvents can be classified into several categories: (1) as a solvent (or a co-solvent) in a reaction with fluorous reagents and/or fluorous substrates because of high solubility of fluorous materials in fluorinated solvents, (2) as an acidic catalyst or reagent for rearrangements, condensations or cleavage reactions, (3) as alternative "green" reaction media for harmful or toxic chlorinated solvents, (4) as amphiphilic solvents easy to remove with biphasic workup, (5) as a liquid phase for immobilizing catalysts and (6) as a liquid membrane. While traditional fluorinated solvents are often used as category (2)-type solvents, new fluorous solvents are employed in the other categories. In particular, categories (5) and (6) are novel uses for these solvents and are expected to be increasingly widely used [20].

References

1 Leading reviews: I.T. Horváth, *Acc. Chem. Res.* **1998**, *31*, 641; D.P. Curran, *Angew. Chem., Int. Ed. Engl.* **1998**, *37*, 1175; L.P. Barthel-Rosa, J.A. Gladysz, *Coord. Chem. Rev.* **1999**, *190–192*, 587; B. Cornils, *Angew. Chem., Int. Ed. Engl.* **1997**, *36*, 2057; E. de Wolf, G. van Koten, B.-J. Deelman, *Chem. Soc. Rev.* **1999**, *28*, 37; T. Kitazume, *J. Fluorine Chem.* **2000**, *105*, 265; G.G. Furin, *Russ. Chem. Rev.* **2000**, *69*, 491; D.P.Curran, *Stimulating Concepts in Chemistry*, F. Vögtle, J. F.Stoddard, M. Shibasaki, eds., Wiley-VCH, New York, **2000**, *Handbook of Fluorous Chemistry*, J. A. Gladysz, D. P. Curran, I. Horváth, eds., Wiley-VCH, Weinheim, **2004**.

2 For historical views and definitions of fluorous chemistry, see: Symposium-in-Print on fluorous chemistry, J. Gladysz, D.P. Curran, *Tetrahedron* **2002**, *58*, 3823.

3 L. A. Paquette, *Encyclopedia of Reagents for Organic Synthesis*, Vol. 7, J. Wiley & Sons, Chichester, **1995**, p. 5131.

4 Jpn. Kokai Tokkyo Koho , **1991**, 03276065; Jpn. Kokai Tokkyo Koho, **1981**, 56053458; Jpn. Kokai Tokkyo Koho, **1981**, 56012551.

5 For recent works on the use of HFIP as a reaction medium, see: F. Fache, O. Piva, *Synlett*, **2002**, 2035; J. Legros, B. Crousse, J. Bourdon, D. Bonnet-Delpon, J.-P. Bégué, *Tetrahedron Lett.* **2001**, *42*, 4463; J. Legros, B. Crousse, D. Bonnet-Delpon, J.-P. Bégué, *Tetrahedron* **2002**, *58*, 3993; C. Bolm, M. Martin, O. Simic, M. Verrucci, *Org. Lett.*, **2003**, *5*, 427.

6 I. T. Horváth, J. Rábai, *Science*, **1994**, *266*, 72.

7 D. P. Curran, *Angew. Chem., Int. Ed. Engl.* **1998**, *37*, 1175.

8 A. Ogawa, D.P.Curran, *J. Org. Chem.* **1997**, *62*, 450.

9 H. Matsubara, S. Yasuda, H. Sugiyama, I. Ryu, Y. Fujii, K. Kita, *Tetrahedron*, **2002**, *58*, 4071.

10 K. Czifrák, L. Somsák, *Tetrahedron Lett.* **2002**, *43*, 8849.

11 T. Fukuyama, M. Arai, H. Matsubara, I. Ryu, *J. Org. Chem.* **2004**, *69*, 8105.

12 H. Matsubara, R. Maeda, I. Ryu, Submitted to JP Patent, **2004**, 65754.

13 B. Richter, B. A. L. Spek, G. van Koten, B.-J.Deelman, *J. Am. Chem. Soc.* **2000**, *122*, 3945.

14 S. M. Pereira, G. P. Savage, G. W.Simpson, *Syn. Commun.* **1995**, *25*, 1023–1026.

15 I. Ryu, H. Matsubara, S. Yasuda, H. Nakamura, D.P. Curran, *J. Am. Chem. Soc.* **2002**, *124*, 12946.

16 H. Matsubara, S. Yasuda, I. Ryu, *Synlett* **2003**, 247.

17 H. Nakamura, B. Linclau, D.P. Curran, *J. Am. Chem. Soc.*, **2001**, *123*, 10119; Z. Luo, S. M. Swaleh, F. Theil, D.P. Curran, *Org. Lett.*, **2002**, *4*, 2585.

18 J. Iskra, S. Staver, M. Zupan, *Chem. Commun.* **2003**, 2497.

19 H. Nakamura, T. Usui, H. Kuroda, I. Ryu, H. Matsubara, S. Yasuda, D.P. Curran, *Org. Lett.* **2003**, *5*, 1167.

20 Recent variations include "extractive PV method", see: N. K. Jana, J. G. Verkade, *Org. Lett.*, **2003**, *5*, 3787; D.P. Curran, S. Werner, *Org. Lett.*, **2004**, *6*, 1021.

3.5.3

**Ionic Liquids: Structure, Properties and Major Applications in Extraction/
Reaction Technology**

Jairton Dupont

3.5.3.1
Introduction

The main goal of green chemistry is the development of chemical products and
processes that reduce or eliminate the use and generation of hazardous substances
[1]. It is evident that the ideal chemical reactions or processes are those that do not
involve the use of solvents [2]. However, most chemical reactions and processes take place in solution, and therefore solvents constitute the "blood" of chemistry. Water is the most widespread solvent in nature (in terms of quantity) and without
question the most important for humankind [3]. However, most reactions and chemical processes employ organic solvents that are usually volatile, flammable, toxic
and hazardous, i.e. incompatible with the aims of green chemistry. The search for
replacements for classical organic solvents is therefore one of the most important
and active fields of contemporary chemistry. Various fluids, including supercritical fluids (Section 3.5.1), perfluorinated liquids (Section 3.5.2) and ionic liquids
(ILs), are being investigated as green alternatives to classical organic solvents. All
these fluids possess advantages and disadvantages in the replacement of the classical volatile or semi-volatile organic compounds (VOCs) that are generally used in
chemical reactions and processes. In this section the structure, properties and
main applications of a new class of fluid materials: ILs will be outlined.

3.5.3.2
Ionic Liquids: Overview

Ionic liquids are a special class of molten salts. It is well accepted that any liquid
electrolyte composed entirely of ions is denominated a molten salt or a fused salt
[4]. Classical examples of molten salts are: pure inorganic compounds (sodium
chloride, mp 801°C), salts of organic compounds (tetrabutylphosphonium chloride, mp 80°C) or even eutectic mixtures of inorganic salts (such as lithium chlori-
de/potassium chloride, 6/4, mp 352°C) or organo-mineral (triethylammonium
chloride/copper chloride, 1/1, mp 25°C). Ionic liquids can be arbitrarily defined as
"non-corrosive" molten salts that are fluid below 100°C and possess relatively low
viscosity. Although molten salts of this type have been known since the beginning
of the 20th century, such as ethylammonium nitrate [5], it was not until 1951, with
the synthesis of air- and water-sensitive salts that are fluid down to −87°C, resulting from the combination of 1-ethyl-3-methylimidazolium chloride or *N*-alkylpy-
ridinium chloride with aluminium trichloride, that ILs have definitively entered
the scene as chemical solvents, in particular for electrochemistry [6] and Zieg-

ler–Natta type biphasic catalysis [7]. The advent in the 1990s of less air- and water-sensitive room temperature ILs such as those based on the 1-alkyl-3-methylimidazolium cations associated with tetrafluoroborate, hexafluorophosphate and trifluoromethane sulfonate anions, has engendered a new impulse in their use as solvents in various domains of science [8–11]. Although the list of new ILs grows very rapidly, most of them are based on cations of tetraalkyl ammonium and phosphonium salts and hetero-aromatics (Fig. 3.5-4) typically associated with inorganic and organic anions such as BF_4, PF_6, $N(CF_3SO_2)_2$, CF_3SO_3, RCO_2, NO_3, ClO_4, etc.

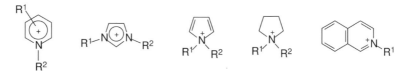

Fig. 3.5-4 Basic skeleton of the nitrogen-containing cations that can generate ILs: *N*-alkylpyridinium; 1,3-dialkylimidazolium; 1,1'-dialkylpyrolinium, 1,1'-dialkylpyrolidinium and *N*-alkylisoquinolinium, respectively.

By far the most used and investigated ILs are those based on 1,3-dialkylimidazolium cations. These liquids possess many physical-chemical properties, that qualifies them as "tailor-made" green solvents for various reactions and processes.

- They are effectively non-volatile (most of them exhibit non-measurable vapor pressure).
- They are usually liquids over a wide range of temperatures, and have relatively lower viscosities and higher densities than most classical organic solvents.
- They usually possess higher thermal, electrochemical and chemical stabilities compared with classical organic solvents.
- They dissolve a very broad spectrum of organic and inorganic compounds and polymeric materials, and their miscibility with these substances can be finely tuned by changing the nature of the cation and/or anion.
- They are typically non-coordinating solvents and their hydrophobicity can also be modulated by the proper choice of the cation and/or anion or by changing the temperature of the process.
- They are easily prepared from commercially available reagents through classical synthetic procedures and several of these liquids are now commercially available.

Not surprisingly, these materials are very popular and enjoy a plethora of applications in various domains of the physical sciences, and an impressive number of specialized reviews and books has appeared dealing with their synthesis, physico-chemical properties and applications in synthesis, catalysis and separation processes [12–26]. This section does not intend to be comprehensive on the vast area of synthesis and applications of ILs; rather it will attempt to provide a critical update of the basic principles and latest developments on the structure and properties of ILs (mainly those based on the 1,3-dialkylimidazolium cation), and their major applications and potentialities in reaction/extraction technologies. For the sake of clarity a limited number of examples has been chosen and readers interested in a more specific type and/or applications of ILs are invited to consult the respective articles and specialized reviews cited in the text. The basic construction of this section is to present and discuss the main physico-chemical properties and structural aspects of the ILs based on the 1,3-dialkylimidazolium cation that may be relevant for their applications as new fluid materials in green reaction/extraction technologies. This will be followed by a presentation and discussion of their use and potential in extraction, multiphase liquid–liquid synthesis and catalytic processes.

3.5.3.3
Preparation and Some Physico-Chemical Properties of 1,3-Dialkylimidazolium ILs

As already pointed out, of the various known ILs, those derived from the combination of the 1,3-dialkylimidazolium cation with various anions are the most popular and investigated class (Scheme 3.5-9). This is most probably due to their facility of synthesis; stability and the possibility of fine-tuning their physico-chemical properties by the simple choice of the N-alkyl substituents and/or anions (Table 3.5-3). The preparation of these salts is usually performed by a simple N-alkylation of 1-methylimidazole generating a 1-alkyl-3-methylimidazolium cation followed by anion metathesis [27]. The synthesis can also start with imidazole, which is consecutively alkylated with alkyl halides (Scheme 3.5-9). The 1,3-dialkylimidazolium cations will be abbreviated throughout this section as $[C_xC_yIm]$, where Im stands for imidazolium and x and y are the number of the carbons of the alkyl chains.

Scheme 3.5-9 R^1, R^2 = alkyl groups; X = Cl, Br, I; Y = BF_4, PF_6, CF_3CO_2, RCO_2, CF_3SO_3, $N(CF_3SO_3)_2$, NO_3, ClO_4, etc.

These methods generate a large variety of 1,3-dialkylimidazolium-based ILs of good quality. The main contaminant is usually residual chloride, which can be detected by a $AgNO_3$ test (limit of 1.4 mg l^{-1}), ion chromatography (below 8 ppm) [28] or more conveniently by cyclic voltammetry (ppb) [29,30]. The water content can be determined by Karl–Fischer titration or by cyclic voltammetry [31]. The presence and quantification of these impurities is essential in many applications since the physico-chemical properties of the ILs can vary significantly depending on their water and chloride contents [32]. Alternatively, chloride-free 1,3-dialkylimidazolium ILs can be prepared from the five-component reaction (glyoxal, formaldehyde, two different amines and acids) [33] and those containing methyl or ethyl sulphate anions by simple alkylation of 1-alkylimidazol with dimethyl or diethyl sulphate, respectively [34].

Table 3.5-3 Selected physical chemical data of some 1-alkyl-3-methylimidazolium-based ILs.

R	X	$T_g{}^a$ (°C)	$T_m{}^b$ (°C)	$T_d{}^c$ (°C)	η (mPa s)^d	d (g·cm^{-3})^e	σ (m S cm^{-1})^f	E.W.^{g} (V)	Ref
Et	BF$_4$	−92	13	447	37	1.28	14	4.0	[35]
^{n}Pr	BF$_4$	−88	−17	435	103	1.24	5.9	4.0	[35]
^{n}Bu	BF$_4$	−85	none	435	180	1.21	3.5	5.5	[35,36]
^{n}Bu	PF$_6$	−61	10	–	219	1.37	1.6	6.4	[36,37]
^{n}Bu	AlCl$_4$	−88	none	–	(294)	1.23	(24.1)	3.0	[38]
^{n}Bu	InCl$_4$	−6	–	–	26	1.55	(9.4)	1.3	[39]
^{n}Bu	CF$_3$SO$_3$	–	16	–	90	1.22	3.7		[11]
^{n}Bu	N(Tf)$_2$	–	−4	>400	69	1.43	3.9		[11,40]
^{n}Bu	CF$_3$CO$_2$	−30	none	–	73	1.21	3.2	4.5	[11,41]

a Glass transition temperature. **b** Melting point. **c** Decomposition temperature. **d** Viscosity at 25°C (30°C). **e** Density at 25°C. **f** Conductivity at 25°C (60°C). **g** Electrochemical window (vitreous carbon electrode).

As can be easily observed from the data presented in Table 3.5-3 the ILs based on 1,3-dialkylimidazolium cations possess relatively low viscosity (as low as 26 mPa s) and high density (usually between 1.2 and 1.5 g cm^{-3}). They are liquid over a large range of temperatures (down to −80°C) have a high thermal stability (starting to decompose at temperatures over 400°C) and display a large electrochemical window (up to 7 V).

3.5.3.4
Chemical Stability and Toxicity of 1,3-Dialkylimidazolium Ionic Liquids

Although most of the 1,3-dialkylimidazolium ILs are stable towards organic and inorganic substances, both the cation and anion can undergo "undesirable" transformations. The reactivity of the cation is mainly related to the relatively higher acidity of the H2 hydrogen of the imidazolium nucleus ($pK_a = 23.0$ for the 1,3-dimethyl imidazolium cation) [42]. Deprotonation of the C2 position of the imidazolium salt generates N-heterocyclic carbene ligands [43] that have been particularly observed in Pd-catalyzed Heck-type reactions performed in ILs [44–46]. Moreover, organic adducts have been also isolated in the presence of organic electrophiles (Scheme 3.5-10) [47].

Under "neutral" conditions the C2–H bond of the imidazolium nucleus can add oxidatively to electron-rich Ni(0) or Pd(0) complexes to generate stable carbene–metal-hydride compounds [48]. Dealkylation of the imidazolium nucleus ("Hoffman elimination") has been observed in the catalytic hydrodimerization of butadiene by Pd(II) compounds immobilized in ILs (Scheme 3.5-11) [49].

In some cases the anions of imidazolium ILs can easily undergo hydrolysis, particularly those containing AlCl$_4$ and PF$_6$ anions. In the case of the hexafluorophosphate anion, phosphate and HF are formed and 1,3-dialkylimidazolium phosphates and fluoride have been isolated during reactions [50] or purification proce-

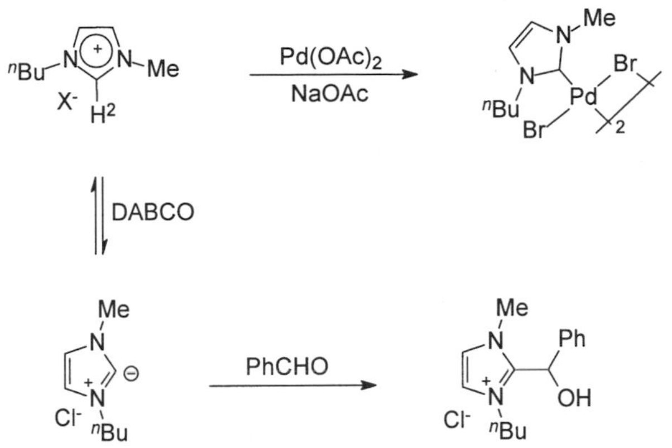

Scheme 3.5-10 Reactions of the imidazolium cation under basic conditions.

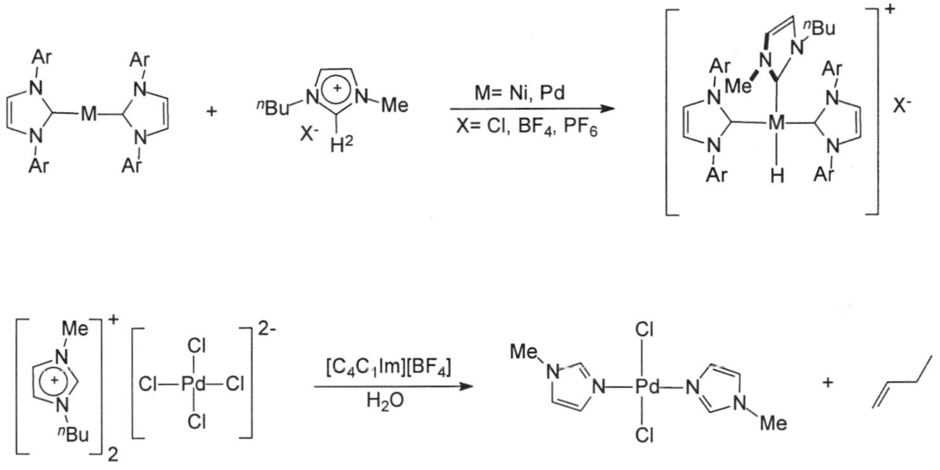

Scheme 3.5-11 Reactions of the imidazolium cation under "neutral" conditions.

dures [51]. Cation metathesis has also been observed with highly negatively charged complexes such as $Na_3Co(CN)_5$ [52] and $Na_2[\{(UO_2)(NO_3)_2\}2(\mu_4\text{-}C_2O_4)]$ [53] dissolved in ILs.

Although it is well known that most of the classical quaternary ammonium salts possess bactericidal activity [54] and have been used as disinfectants, little was known until very recently about the toxicity of new ILs [55]. It is evident that use of these liquids as green materials requires the consideration of their toxicity and their minimum inhibitory concentration values [56]. This situation is now rapidly changing and most of the data available indicate that 1,3-dialkylimidazolium ILs display antimicrobial activity and that their biological activity is dependent on the N-alkyl chain lengths [57]. The relatively low values of acute toxicity of 3-hexyloxymethyl-1-methylimidazolium tetrafluoroborate (LD_{50} = 1400 mg kg^{-1} for female and LD_{50} = 1370 mg kg^{-1} for male Wistar rats) are one of the first indications that tetrafluoroborate salts may be safely used [58]. It is also noticeable that biotransformations can be performed to great advantage (compared with the classical systems) in ILs and these materials can be used for embalming and tissue preservation [59]. The environmental risk of ILs may be assessed by a combination of structure–activity relationships, toxicological and ecotoxicological tests and modeling [60].

3.5.3.5
"Solvent" Properties and Structure of Imidazolium ILs

Several attempts have been made to apply to ILs the classical descriptors generally employed for the description of molecular solvents, such as "polarity" (dielectric constant) and coordinating properties. These studies include the use of chemical probes (Reichardt's dye, pyrene, dansylamide, Nile Red, or 1-pyrenecarbaldehyde) and chromatographic methods [24] but they have so far met with limited success. It has been demonstrated that interaction of the IL with the solute occurs through high dipolar and dispersion forces and hydrogen bonds. The dispersion forces are nearly constant, whereas the dipolar forces and hydrogen bonds varied with the type of IL [61]. Nonetheless, a complementary scale of hydrogen-bond acidity and basicity, and polarizability/dipolarity has been constructed for some ILs [62]. The dipolarity/polarity effects are relatively higher than those usually encountered for classical organic solvents and, as expected, the type of cation largely dominates the hydrogen-bond acidity while the hydrogen acidity is determined by the anion nature. The lack of success achieved using the classical approach is most probably related to the assumption that ILs are "homogenous" liquids. In fact there are now several indications that the mixture of ILs can generate nano-inhomogeneities similar to those encountered in some liquid crystals and concentrated surfactant media [63].

The model of "non-homogeneous" liquids for ILs may be more helpful not only for the rationalization of many of their properties but also in allowing the design of new IL materials with tailor-made properties. In order to develop this model a close look at the structure of pure ILs in the solid, liquid and gas phases is necessary, followed by analysis of the interaction of other substances with the ILs through changes in the physico-chemical properties.

An over-view of the X-ray studies reported on the structure of 1,3-dialkylimidazolium salts (see Table 3.5-3 and Scheme 3.5-9 for examples) reveals a typical trend: in the solid state they form an extended network of cations and anions connected together by hydrogen bonds. The monomeric unit is always constituted of one imidazolium cation surrounded by at least three anions and in turn each anion is surrounded by at least three imidazolium cations (Fig. 3.5-5).

The strongest hydrogen bond always involves the most acidic H2 of the imidazolium cation followed by the other two hydrogens (H4 and H5) of the imidazolium nucleus and/or the hydrogens of the *N*-alkyl radicals. These bonds possess the

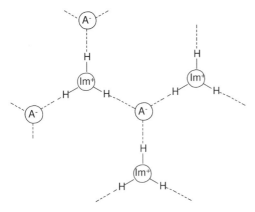

Fig. 3.5-5 Simplified two-dimensional model of the polymeric supramolecular structure of 1,3-dialkyl imidazolium ILs showing the hydrogen bonds between the imidazolium cation (Im⁺) and the anions (A⁻) (one cation is surrounded by three anions and vice-versa).

properties of weak to moderate hydrogen bonds: they are mostly electrostatic in nature (H$\cdots$X bond lengths > 2.2 Å; C–H$\cdots$X bond angles between 100 and 180°) [64].

Although the number of anions that surround the cation (and vice versa) can change depending on the anion size and type of the N-alkyl imidazolium substituents, the structural trend of one imidazolium hydrogen bonded to at least three anions and one anion hydrogen bonded to at least three cations is a general feature in 1,3-dialkylimidazolium salts. The three-dimensional arrangement of the imidazolium ILs is generally formed through chains of imidazolium rings or alternating with the anions (Fig. 3.5-6).

In some cases there are typical π–π stacking interactions among the imidazolium rings and in the case of 1-alkyl-3-methylimidazolium salts relatively weak C–H$\cdots$

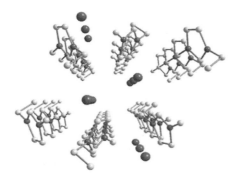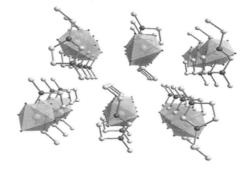

Fig. 3.5-6 A three-dimensional simplified schematic view of the arrangements of 1,3-dialkylimidazolium cations showing: the channels in which the "spherical" anions are accommodated (left) and the arrangements of 1,3-dialkylimidazolium cations alternating with octahedral anions (right).

π interactions via the methyl group and the imidazolium ring-π system can also be found. This molecular arrangement can generate channels in which the anions are accommodated as chains.

This structural pattern depends on the anion type, and the internal arrangements along the imidazolium columns vary with the type of the *N*-alkyl substituents and can be used for the generation of fluid self-organized columnar ILs that exhibits one-dimensional ionic conductivities. ILs that exhibit hexagonal columnar phases over a wide range of temperatures including room temperature have been constructed and their one-dimensional anisotropic ionic conductivities for the macroscopically ordered columnar assemblies determined [65].

It is important to note that effects other than π–π stacking, such as the entropy effect and electrostatic interactions, may not favor the formation of structures of the type shown in Fig. 3.5-6. It can be proposed that the best representation for the imidazolium salts in the solid phase is $[(DAI)_x(X)_{x-n})]^{n+}[(DAI)_{x-n}(X)_x)]^{n-}$, where DAI is the 1,3-dialkylimidazolium cation and X is the anion.

There are several indications that 1,3-dialkylimidazoliums do indeed possess analogous structural patterns in both solid and liquid phases, and to some extent even in the gas phase. Note that although significant randomness in organization is necessary to describe the structure of a liquid, in most cases there is only a 10–15% volume expansion on going from the crystalline to the liquid state and the ion–ion or atom–atom distances are similar in both the solid and the liquid states. Furthermore, while long-range order is lost on going from the crystal to the liquid, similarities remain as a consequence of the Coulombic forces between the cations and anions of the imidazolium ionic liquids [66]. It is clear that the long-range Coulomb interactions in ionic organic liquids can lead to more extensive spatial correlations than those in comparable van der Waals organic liquids [67].

The IR spectra of 1,3-dialkylimidazolium ILs in both solid and liquid phases are quite similar and show the characteristic C–H $\cdots$ X hydrogen-bonding bands in the 3100–3200 cm^{-1} region pertaining to the interaction of H2, H4 and H5 of the imidazolium cations with the anions. The same result was observed by Raman spectroscopy [68] and neutral diffraction [69] studies, indicating that the three-dimensional structure found in the solid state is maintained in the liquid phase. Similar observations have been obtained from various X-ray and neutron-scattering studies of solid and liquid NaCl, i.e. the structural organization observed in the crystal exists in the liquid phase [70]]. The hydrogen-bonding system between the cation and the anion has been also evidenced by NMR studies of pure imidazolium ILs [71, 72], including evidence of spin diffusion from H,H-NOESY experiments [73]. Optical heterodyne-detected optical Kerr effect experiments used to study the orientation dynamics of 1-ethyl-3-methylimidazolium nitrate indicated that the nature of the interactions in this IL is very different from those occurring in van der Waals organic liquids. These intermolecular interactions are stronger than those in the van der Waals liquids and in liquid crystals [74].

This structural pattern is even maintained in the gas phase and evidence for the clustering – supramolecules formed through association of the 1,3-dialkylimidazolium cations with the anions – has been obtained by mass spectrometry experiments. Clusters of the type $[(DAI)_2(AlCl_4)]^+$ (DAI = 1,3-dialkylimidazolium cation) have been observed in organo-aluminate molten salts under conditions of fast atom bombardment mass spectrometry (FAB-MS) [75]. Moreover, electron spray mass spectrometry experiments in both positive and negative modes (ESI(+) and ESI(−)) of various 1,3-dialkylimidazolium cations associated with different anions also show the formation of supramolecular structures of the type $[(DAI)_x(X)_{x-n})]^{n+}$ $[(DAI)_{x-n}(X)_x)]^{n-}$ ($x, n = 1, 2, 3, ...; x > n$) [76]. Notably, stable supramolecular aggregates with magic mass numbers such as $[(DAI)_2(BF_4)_3]^-$ and $[(DAI)_5(BF_4)_4]^+$ were found in the case of the tetrafluoroborate anion. It was possible to produce and isolate mixed clusters of the type $[(X^1)(DAI)(X^2)]^-$ ($X^1, X^2 = CF_3SO_3, PF_6, BF_4$ and BPh_4) via mass selection and then dissociate by collision activation to determine the intrinsic strength of the hydrogen bonds between imidazolium cations and anions. The strength of the hydrogen bonds follows the order $CF_3CO_2^- > BF_4^- > PF_6^- > BPh_4^-$, which shows the same trend as those obtained earlier based on IR studies [36].

It is clear that "pure" 1,3-dialkyimidazolium ILs in the solid, liquid and gas phases can be described as well-organized hydrogen-bonded polymeric supramolecules. Moreover, there is now much evidence indicating that this polymeric nature is maintained to a great extent when they are mixed with other substances. 1,3-Dialkylimidazolium ILs–aromatic mixtures form liquid clathrates and in the case of a $[C_1C_1Im]PF_6$–benzene mixture the inclusion compound $[([C_1C_1Im]PF_6)_2(benzene)]_n$ could be trapped and its X-ray structure determined [77]. The conductivity of 1,3-dialkylimidazolium liquids initially decreases with an increase in concentration and then increases with a further increase in salt concentration indicating the formation of triple ions of the type $[(DAI)_2(X)]^+[(DAI)(X)_2]^-$ [78]. The formation of hydrogen-bonded aggregates through disruption of the polymeric structure can also explain the augmentation of the $BMI \cdot PF_6$ conductivity when saturated with carbon dioxide (a molecule with a relatively large quadrupole moment). Indeed, an increment of greater than 3 times (from 3×10^3 to 10×10^3 S cm^{-1}) in the conductivity was observed by the addition of 60 bar of CO_2 at 50°C [79]. Moreover, an irregular ionic lattice model has been successfully used to predict the CO_2 solubilities with $BMI \cdot PF_6$ [80]. The formation of such floating aggregates was also observed by the changes of solution enthalpies of these ILs in classical organic solvents [75].

Multinuclear NMR spectroscopy experiments of various 1,3-dialkylimidazolium ILs dissolved in organic solvents have also pointed to the formation of floating aggregates through hydrogen bonds [81–84]. In particular, it has been demonstrated by heteronuclear NMR experiments on $[C_4C_1Im]BF_4$ that contact ion pairs exist in the presence of small amounts of water and even in dimethyl sulfoxide (DMSO) solution [85].

It has also been demonstrated that the diffusion coefficients of both neutral and charged molecules in $[C_4C_1Im]BF_4$ and $[C_4C_1Im]PF_6$ increase with the addition of controlled amounts of water. The effect of water on the diffusion coefficient for neutral and ionic species suggests that wet ILs may not always be regarded as homogeneous solvents, but have to be considered as nano-structures with polar and non-polar regions [86]. The presence of nano-structures in the wet IL may allow neutral molecules to reside in less polar regions and ionic species to undergo faster diffusion in the more polar or wet regions. The presence of these hydrogen-bonded nano-structures with polar and non-polar regions may be responsible for the stabilization of enzymes supported in ILs that can maintain their functionality under very extreme denaturative conditions. It is well known that the thermal stability of enzymes is enhanced in both aqueous and anhydrous media containing polyols as a consequence of the increase in hydrogen bond interactions. Thus, both the solvophobic interactions essential to maintain the native structure and the water shell around the protein molecule are preserved by the "inclusion" of the aqueous solution of free enzyme into the IL network, resulting in a clear enhancement of the enzyme stability [87]. Moreover, the presence of these nano-structures provides an excellent environment for the formation and stabilization of nano-materials [88, 89].

In summary pure 1,3-dialkylimidazolium ILs can be described as hydrogen-bonded polymeric supramolecules of the type $[(DAI)_x(X)_{x-n})]^{n+}[(DAI)_{x-n}(X)_x)]^{n-}$ where DAI is the 1,3-dialkylimidazolium cation and X is the anion. This structural pattern is a general trend for both the solid and the liquid phase and is apparently maintained to a great extent even in the gas phase. The introduction of other molecules and macromolecules occurs with a disruption of the hydrogen bond network, generating nano-structures with polar and non-polar regions where inclusion-type compounds can be formed. These inclusion compounds can involve molecules (such as arenes) [74], ions (such as charged transition metal complexes) [90], macromolecules (such as enzymes [22] or cellose [91]) and nano-particles (such as transition metal nano-clusters) [92], and the stabilization of this process is mainly due to the electronic and steric effects provided by the nano-structures of the type $[(DAI)_x(X)_{x-n})]^{n+}[(DAI)_{x-n}(X)_x)]^{n-}$. When they are infinitely diluted in other molecules they can form solvent-separated ion pairs. With an increase of the concentration of the imidazolium salt, these collapse to form contact ion pairs – through hydrogen bonds involving the cation with the anion – and with a further salt concentration increment they form triple ions, etc,

3.5.3.6
Solubility of Ionic Liquids

As already pointed out (Section 3.5.3.5) ILs can be described as polymeric supramolecules formed from the association of imidazolium cations with the anions through hydrogen bonds. The introduction of other substances occurs with the disruption of these extended networks of anions and cations, and ILs can thus act as both hydrogen bond acceptors (anion) and donors (imidazolium cation) and as expected they interact with substances with both accepting and donating sites. Indeed, 1,3-dialkylimidazolium ILs can dissolve a plethora of classical polar and non-polar compounds. Moreover, in some cases the mixture of 1,3-dialkylimidazolium ILs with other substances can generate nano-structures with "polar" and "non-polar" regions leading to a dual behavior.

All 1,3-dialkylimidazolium ILs reported to date are hygroscopic, and their miscibility with water is largely controlled by the nature of the anion. While those salts
containing the nitrate, chloride and perchlorate anions are usually miscible with
water in all proportions, those associated with hexafluorophosphate and bis(trifluoromethane) sulfonylamidate anions are almost completely immiscible with
water [93]. Interestingly, the miscibility with water of those containing the tetrafluoroborate anion is temperature dependent (Fig. 3.5-7) [36]. It is also known that an
increase of the N-alkyl chain lengths increases the hydrophobicity for a series of 1-
alkyl-3-methylimidazolium hexafluorophosphate ILs [94] The miscibility of water
with ILs can be increased by the addition of short-chain alcohols [95] or diminished
by the addition of salts (salting-out effect) [96].

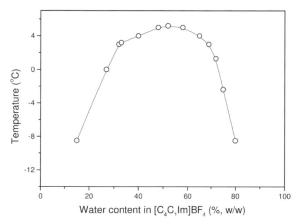

Fig. 3.5-7 Miscibility of water in $[C_4C_1Im]BF_4$ as a function of the
temperature.

The solubility of 1,3-dialkylimidazolium hexafluorophosphate ILs decreases
with an increase of the molecular weight of alcohols and is higher in secondary
than in primary alcohols [97-100]. In these cases, alcohols are most probably stabilizing the hydrogen-donor sites since they form hydrogen-bonded structures with
both high enthalpies and high constants of association. The solubility of saturated
hydrocarbons is usually very low in 1,3-dialkylimidazolium ILs and increases with
the augmentation of the length of the alkyl substituents at the imidazole ring,
being in the order of 0.05 mole fraction for hexane and cyclohexane in $[C_4C_1Im]PF_6$
at room temperature [101]. Unsaturated hydrocarbons are, however, more soluble,
and within this class of hydrocarbons dienes are more soluble than alkenes. For
example, the solubility of butadiene in $[C_4C_1Im]BF_4$ and $[C_4C_1Im]PF_6$ is 0.16 and
0.11 mole fraction, respectively, whereas this drops to 0.05 mole fraction for 1-bu-
tene in both ILs [102]. This is probably related to the relatively higher hydrogen
bond accepting properties of the diene, which can interact with the imidazolium
cation, [103] compared to the monoene. Since aromatic hydrocarbons can form "inclusion" compounds through $C–H\cdots\pi$ interactions with the IL they display high
solubility and can attain 0.35 mole fraction for benzene in $[C_4C_1Im]PF_6$ at room

temperature [104], for example. The solubility in ILs of aromatic hydrocarbons decreases with an increase of the molecular weight of the hydrocarbon and the differences of solubilities of *o-*, *m-* and π–xylenes are not significant. Other aromatics such as naphthalene, mono-, di- and trichloro benzenes are also very soluble in imidazolium ILs [105].

The solubility of various gases such as carbon dioxide, ethylene, ethane, methane, argon, oxygen, carbon monoxide, hydrogen and nitrogen in $[C_4C_1Im]PF_6$ have been determined [106]. Carbon dioxide has the highest solubility and strongest interactions with the IL, followed by ethylene and ethane. The solubility of carbon dioxide in $[C_4C_1Im]PF_6$ can reach 0.6 mole fraction at 8 Mpa [107]. Argon and oxygen have very low solubilities and immeasurably weak interactions. Carbon monoxide, hydrogen and nitrogen all have very low solubilities, which could not be measured by classical methods. Nonetheless, the solubility of hydrogen in some ILs could be estimated by high-pressure NMR experiments [108]. It was also estimated that hydrogen is at least four times more soluble in $[C_4C_1Im]BF_4$ than in $[C_4C_1Im]PF_6$ at the same pressure and temperature [109].

3.5.3.7
Extraction/Separation Processes Involving Ionic Liquids.

ILs, as already pointed out, are largely used in synthesis and as alternative "solvents" to classical organic solvents and various excellent reviews on the subject have recently been published [12–26]. In various cases these liquids could be used with some advantages, such as product yields, selectivity and recycling of the IL, compared with the traditional procedures. In some cases these processes can be considered "clean" when the products are sufficiently immiscible with the IL phase and can be separated by decantation, or are relatively volatile and can be recovered by bulk distillation procedures. Unfortunately this is not always the case and other separation methods must be applied. On a laboratory scale, liquid–liquid extraction is the preferred method mainly because of its simplicity and practicality, but it usually involves the use of molecular organic solvents (back to the starting problem!) resulting in cross-contaminations. There are other approaches that have been proposed to circumvent this problem such as per-evaporation and extraction with supercritical carbon dioxide (another "green" fluid).

It has been demonstrated that with the appropriate choice of membrane (hydrophobic or hydrophilic polymers) the per-evaporation procedure can be efficiently applied for the quantitative and selective recovery of organic solutes, such as naphthalene, water, ethyl hexanoate and chlorobutane, from $[C_4C_1Im]PF_6$ IL [110]. It was also reported that the same ILs can be used as supported liquid membranes for the selective transport of secondary amines over tertiary amines with similar boiling points, and this was attributed to the higher hydrogen bond affinity of the secondary derivative with the imidazolium cation [111,112].

As mentioned before, carbon dioxide is highly soluble in $[C_4C_1Im]PF_6$, although the ionic liquid does not dissolve in carbon dioxide [103]. These properties can facilitate the extraction of solutes from the IL without contamination of the non-ionic phase. Supercritical carbon dioxide has thus been used for the extraction of various volatile and relatively non-volatile compounds from ILs without any extraction of the ILs themselves [113,114]. The carbon dioxide can be also used as a modifier for an ILs/organic compounds mixture to induce the formation of three phases and facilitate the separation procedure [115].

The use of ILs as media for liquid–liquid extractions is growing rapidly since their hydrophobic or hydrophilic nature can be modulated by modifications in both cation and anion. The portioning of various charged and non-charged aromatic compounds between $[C_4C_1Im]PF_6$ IL and water is similar to their portioning in traditional water–molecular organic solvent systems [116]. Various ILs have been tested for the extraction of sulfur from *n*-dodecane (model of diesel oil) and those containing higher Lewis acid properties such as $[C_4C_1Im]AlCl_4$ gave superior results in the desulfurization process of "real" diesel samples [117].

Traditional crown ether extractants in $[C_nC_1Im]PF_6$ ($n = 4$, 6 and 8) ILs have been investigated in the extraction of Na^+, Cs^+ and Sr^{2+} from aqueous solutions as a result of metal ion portioning, which is dependent on the crown ether hydrophobicity and the aqueous phase composition. Although some interesting results were obtained in comparison to the traditional solvent extraction process, the portioning behavior suggested a complicated mechanism [118]. More interestingly, imidazolium ionic liquids containing specific functionality (for example complexant groups covalently bound, usually to the cation) can generate better liquids for the dissolution or extraction of metal ions. Using this approach, a new family of 1-alkyl-3-methylimidazolium ILs with ether and alcohol functional groups on the alkyl side-chain ($[C_nO_mC_1Im]X$) has been prepared. These new ILs present a considerably lower viscosity and an increased ability to dissolve $HgCl_2$ and $LaCl_3$ (up to 16 times higher) than the non-oxygenated analogues ($[C_nC_1Im]X$) [119]. A series of hydrophobic imidazolium ILs containing urea-, thiourea- and thioether-substituted alkyl groups were also prepared and their properties for the extraction of Hg^{2+} and Cd^{2+} from aqueous have been also reported. The urea- and thiourea-containing imidazolium ILs yielded the highest distribution ratios, which are higher for Hg^{2+} than for Cd^{2+}. An aqueous-phase pH change did not promote the stripping of the metal ions to the ILs phase [120]. However, an IL containing an ethylene glycol spacer between two methylimidazolium cations (Fig. 3.5-8) is highly selective for the extraction of Hg^{2+} in the presence of Cs^{2+} from aqueous solution by simple aqueous-phase pH change [121].

ILs-separation processes have also been applied for actinides in aqueous solution by the incorporation of *n*-octyl(phenyl)-N,N-diisobutylcarbamoylmethyl phosphine oxide as the extractant. The portioning distribution ratios for UO_2^{2+}, Pu^{4+}, Th^{4+} and Am^{3+} are higher in $[C_nC_1Im]PF_6$ than those obtained using dodecane as the extracting phase [89].

The dual-nature behavior of ILs, i.e. the ability to separate both polar and nonpolar compounds, has been explored for the development of a new class of stationary phases for gas–liquid chromatography (GLC). The popular $[C_4C_1Im]PF_6$ IL was one of the first to be tested as stationary phase for GLC giving highly interesting results, acting as a low-polarity stationary phase for nonpolar compounds whi-

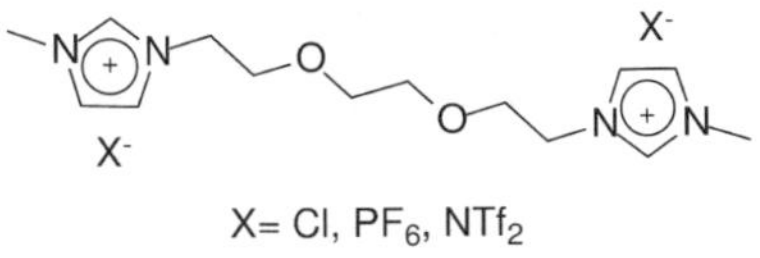

X= Cl, PF$_6$, NTf$_2$

Fig. 3.5-8 Structure of substituted ethylene glycol imidazolium ILs.

le retaining more polar molecules tenaciously. It has been shown that the nature of the anion can have a significant effect on both the solubilizing ability and the selectivity of IL stationary phases [122]. Although many imidazolium IL stationary phases have low maximum operating temperatures and produce low peak efficiencies for certain analytes, new ILs that circumvent these problems have been reported. For example, the IL l-3-methylimidazolium trifluoromethane sulfonate possesses a higher bleed temperature (around 250°C) than commercial GLC stationary phases, producing faster separations of complex mixtures of analytes that are generally different from those observed with such phases[123].

Imidazolium ILs easily form micro-emulsions using different surfactants such as long-chain alcohols and the properties of the new micelle–ionic liquid solutions can be explored in inverse gas chromatography processes [124]. Moreover, ILs have been used as run buffer additives in capillary electrophoresis [125] and as ultra-low-volatility liquid matrixes for matrix-assisted laser desorption/ionization mass spectrometry [126].

3.5.3.8
Multiphase Catalysis Employing Ionic Liquids

Catalysis plays a major role in the chemical industry, and heterogeneous processes are preferred because of the problems (product separation, catalyst recycling and the use of organic solvents) associated with homogeneous catalysis. In terms of atom economy the ideal process is a simple addition A + B giving only C, with all other components only required in catalytic amounts [127]. Multiphase catalysis, in particular liquid–liquid biphasic catalysis involving two immiscible phases, may offer the possibility of circumventing these problems and generate "greener" catalytic process [128,129]. The concept of this system requires that the catalyst is soluble in only one phase whereas the substrates/products remain in the other phase. The reaction can take place in one (or both) of the phases or at the interface. In most cases, the catalyst phase can be reused and the products/substrates are simply removed from the reaction mixture by decantation. Various fluids, such as pefluorinated solvents, [130] water [124], supercritical fluids (mainly carbon dioxide) [131] and ionic liquids [14–17,19,20] are currently being investigated as green media for multiphase catalytic processes.

Multiphase catalysis performed in ILs can lead to various phase systems in which the catalyst should reside in the IL. Before the reaction starts, and where there is no involvement of gaseous reactants, two systems can usually be formed: a single phase, in which the substrates are soluble in the ionic liquid or biphasic, in which one or all of the substrates reside preferentially in an organic phase. If a gaseous reactant is involved, two-phase and three-phase systems can be formed. At the end of the reaction three systems can be formed: a single-phase system; a two-phase system in which the residual substrates are soluble in the ionic catalytic solution and the products reside preferentially in the organic phase; a three-phase system, formed, for example, by an ionic catalytic solution, an organic phase containing the desired product and a third phase containing the byproducts. In most cases, catalysis performed in ILs forms two-phase systems (before and after catalysis) as shown schematically in Fig. 3.5-9.

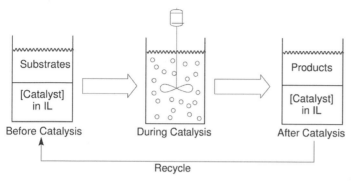

Fig. 3.5-9 Schematic view of a typical liquid–liquid biphasic catalytic process with catalysts immobilized in ILs.

Various catalytic processes can be "directly" transposed in ionic liquids, such as those based on homogeneous transition metal catalyst precursors [14–17,19,20] and colloids [49], with great advantages compared with those performed in organic solvents or in water. In particular the classical transition metal catalyst precursors are, in most of the cases, "soluble" in imidazolium ILs and they are not removed from the ionic solution by a great many organic compounds. Thus a legion of transition metal catalyzed reactions, such as hydrogenations, oxidations, carbonylations, C–C coupling, etc., have been performed in ionic liquids, and excellent reviews on the subject are available [14-17,19,20]. This is probably one of the great advantages of ILs in organometallic catalysis, i.e. it allows direct transposition of well-known homogeneous processes for liquid–liquid biphasic conditions without the use of the specially designed ligands/complexes that are necessary for the aqueous, perfluorinated or supercritical fluids catalytic processes [125]. Moreover, in these IL multiphase processes it is possible to extract the primary products during the reaction and thus modulate the product selectivity (modifying the solubility of the different substrates and reaction products in the catalyst-containing phase). This approach can constitute a suitable method for avoiding consecutive reactions of primary products and it has been exploited to some extent in an IL catalytic process for the selective hydrogenation of dienes to monoenes [98] and benzene to cyclohexene [132]. In the cases where the catalyst is removed from the IL catalytic solution by the products, catalyst leaching can be avoided by the use of modified ligands containing anionic or cationic groups such as sulfonic and quaternary ammonium and phosphonium (Fig. 3.5-10) [133–135].

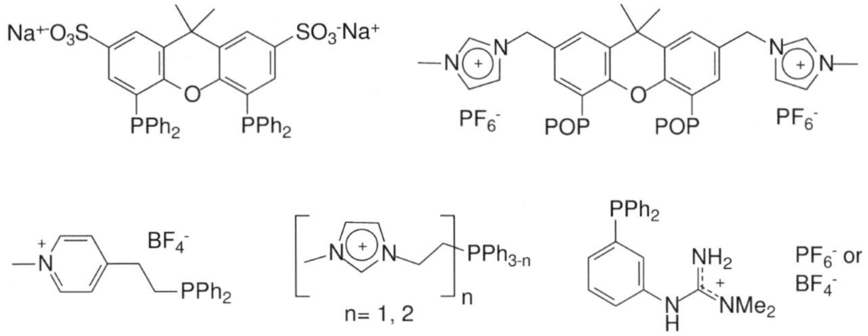

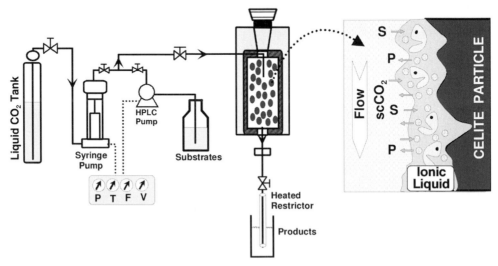

Fig. 3.5-10 Examples of ionic phosphine ligands employed for the immobilization of transition metal catalyst precursors in ILs.

The separation of the products from the IL catalytic mixture can be performed in various cases by simple decanting and phase separation or by product distillation. In this respect, a continuous-flow process using toluene as extractant has been applied for the selective Pd-catalyzed dimerization of methyl acrylate in ILs [136]. However, in cases where the products are retained in the IL phase, extraction with supercritical carbon dioxide can be used instead of classical liquid–liquid extractions that necessitate the use of organic solvents, which may result in cross-contamination of products. This process was successfully used in catalyst recycling and product separation for the hydroformylation of olefins employing a continuous-flow process in supercritical carbon dioxide–IL mixtures [137]. Similarly, free and immobilized Candida antarctica lipase B dispersed in ILs were used as catalyst for the continuous kinetic resolution of rac-1-phenylethanol in supercritical carbon dioxide at 120°C and 150°C and 10 Mpa with excellent catalytic activity, enzyme stability and enantioselectivity levels (Fig. 3.5-11).

Fig. 3.5-11 Experimental setup of the bioreactor with ILs-scCO$_2$ (S = substrates and P = products). The diagram was provided by Prof. P. Lozano (University of Murcia, Spain).

The potential of ILs in liquid–liquid biphasic catalysis for the generation of clean technology has became a reality with the announcement of a commercial process for the dimerization of butenes to isooctenes (Difasol process) by IFP (France) based on nickel complexes immobilized in organo-aluminate imidazolium ILs [138]. This new process provides significant benefits over the existing homogeneous Dimersol X process, which is currently in operation in five industrial plants, producing nearly 200 000 tonnes per year of isooctenes [139].

3.5.3.9
Conclusions and Perspectives

It is clear that ILs represent one of the most important classes of alternative "solvents" for the substitution of volatile or semi-volatile organic compounds in many extraction/reaction processes. Ionic liquids possess unique structures in the liquid phase (similar to the solid phase) and solvation characteristics (dual-nature) that confer on this class a large potential for the development of new tailor-made materials and for new applications. Moreover, the association of ILs with other green fluids such as supercritical carbon dioxide will certainly boost clean extraction/reaction processes. Although knowledge about their structure and properties is growing very rapidly (more than 600 papers dealing solely with ILs in 2003) ILs still represent a wide field. However, there is no doubt that are establishing themselves alongside van der Waals and water-like solvents, and, like these, will frequently be the best solvents or a modifiers for a given preparation/extraction technologies.

References

1 P.T. Anastas, M.M. Kirchhoff, *Acc. Chem. Res.* **2002**, *35*, 686–694.

2 For a recent review of solventless reactions, see: G.W.V. Cave, C.L. Raston, J.L. Scott, *Chem. Commun.* **2001**, 2159–2169.

3 F. Franks, *Chem. Unserer Zeit* **1986**, *20*, 146–155.

4 G. Mamantov, C. Mamantov, eds., *Advances in Molten Salt Chemistry*, Vols 5 and 6, Elsevier, New York, **1983**.

5 P. Walden, *Bull. Acad. Imper. Sci. (St. Petersburg)* **1914**, 1800.

6 F.H. Hurley, T.P. Wier, *J. Electrochem. Soc.* **1951**, *98*, 203–206.

7 Y. Chauvin, B. Gilbert, I. Guibard, *J. Chem. Soc., Chem. Commun.* **1990**, 1715–1716.

8 J.S. Wilkes, M.J. Zaworotko, *J. Chem. Soc. Chem. Commun.* **1992**, 965–967.

9 Y. Chauvin, L. Mussmann, H. Olivier, *Angew. Chem., Int. Ed. Engl.* **1996**, *34*, 2698–2700.

10 P.A.Z. Suarez, J.E.L. Dullius, S. Einloft, R.F. deSouza, J. Dupont, *Polyhedron* **1996**, *15*, 1217–1219.

11 P. Bonhote, A.P. Dias, N. Papageorgiou, K. Kalyanasundaram, M. Gratzel, *Inorg.Chem.* **1996**, *35*, 1168–1178.

12 Y. Chauvin, *Act. Chim.* **1996**, 44–46.

13 K.R. Seddon, *J. Chem. Technol. Biotechnol.* **1997**, *68*, 351–356.

14 T. Welton, *Chem. Rev.* **1999**, *99*, 2071–2083.

15 P. Wasserscheid, W. Keim, *Angew. Chem., Int. Ed.* **2000**, *39*, 3773–3789.

16 R. Sheldon, *Chem.Commun.* **2001**, 2399–2407.

17 C.M. Gordon, *Appl. Catal. A-Gen.* **2001**, *222*, 101–117.

18 P.J. Dyson, *Transition Met. Chem.* **2002**, *27*, 353–358.

19 H. Olivier-Bourbigou, L. Magna, *J. Mol. Catal. A-Chem.* **2002**, *182*, 419–437.

20 J. Dupont, R.F. de Souza, P.A.Z. Suarez, *Chem. Rev.* **2002**, *102*, 3667–3691.

21 C. Baudequin, J. Baudoux, J. Levillain, D. Cahard, A.C. Gaumont, J.C. Plaquevent, *Tetrahedron-Asymm.* **2003**, *14*, 3081–3093.

22 S. Park, R. Kazlauskas, *Curr. Opin. Biotech.* **2003**, *14*, 432–437.

23 C.E. Song, *Chem. Commun.* **2004**, 1033–1043.

24 C.F. Poole, *J. Chromat.-A* **2004**, *1037*, 49–82.

25 P. Wasserscheid, T. Welton, *Ionic Liquids in Synthesis*, VCH-Wiley, Weinheim, **2002**.

26 R.D. Rogers, K.R. Seddon, *Ionic Liquids; Industrial Applications for Green Chemistry*, ACS Symposium Series 818, American Chemical Society, Washington, DC, 2002.

27 J. Dupont, P.A.Z. Suarez, C.S. Consorti, R.F. de Souza, *Org. Synth.* **2002**, *79*, 236–243.

28 C. Villangran, M. Deetlefs, W.P. Pitner, C. Hardacre, *Anal. Chem.* **2004**, *76*, 2118–2123.

29 C. Villangran, C.E. Banks, C. Hardacre, R.G. Compton, *Anal. Chem.* **2004**, *76*, 1998–2003.

30 B.K. Sweeny, D.G. Peters, *Electrochem. Commun.* **2001**, *3*, 712–715.

31 V. Gallo, P. Mastrorilli, C.F. Nobile, G. Romanazzi, G.P. Suranna, *J. Chem. Soc., Dalton Trans.* **2002**, 4339–4342.

32 K.R. Seddon, A. Stark, M.J. Torres, *Pure Appl. Chem.* **2000**, *72*, 2275–2287.

33 R.F. de Souza, V. Rech, J. Dupont, *Adv. Synth. Catal.* **2002**, *344*, 153–155.

34 J.D. Holbrey, W.M. Reichert, R.P. Swatloski, G.A. Broker, W.R. Pitner, K.R. Seddon, R.D. Rogers, *Green Chem.* **2002**, *4*, 407–413.

35 T. Nishida, Y. Tashiro, M. Yamamoto, *J. Fluor. Chem.* **2003**, *120*, 135–141.

36 P.A.Z. Suarez, S. Einloft, J.E.L. Dullius, R.F. de Souza, J. Dupont, *J. Chim. Phys. Phys. -Chim. Biol.* **1998**, *95*, 1626–1639.

37 J.G. Huddleston, A.E. Visser, W.M. Reichert, H.D. Willauer, G.A. Broker, R.D. Rogers, *Green Chem.*, **2001**, *3*, 156–164.

38 A. Fannin Jr., D.A. Floreani, L.A. King, J.S. Landers, B.J. Piersma, D.J. Stech, R.L. Vaughn, J.S. Wilkes, J.L. Williams, *J. Phys. Chem.* **1984**, *88*, 2609–2614.

39 A.S. Neto, G. Ebeling, R.S. Gonçalves, F.C. Gozzo, M.N. Eberlin, J. Dupont, *Synthesis* **2004**, 1155–1158.

40 P.J. Dyson, G. Laurenczy, C.A. Ohlin, J. Vallance, T. Welton, *Chem. Commun.*, **2003**, 2418–2419.

41 P.A.Z. Suarez, C.S. Consorti, R.F. de Souza, J. Dupont, R.S. Goncalves, *J. Braz. Chem. Soc.* **2002**, *13*, 106–109.

42 T.L. Amyes, S.T. Diver, J.P. Richard, F.M. Rivas, K. Toth, *J. Am. Chem. Soc.* **2004**, *126*, 4366–4374.

43 J. Arduengo III, *Acc. Chem. Res.* **1999**, *32*, 913–921.

44 J. Carmichael, M.J. Earle, J.D. Holbrey, P.B. McCormac, K.R. Seddon, *Org. Lett.* **1999**, *1*, 997–1000.

45 L.J. Xu, W.P. Chen, J.L. Xiao, *Organometallics* **2000**, *19*, 1123–1127.

46 J. Mathews, P.J. Smith, T. Welton, A.J.P. White, D.J. Williams, *Organometallics* **2001**, *20*, 3848–3850.

47 V.K. Aggarwal, I. Emme, A. Mereu, *Chem. Commun.* **2002**, 1612–1613.

48 N.D. Clement, K.J. Cavell, C. Jones, C.J. Elsevier, *Angew. Chem., Int. Ed.*, **2004**, *43*, 1277–1279.

49 J.E.L. Dullius, P.A.Z. Suarez, S. Einloft, R.F. de Souza, J. Dupont, J. Fischer, A. De Cian, *Organometallics* **1998**, *17*, 815–819.

50 G.S. Fonseca, A.P. Umpierre, P.F.P. Fichtner, S.R. Teixeira, J. Dupont, *Chem. Eur. J.* **2003**, *9*, 3263–3269.

51 R.P. Swatloski, J.D. Holbrey, R.D. Rogers, *Green Chem.* **2003**, *5*, 361–363.

52 P.A.Z. Suarez, J.E.L. Dullius, S. Einloft, R.F. deSouza, J. Dupont, *Inorg. Chim. Acta* **1997**, *255*, 207–209.

53 E. Bradley, J.E. Hatter, M. Nieuwenhuyzen, W.R. Pitner, K.R. Seddon, R.C. Thied, *Inorg. Chem.* **2002**, *41*, 1692–1694.

54 G. Domagk, *Dtsch. Med. Wochenschr.* **1935**, *61*, 829–832.

55 J. Pernak, J. Rogoza, I. Mirska, *Eur. J. Med. Chem.* **2001**, *36*, 313–320

56 W.M. Nelson, *ACS Symposium Series* **2002**, *818*, 30–41.

57 J. Pernak, K. Sobaszkiewicz, I. Mirska, *Green Chem.* **2003**, *5*, 52–56.

58 J. Pernak, A. Czepukowicz, R. Pozniak, *Ind. Eng. Chem. Res.* **2001**, *40*, 2379–2383.

59 P. Majewski, A. Pernak, M. Grzymislawski, K. Iwanik, J. Pernak, *Acta Histochem.* **2003**, *105*, 135–142.

60 B. Jastorff, R. Stormann, J. Ranke, K. Molter, F. Stock, B. Oberheitmann, W. Hoffmann, J. Hoffmann, M. Nuchter, B. Ondruschka, J. Filser, *Green Chem.* **2003**, *5*, 136–142.

61 J.L. Anderson, J. Ding, T. Welton, D.W. Armstrong, *J. Am. Chem. Soc.* **2002**, *124*, 14247–14254.

62 L. Crowhurst, P.R. Mawdsley, J.M. Perez-Arlandis, P.A. Salter, T. Welton, *Phys. Chem. Chem. Phys.* **2003**, *5*, 2790–2794.

63 G.S. Attard, P.N. Bartlett, N.R.B. Coleman, J.M. Elliott, J.R. Owen, J.H. Wang, *Science* **1997**, *278*, 838–840.

64 G.A. Jeffrey, *An Introduction to Hydrogen Bonding*, Oxford University Press, Oxford, **1997**.

65 M. Yoshio, T. Mukai, H. Ohno, T. Kato, *J. Am. Chem. Soc.* **2004**, *126*, 994–995.

66 J.D. Martin, *ACS Symp. Ser.* **2002**, *818*, 413–427.

67 H. Cang, J. Li, M.D. Fayer, *J. Chem. Phys.* **2003**, *119*, 13017–13023.

68 S. Hayashi, R. Ozawa, H. Hamaguchi, *Chem. Lett.* **2003**, *32*, 498 499; R. Ozawa, S. Hayashi, S. Saha, A. Kobayashi, H. Hamaguchi, *Chem. Lett.* **2003**, *32*, 948–949.

69 C. Hardacre, J.D. Holbrey, S.E.J. McMath, D.T. Bowron, A.K. Soper, *J. Chem. Phys.* **2003**, *118*, 273–278.

70 S. Biggin, J.E. Enderby, *J. Phys. C-Solid State Phys.* **1982**, *15*, L305-L309; M. Rovere, M.P. Tosi, *Rep. Prog. Phys.* **1986**, *49*, 1001–1081.

71 J.S. Wilkes, J.S. Frye, G.F. Reynolds *Inorg. Chem.* **1983**, *22*, 3870–3872.

72 A. Mele, C.D. Tran, S.H.D. Lacerda, *Angew. Chem. Int. Ed.* **2003**, *42*, 4364–4366.

73 N.E. Heimer, R.E. Del Sesto, W.R. Carper, *Magn. Reson .Chem.* **2004**, *42*, 71–75.

74 H. Cang, J. Li, M.D. Fayer, *J. Chem. Phys.* **2003**, *119*, 13017–13023.

75 K. AbdulSada, A.E. Elaiwi, A.M. Greenway, K.R. Seddon, *Eur. Mass Spect.* **1997**, *3*, 245–247.

76 C. Gozzo, C.S. Consorti, J. Dupont, M.N. Eberlin, *Chem. Eur. J.* **2004**, *10*, 6187–6193.

77 J.D. Holbrey, W.M. Reichert, M. Nieuwenhuyzen, O. Sheppard, C. Hardacre, R.D. Rogers, *Chem. Commun.* **2003**, 476–477.

78 S. Consorti, P.A.Z. Suarez, R.F. de Souza, R.A. Burrow, W. Loh, L.H.M. da Silva, J. Dupont, *J. Phys. Chem. B*, **2005**, in press.

79 J.M. Zhang, C.H. Yang, Z.S. Hou, B.X. Han, T. Jiang, X.H. Li, G.Y. Zhao, Y.F. Li, Z.M. Liu, D.B. Zhao, Y. Kou, *New J. Chem.* **2003**, *27*, 333–336

80 M.R. Ally, J. Braunstein, R.E. Baltus, S. Dai, D.W. DePaoli, J.M. Simonson, *Ind. Eng. Chem. Res.* **2004**, *43*, 1296–1301.

81 G. Avent, P.A. Chaloner, M.P. Day, K.R. Seddon, T. Welton, *J. Chem. Soc., Dalton Trans.* **1994**, 3405–3413.

82 J. Dupont, P.A.Z. Suarez, R.F. de Souza, R.A. Burrow, J.P. Kintzinger, *Chem. Eur. J.* **2000**, *6*, 2377–2381.

83 J.-F. Huang, P.-Y. Chen, I.-W. Sun, S.P.Wang, *Inorg. Chim. Acta* **2001**, *320*, 7–11.

84 J.F. Huang, P.Y. Chen, I.W. Sun, S.P. Wang, *Spectrosc. Lett.* **2001**, *34*, 591–603.

85 A. Mele, C.D. Tran, S.H.P. Lacerda, *Angew. Chem., Int. Ed.* **2003**, *42*, 4364–4366.

86 U. Schroder, J.D. Wadhawan, R.G. Compton, F. Marken, P.A.Z. Suarez, C.S. Consorti, R.F. de Souza, J. Dupont, *New J. Chem.* **2000**, *24*, 1009–1015.

87 See for example: P. Lozano, T. de Diego, D. Carrie, M. Vaultier, J.L. Iborra, *Biotech. Progr.* **2003**, *19*, 380–382.

88 J. Carmichael, C. Hardacre, J.D. Holbrey, M. Nieuwenhuyzen, K.R. Seddon, *Mol. Phys.* **2001**, *99*, 795–800.

89 J. Dupont, G.S. Fonseca, A.P. Umpierre, P.F.P. Fichtner, S.R. Teixeira, *J. Am. Chem. Soc.* **2002**, *124*, 4228–4229.

90 See for example: R.P.J. Bronger, S.M. Silva, P.C.J. Kamer, P.W.N.M. van Leeuwen, *Chem. Commun.*, **2002**, 3044–3045.

91 R.P. Swatloski, S.K. Spear, J.D. Holbrey, R.D. Rogers, *J. Am. Chem. Soc.* **2002**, *124*, 4974–4975.

92 W. Scheeren, G. Machado, J. Dupont, P.F.P. Fichtner, S.R. Teixeira, *Inorg. Chem.* **2003**, *42*, 4738–4742.

93 See for example: A.E. Visser, R.D. Rogers, *J. Solid State Chem.* **2003**, *171*, 109–113 and references therein.

94 J.G. Huddleston, A.E. Visser, W.M. Reichert, H.D. Willauer, G.A. Broker, R.D. Rogers, *Green Chem.* **2001**, *3*, 156–164.

95 R.P. Swatloski, A.E. Visser, W.M. Reichert, G.A. Broker, L.M. Farina, J.D. Holbrey, R.D. Rogers, *Green Chem.* **2002**, *4*, 81–87.

96 K.E. Gutowski, G.A. Broker, H.D. Willauer, J.G. Huddleston, R.P. Swatloski, J.D. Holbrey, R.D. Rogers, *J. Am. Chem. Soc.* **2003**, *125*, 6632–6633.

97 U. Domanska, A. Marciniak, *J. Phys. Chem. B* **2004**, *108*, 2376–2386.

98 U. Domanska, E. Bogel-Lukasik, R. Bogel-Lukasik, *J. Phys. Chem. B* **2003**, *107*, 1858–1863.

99 U. Domanska, E. Bogel-Lukasik, R. Bogel-Lukasik, *Chem. Eur. J.* **2003**, *9*, 3033–3041.

100 U. Domanska, E. Bogel-Lukasik, *Ind. Eng. Chem. Res.* **2003**, *42*, 6986–6992.

101 U. Domanska, A. Marciniak, *J. Chem. Eng. Data* **2003**, *48*, 451–456.

102 S. Consorti, A.P. Umpierre, R.F. de Souza, J. Dupont, P.A.Z. Suarez, *J. Braz. Chem. Soc.* **2003**, *14*, 401–405.

103 V. K. Aggarwal, N.L. Lancaster, A.R. Sethi, T. Welton, *Green Chem.* **2002**, *4*, 517–520.

104 L.A. Blanchard, J.F. Brennecke, *Ind. Eng. Chem. Res.* **2001**, *40*, 287–292.

105 J.G. Huddleston, H.D. Willauer, R.P. Swatloski, A.E. Visser, R.D. Rogers, *Chem. Commun.* **1998**, 1765–1766.

106 J.L. Anthony, E.J. Maginn, J.F. Brennecke, *J. Phys. Chem. B* **2002**, *106*, 7315–7320.

107 L.A. Blanchard, D. Hancu, E.J. Beckman, J.F. Brennecke, *Nature* **1999**, *399*, 28–29.

108 P.J. Dyson, G. Laurenczy, C.A. Ohlin, J. Vallance, T. Welton, *Chem. Commun.* **2003**, 2418–2419.

109 A. Berger, R.F. de Souza, M.R. Delgado, J. Dupont, *Tetrahedron-Asymm.* **2001**, *12*, 1825–1828.

110 T. Schafer, C.M. Rodrigues, C.A.M. Afonso, J.G. Crespo, *Chem. Commun.* **2001**, 1622–1623.

111 L.C. Branco, J.G. Crespo, C.A.M. Afonso, *Angew. Chem., Int. Ed.* **2002**, *41*, 2771–2773.

112 L.C. Branco, J.G. Crespo, C.A.M. Afonso, *Chem. Eur. J.* **2002**, *8*, 3865–3871.

113 L.A. Blanchard, Z.Y. Gu, J.F. Brennecke, *J. Phys. Chem. B* **2001**, *105*, 2437–2444.

114 L.A. Blanchard, J.F. Brennecke, *Ind. Eng. Chem . Res.* **2001**, *40*, 287–292.

115 M. Scurto, S.N.V.K. Aki, J.F. Brennecke, *J. Am. Chem. Soc.* **2002**, *124*, 10276–10277.

116 J.G. Huddleston, H.D. Willauer, R.P. Swatloski, A.E. Visser, R.D. Rogers, *Chem. Commun.* **1998**, 1765–1766.

117 Bosmann, L. Datsevich, A. Jess, A. Lauter, C. Schmitz, P. Wasserscheid, *Chem. Commun.* **2001**, 2494–2495.

118 E. Visser, R.P. Swatloski, W.M. Reichert, S.T. Griffin, R.D. Rogers, *Ind. Eng. Chem. Res.* **2000**, *39*, 3596–3604.

119 L.C. Branco, J.N. Rosa, J.J.M. Ramos, C.A.M. Afonso, *Chem. Eur. J.* **2002**, *8*, 3671–3677.

120 E. Visser, R.P. Swatloski, W.M. Reichert, R. Mayton, S. Sheff, A. Wierzbicki, J.H. Davis, R.D. Rogers, *Environ. Sci. Technol.* **2002**, *36*, 2523–2529.

121 J.D. Holbrey, A.E. Visser, S.K. Spear, W.M. Reichert, R.P. Swatloski, G.A. Broker, R.D. Rogers, *Green Chem.* **2003**, *5*, 129–135.

122 W. Armstrong, L.F. He, Y.S. Liu, *Anal. Chem.* **1999**, *71*, 3873–3876.

123 J.L. Anderson, D.W. Armstrong, *Anal. Chem.* **2003**, *75*, 4851–4858.

124 J.L. Anderson, V. Pino, E.C. Hagberg, V.V. Shereas, D.W. Armstrong, *Chem. Commun.*, **2003**, 2444–2445.

125 G. Yanes, S.R. Gratz, M.J. Baldwin, S.E. Robison, A.M. Stalcup, *Anal. Chem.* **2001**, *73*, 3838–3844.

126 S. Carda-Broch, A. Berthod, D.W. Armstrong, *Rapid Commun. Mass Spectrom.* **2003**, *17*, 553–560.

127 M. Trost, *Acc. Chem. Res.* **2002**, *35*, 695–705.

128 B. Cornils, W.A. Herrmann, eds., *Aqueous-Phase Organometallic Catalysis, Concepts and Applications*, Wiley-VCH, Weinheim, **1996**.

129 W. Keim, *Green Chem.* **2003**, *5*, 105–111.

130 T. Horvath, *Acc. Chem. Res.* **1998**, *31*, 641–650.

131 W. Leitner, *Acc. Chem. Res.* **2002**, *35*, 746–756.

132 T. Silveira, A.P. Umpierre, L.M. Rossi, G. Machado, J.Morais, G.V. Soares, I.J.R. Baumvol, S.R. Teixeira, P.F.P. Fichtner, J. Dupont, *Chem. Eur. J.* **2004**, 10, 3734–3740.

133 F. Favre, H. Olivier-Bourbigou, D. Commereuc, L. Saussine, *Chem. Commun.* **2001**, 1360–1361.

134 C. Brasse, U. Englert, A. Salzer, H. Waffenschmidt, P. Wasserscheid, *Organometallics* **2000**, *19*, 3818–3823.

135 R.P.J. Bronger, S.M. Silva, P.C.J. Kamer, P.W.N.M. van Leeuwen, *Chem. Commun.* **2002**, 3044–3045.

136 M. Picquet, S. Stutzmann, I. Tkatchenko, I. Tommasi, J. Zimmermann, P. Wasserscheid, *Green Chem.* **2003**, *5*, 153–162.

137 M.F. Sellin, P.B. Webb, D.J. Cole-Hamilton, *Chem. Commun.* **2001**, 781–782.

138 S. Einloft, Y. Chauvin, H. Olivier, Patents US 55503041996–08–27 and US 55503061996–08–27; H. Olivier-Bourbigou, P. Travers, J.A. Chodorge, *Petr. Techn. Quat.*, **1999**, 141–149.

139 M. Freemantle, *Chem. Eng. News* **1998**, *76*, 32–37.

3.6
Membrane Processes

3.6.1
Pressure-driven Membrane Processes

Ivo F.J. Vankelecom and Lieven E.M. Gevers

3.6.1.1
Introduction

Ranging from the very common coffee filter to the huge desalination plants in the Middle East producing drinking water for millions, pressure-driven membrane processes span a wide and diverse range of applications. In all cases, the driving force that delivers the energy to separate the feed molecules is a pressure gradient, in contrast to other membrane separations where an electrical potential, temperature, concentration or partial pressure gradient maintains the separation. Pressure-driven processes are generally classified as microfiltration (MF), ultrafiltration (UF) and hyperfiltration (HF), with the latter mostly subdivided into reverse osmosis (RO) and nanofiltration (NF). However, this distinction is not always very sharp as seen from Table 3.6-1, summarizing the main characteristics of the different membrane processes. Moreover, this nomenclature has not been used systematically in literature, and NF for instance is sometimes also referred to as dense UF or low-pressure RO.

It is mainly the size of the particle or molecule to be separated, in addition to its chemical properties, that will determine the structure of the required membrane (porous or dense, mean pore size, pore-size distribution, ...), while the feed solvent, the cleaning method, the applied pressure and the operating temperature will all co-determine the membrane material.

In this major section the fundamentals and the main problems of these pressure-driven membrane processes will be summarized. A short overview will then be given of the membrane preparation methods and the most common materials used. Their application in organic media will then be reviewed in more detail, followed by comments on the current commercial membrane market and some perspectives for the future.

Green Separation Processes. Edited by C. A. M. Afonso and J. G. Crespo
Copyright © 2005 WILEY-VCH Verlag GmbH & Co. KGaA, Weinheim
ISBN 3-527-30985-3

Table 3.6-1 Summary of the main properties of the pressure-driven separation processes (adapted from Ref. [1]).

	MF	UF	HF
Typical separation	Particles (bacteria, yeasts)	Macromolecules (proteins)	Low MW solutes (salt, glucose)
Osmotic pressure	Negligible	Negligible	High (5–25 bar)
Applied pressure	Low (< 2 bar)	Low (1–10 bar)	High (10–100 bar)
Structure	(mostly) Symmetric	Asymmetric	Asymmetric
Separating layer thickness	10–150 μm	0.1–1.0 μm	0.1–1.0 μm
Pore size	0.05–10 μm	1–100 nm	< 2 nm
Main separation principle	Size exclusion	Size exclusion	Solution-diffusion + Convective flow

3.6.1.2
Fundamentals of the Processes

3.6.1.2.1 Microfiltration

Microfiltration (MF) is the most common pressure-driven membrane separation process and also the easiest to understand [1, 2]. It is simply "conventional coarse filtration" running at very low pressures (typically below 2 bar) owing to the open structure of the membranes. Darcy's law is applicable showing a proportionality between the applied pressure difference ΔP and the flux J through the membrane.

The proportionality factor A can be further specified depending on the shape of the pores.

$$J = A \cdot \Delta P \tag{1}$$

For membranes with straight capillaries as pores the Hagen–Poiseuille equation (2) applies, while the Karman–Kozeny equation (3); various forms of this equation have been applied) is valid for transport through membranes with a more nodular structure.

$$J = \frac{\varepsilon.r^2}{8\eta\tau} \cdot \frac{\Delta P}{\Delta x} \tag{2}$$

$$J = \frac{\varepsilon^2}{K\eta S^2.(1-\varepsilon)^2} \cdot \frac{\Delta P}{\Delta x} \tag{3}$$

where ε = porosity; η = viscosity; τ = tortuosity; K = Kozeny constant (often calculated in terms of hydraulic radius); S = total surface area; r = pore radius).

Both laws relate flux to viscosity and some simple structural parameters, such as pore radius and porosity. Only membrane parameters, combined with some solvent parameters, are involved. It follows that membrane transport is clearly – at least in the ideal case – independent of the solute. The laws also indicate that optimized MF-membranes combine the highest possible (surface) porosity to allow large fluxes with the narrowest possible pore-size distribution to avoid "leakage" of the solute through the membrane.

3.6.1.2.2 Ultrafiltration

As a process (formulas, separation principle, ...), UF is very similar to MF [1, 2]. Ultrafiltration typically retains macromolecules with MW > 1000 Da. In order to retain such small molecules, the membrane pore size is smaller than in MF and the hydrodynamic resistance of the layer is thus higher. That is why the thickness of the selective layer should be minimized, commonly to less than 1 μm. Such thicknesses can be realized by preparing composite membranes, in which the different layers are prepared from different materials. Alternatively, integral asymmetric membranes with a single chemical composition can be prepared (Fig. 3.6-1), meaning that the pore sizes at the top of the membrane in a cross-section are different from those at the bottom. Optimized UF-membranes show a good resistance to solvents, pH extremes and high temperatures. Above all, they should limit fouling as much as possible (see below).

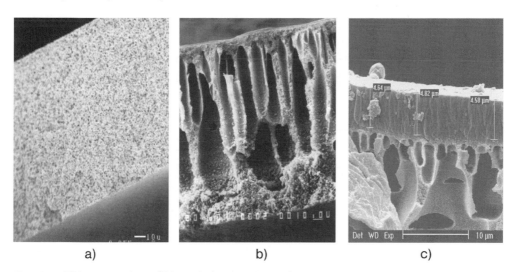

a)   b)   c)

Fig. 3.6-1 SEM-cross-sections of (a) a typical symmetric membrane, (b) an integral asymmetric and (c) a composite membrane.

3.6.1.2.3 Hyperfiltration

Hyperfiltration, particularly RO, was the first membrane process to be run on an industrial scale, as early as the 1960s [1,3]. The great breakthroughs here were the invention in the early sixties by Loeb and Sourirajan [4] of asymmetric membranes prepared via phase inversion and the development of membranes prepared via interfacial polymerization [5]. The membranes applied are densified even more than those for UF and a limit is reached: membranes may get so dense that the pores can no longer be visualized by any means and certain membranes (especially those useful in the high-pressure range) are simply to be considered as "dense", with molecular transport taking place through the free volume elements of the membranes and no longer through pores that are permanently present [6]. Hyperfiltration is typically used to retain small organic molecules or inorganic salts. Good rejections of such compounds can induce very significant osmotic pressures over the membrane, thus drastically decreasing the actual driving force. To realize high fluxes and selectivities, selection of the HF-membrane material is very crucial: it should show a high affinity for the solvent (e.g. made from a hydrophilic material in the case of water treatment) but a low affinity for the solute.

3.6.1.2.4 Diafiltration

Sometimes considered as another pressure-driven membrane process, diafiltration (dilution mode) is in fact nothing other than an improved design for a more enhanced purification (Fig. 3.6-2) [1, 5]. It has, for instance, applications in the pharmaceutical, biotech and food industries where a complete separation of high MW compounds from low MW compounds is required. After a first separation, the retentate is diluted again before the next separation purifies the stream further. This process can be repeated as often as desired. Without the dilution, the filtration would stop at a certain point when still too much of the unwanted compound would be present because of fouling or an excessive increase in osmotic pressure.

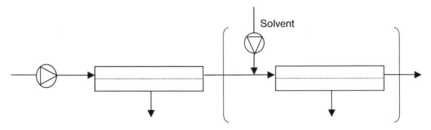

Fig. 3.6-2   Scheme of a diafiltration type of operation.

3.6.1.3
Main Problems

3.6.1.3.1 Introduction

The main problem with pressure-driven membrane processes is the flux decline as a function of filtration time (Fig. 3.6-3) due, most importantly, to concentration polarization (remaining constant once established) and membrane fouling (worsening as a function of time). These cause extra resistances on top of the membrane resistance and thus slow down the transport. This reduction can be as severe as 99% of the initial flux value in MF. Reviews are available on these matters [7], some focusing in more detail on in situ monitoring techniques [8], some only on concentration polarization [9] others only on fouling [10–12].

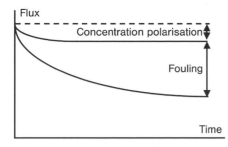

Fig. 3.6-3 Flux decrease as a function of time due to the combined effect of fouling and concentration polarisation.

3.6.1.3.2 Membrane Fouling

Fouling refers to the accumulation of retained molecules or particles in the pores of the membrane or at the membrane surface (Fig. 3.6-4). The retained species can build up a gel or cake layer, or can block pores. The latter effect may be caused by adsorption of the solutes on the pore walls, or may be a consequence of a non-uniform pore-size distribution, for instance when pores narrow in the permeation direction. Since species also adsorb on the membrane surface, this makes even dense membranes prone to fouling. Fouling is not a problem of the membrane sheet alone, but also of the module – the operational assembly of different membrane sheets or tubes into one vessel – as a whole. Foulants can be subdivided into three categories, each of them having their own peculiarities that affect their treatment: organic precipitates, inorganic precipitates and suspended solids. Their fouling potential is a complicated function of time, temperature, pH, concentration, ionic strength, specific interactions with the membrane material or among the particles, operating conditions, ...

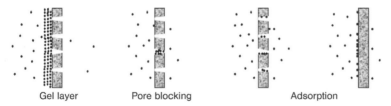

Gel layer   Pore blocking   Adsorption

Fig. 3.6-4 Schematic overview of the different processes leading to membrane fouling.

Even though fouling remains a serious problem, many solutions exist these days to cope with it. Some of them can be easily automated so that fouling does not have to be a real worry once a process is running. An appropriate pre-treatment of the feed (adjustment of pH and T, the use of additives, adsorption, prefiltration, ...) can avoid many problems. At the level of membrane development, special attention should be paid to narrow pore-size distributions and a careful selection of the membrane hydrophobicity and the presence on the membranes of charged or functional groups with specific interactions. On the level of module and process conditions, turbulence promoters and high feed-flow velocities in cross-flow filtration modes are favorable. In certain cases, it might even be worthwhile considering the operation of the membrane at reduced fluxes by lowering the trans-membrane pressure for instance.

Unsuccessful prevention can be cured by appropriate cleaning procedures: mechanical, hydraulic (back flushing, alternating pressurizing and de-pressurizing cycles,...), electrical (to remove charged species from the cake) or chemical. For the latter, the concentrations used and the necessary cleaning times are of utmost important for membrane durability. Commonly used chemicals are: acids and bases, enzymes (proteases, amylases, ...), detergents, complexing agents (EDTA, polyacrylates, ...), disinfectants (H_2O_2, NaOCl, ...) and steam sterilization. More refined concepts have been developed recently, such as the air flush concept, or rotating and vibrating modules. The means of cleaning has implications for the selection of the membrane material, which has to be resistant to the feed solution and also to the cleaning agents applied. The same obviously holds for the module housings and sealing.

3.6.1.3.3 Concentration Polarization

Concentration polarization occurs particularly when high rejections are combined with high fluxes. In the feed, a convective flow exists forcing the solute to the membrane. The solute is then partially rejected and the solute concentration at the membrane surface builds up, initiating a diffusive solute flux from the membrane surface towards the feed. From a certain distance onwards, feed stirring takes over and causes perfect mixing again. The consequences of concentration polarization are difficult to predict. Retentions can be lowered, since a higher solute concentration is built up at the membrane surface, thus increasing the amount of solute that is transported through the membrane either in the form of a diffusive flow for dense membranes or in the form of a convective flow for porous membranes. On the other hand, higher retentions can be induced when larger molecules or particles are being retained that form a kind of "dynamic" membrane creating an extra resistance, mainly for the solute. In all cases, however, fluxes decline.

To avoid concentration polarization, an improved mass transfer should be realized in the feed compartment. Determining parameters are: feed flow velocity (modified through the hydraulic diameter of the feed cell or the pump characteristics), solute diffusion (changed via the feed temperature), feed viscosity (idem), shape and dimensions of the module (introduction of turbulence promoters, use of pulsating flows to break the boundary layer, increased Reynolds numbers, ...).

In general, it can be said that concentration polarization is most severe in MF and UF, where high fluxes are combined with retention of macromolecules that show only low diffusion coefficients. The problem is less severe for HF with typically lower fluxes and higher diffusion coefficients of the retained low MW solutes.

3.6.1.3.4   Membrane wetting

Non-wettability of the membrane material by the filtration solvent can be a problem, overcome easily by pretreating the membrane with a solvent with appropriate surface tension, for instance ethanol. Even though applied only initially, such pretreatments can potentially alter the membrane's performance during the rest of the operation.

3.6.1.4
Membrane Materials and Preparation

3.6.1.4.1   Introduction

Membranes for pressure-driven separations can have either a polymeric or an inorganic nature. The former are mostly prepared via sintering, stretching, track-etching or phase inversion when MF-membranes are involved. Inorganic MF-membranes can be ceramic, metal or glass based and are prepared via sintering, sol–gel processes or anodic oxidation. These inorganic membranes obviously have a higher thermal and chemical stability than polymeric ones. Mostly, they also possess a narrower pore-size distribution, but they are more brittle and more expensive. Polymeric UF and HF membranes can be prepared via phase inversion and the densest ones also via interfacial polymerizetion and coating. In some cases, a prepared membrane can be tuned still further by modifying its surface. When UF and NF-membranes are inorganic, only sol–gel synthesis can lead to pores that are narrow enough [1]. Whenever the performance of the prepared membrane is linked to molecular weight cut-offs (MWCO) and pore diameters, care should be taken in interpreting these numbers as they might have been determined under slightly different conditions (pressures, flow mode, solutes,...), with different definitions (90 or 95% rejection) or obtained via different calculation methods.

3.6.1.4.2   Sintering

During sintering, a powder of particles of a given size is pressurized at elevated temperatures in a preformed shape so that the interface between the particles disappears. Microfiltration membranes can thus be obtained from PTFE (polytetrafluoroethylene), PE (polyethylene), PP (polypropylene), metals, ceramics, graphite and glass, with pore sizes depending on the particle size and the particle-size distribution. Porosities up to 80% for metals and 10–20% for polymeric membranes can be reached with pore sizes varying between 0.1 and 10 µm. Most of these materials have excellent solvent and thermal stability.

3.6.1.4.3   Stretching

Stretched membranes are prepared by stretching an extruded partially crystalline polymer film perpendicular to the direction of extrusion (Fig. 3.6-5). This process creates small ruptures between the crystalline regions in the film that constitute the actual pores. Thermally and chemically very stable PTFE, PP and PE membranes can thus be prepared with pore sizes between 0.1 and 3 µm and porosities up to 90%.

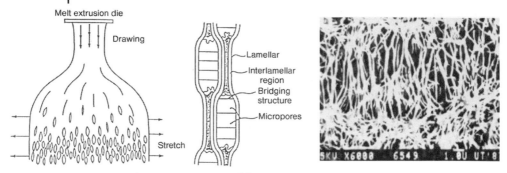

Fig. 3.6-5 Schematic presentation of the preparation of a stretched membranes (left).

A SEM-picture of a typical stretched membrane (right).

3.6.1.4.4 Track-etching

A polymeric film or foil is subjected to high-energy particle radiation (Fig. 3.6-6). This causes a certain damage to the structure along some tracks. By immersing the film in an acid or alkaline bath, the polymer can be etched away to leave parallel cylindrical pores of uniform dimension ranging between 0.02 and 10 μm pore size. Only low porosities (up to 10%) can be reached, but the pores have a very narrow pore-size distribution and the membranes show excellent thermal and chemical stability.

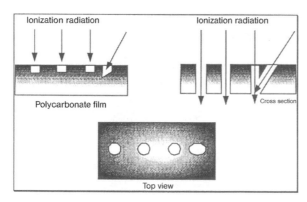
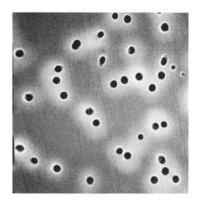

Fig. 3.6-6 Schematic presentation of the preparation of a track-etched membranes (left).

A SEM-picture of a typical track-etched membrane (right).

3.6.1.4.5 Anodic Oxidation

Membranes prepared via anodic oxidation possess a very regular morphology consisting of hexagonally close-packed cells with cylindrical pores. They are formed by applying an aluminum foil as the anode in an acid electrolyte. Depending on the experimental conditions, pore sizes can vary between 0.1 and 0.2 μm (Fig. 3.6-7).

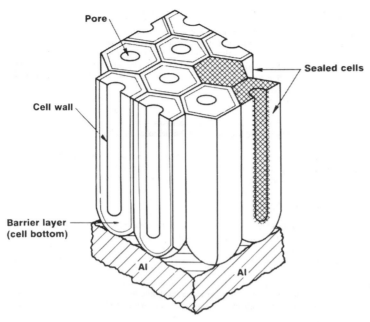

Fig. 3.6-7 Schematic drawing of a typical membrane prepared via anodic oxidation. ([taken from [H.P. Hsieh, Chapter 3 in Inor- ganic Membranes for Separation and Reaction, Elsevier, Amsterdam, 1996.])

3.6.1.4.6 Phase Inversion

Based on the pioneering work of Loeb and Sourirajan [4], membranes prepared according to the phase-inversion technique form the most important group of NF/RO-membranes, together with those prepared via interfacial polymerization [5]. Since phase-inversion membranes are prepared in one step, they are generally cheaper than other membranes and thus more appropriate for low-cost applications, such as water treatment. Cellulose acetate (CA) membranes, for example, can only be prepared via this method. Together with polyamides (PA), which can be prepared via both phase inversion and interfacial polymerization, CA dominates the field of commercial (still mostly aqueous) NF/RO-applications.

"Phase inversion" refers to the controlled transformation of a cast polymeric solution from a liquid into a solid state. During the phase-inversion process, a thermodynamically stable polymer solution is usually subjected to controlled liquid–liquid demixing. This "phase separation" of the cast polymer solution into a polymer-rich and a polymer-lean phase can be induced by immersion in a non-solvent bath ("immersion precipitation"), by evaporating the volatile solvent from a polymer that was dissolved in a solvent/non-solvent mixture ("controlled evaporation"), by lowering the temperature ("thermal precipitation") or by placing the cast film in a vapor phase that consists of a non-solvent saturated with a solvent ("precipitation from vapor phase") [1].

The whole membrane formation process is a complex combination of the thermodynamics and kinetics that finally determine the ultimate membrane morphology [13]. So many parameters and different compounds can be involved that it is often difficult to make general predictions. At the least, a three-component system, consisting of a polymer, a solvent for the polymer and a so-called non-solvent, is normally involved. But the casting solution can also contain one or more co-solvents (often more-volatile solvents), small amounts of non-solvent, additives of an organic (maleic acid, pyrrolidone, ...), inorganic (LiCl, $Mg(ClO_4)_2$, ...) or polymeric (poly(ethylene glycol), ...) nature. Moreover, all can be present in different concentrations. After the membrane casting, solvent evaporation can take place over a certain time period at a given temperature. The composition and temperature of the coagulation medium are the subsequent parameters, while a final post-treatment of the membranes has often proven extremely important for the final membrane performance. Such post-treatment can consist of exposing the membranes to solvents or acids, annealing, for instance in water, cross-linking, drying by exchanges in a series of solvents or treating with conditioning agents like lube oil.

But of prime importance with regard to the final separation process is the nature of the membrane-forming polymer: its hydrophilicity, charge density, polymer structure and molecular weight. Typical polymers used in this phase-separation process are: cellulose esters (most commonly CA), polyamides, poly(amide-hydrazides), polyimides, (sulfonated) polysulfones, poly(phenylene oxide) and (sulfonated) poly(phthalazine ether sulfone ketone).

In preparing membranes via the phase inversion process for applications in pressure-driven processes, the formation of macrovoids should be avoided completely. These finger-like pores of the type present in the substructure of membranes (b) and (c) of Fig. 3.6-1, severely limit the compaction resistance of the membrane. Membranes with a sponge-like structure (Fig. 3.6-1a) are to be preferred.

3.6.1.4.7 Interfacial Polymerization

Interfacial polymerization has become a very important and useful technique for the synthesis of thin-film composite RO and NF membranes [5, 13]. Polymerization occurs at the interface between two immiscible solvents that contain the reactants (Fig. 3.6-8). For instance, a UF membrane is immersed in an aqueous diamine solution. The excess of water is removed, and the saturated support is put in contact with an organic phase that contains an acyl chloride. As a consequence, the two monomers react to form a thin layer (1 to 0.1 μm) of PA on top of the UF membrane.

Polyamides clearly dominate the field of thin-film composites by interfacial polymerization. The composition and morphology of the membranes depend on different parameters, including the concentration of the reactants, their partition coefficients and reactivities, the kinetics and diffusion rates of the reactants, the presence of by-products, competitive side-reactions, cross-linking reactions and post-reaction treatment.

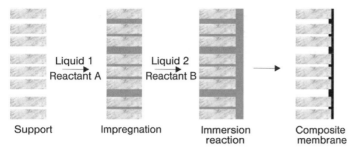

Fig. 3.6-8 Schematic illustration of the interfacial polymerization method.

3.6.1.4.8 Coating

Coated membranes consist of a thick, porous, non-selective supporting layer covered with an ultra-thin barrier layer [13]. The supporting layer is sometimes itself supported by a woven or non-woven fabric, most commonly a polyester, to improve the mechanical handling properties of the membranes. The multilayer approach allows a more flexible optimization of each layer. The support layer should offer maximal mechanical strength and compression resistance, combined with minimal resistance to permeation. The top layer on the other hand should show the desired combination of solvent flux and solute rejection. Before or after drying, a cross-linking can give the membrane more stability and sometimes better separation properties, generally making it less permeable and more selective. The type of cross-linker, its concentration, reaction pH, reaction time, temperature, ... determine the degree of cross-linking. The choice of polymer depends on many parameters, such as the strength and stability of the polymer, its film-forming properties, solubility in solvents, cost price, possibility of cross-linking, ... More-viscous casting solutions and multiple castings result in thicker membranes and thus lower fluxes but normally unchanged rejections. The viscosity of the casting solution depends on the temperature of the solution, the concentration of the solution, additives ... .

Typical membranes prepared according to this synthesis procedure have chitosan, PVA (polyvinylalcohol), PPO (polyphenylene oxide), PDMS (polydimethyl siloxane) or Nafion top-layers. Typical support layers are made of PAN (polyacrylonitrile), PI (polyimide), PVDF (polyvinylidenedifluoride) or PSf (polysulfone).

3.6.1.4.9 Surface Modification of Membranes

Surface modifications of RO or NF membranes are often applied to further enhance the performance of the prepared membranes or they might be needed to further improve the long-term stability of the membrane [13]. Modification techniques can change the pore structure, introduce functional groups, change hydrophilicity, ... Membranes can be modified via a plasma which is generated after ionization of a gas by means of an electrical discharge at high frequencies. Among the classical organic reactions applied to modify membranes, sulfonation, nitration, halogenation and treatment in acid or base are the most common. In polymer grafting, covalent bonds are generated on the polymer segments, often after creation of reactive sites. They can be introduced by means of UV photoinitiation, redox, gamma-ray or plasma initiation. Finally, photochemical and surfactant modification have also been reported.

3.6.1.4.10 Ceramic Membranes

Ceramic membranes normally have an asymmetrical structure composed of at least two, normally three, different porosity levels. Indeed, before applying the active top layer, a mesoporous intermediate layer is often applied in order to reduce the surface roughness. A macroporous support ensures mechanical stability. Ceramic membranes generally show a higher chemical, structural and thermal stability. They do not deform under pressure, do not swell and are cleaned easily [13].

The most common membranes are made of Al, Si, Ti or Zr oxides, with Ti and Zr oxides being more stable than Al or Si oxides. In some less-common cases, Sn or Hf have been used as base elements. Each oxide has a different surface charge in solution. Other membranes are composed of mixed oxides of two of the previous elements, or are stabilized by some additional compounds present in minor concentrations.

Sol–gel processes are very general means of converting a colloidal or polymeric solution into a gelatinous substance. They involve the hydrolysis and condensation reactions of alkoxides or salts dissolved in water or organic solvents. Viscosity modifiers or binders are frequently added to the sol before its deposition as a layer on a porous support via dip or spin coating. This is followed by gelation of the layer on drying. A controlled calcination and/or sintering finally leads to the actual NF-membrane. Post-synthesis, ceramic HF-membranes can be modified by covalent grafting of chelating ligands to reduce the pore size and change the isoelectric point (IEP). Adsorption of alcohols has been used to tailor the pore sizes, to enhance the hydrophobicity and decrease the fouling tendency. A similar chemisorption of water during permeations can lead to a severe initial decline of the fluxes, which can be restored completely after heat treatment.

3.6.1.4.11 Supports

Even though they only give support to the top layer that performs the actual separation, the importance of the support layers cannot be overestimated. Apart from sometimes influencing the charge of the composite membrane, the quality of the support often determines the final quality of the composite membrane itself, since defects and irregularities in the support usually also produce defects in the thin top layer. With respect to wettability during the application of the top layer, constant and homogeneous surface characteristics are required. As the roughness of the support should be as low as possible, intermediate layers with a gradual decrease in thickness and pore size are often applied. Thermal and chemical stability, roughness, non-ageing mechanical strength and compatibility with the other materials in the membrane and the module are all of importance.

3.6.1.5
Applications in Organic Media

3.6.1.5.1 **Introduction**

After decades of successful applications in aqueous media, pressure-driven membrane separation processes have recently received substantial attention in attempts to integrate them in processes taking place in organic solvents. They have already been reported to purify organic solvents, exchange solvents and recover products and catalysts from reaction mixtures. The membrane process is often used to complement other more traditional separation processes such as evaporation, distillation, extraction, ... to form an optimized hybrid system. The main advantages of membrane processes are the low energy consumption and reduced emission of harmful compounds into the environment. For applications allowing only low thermal stress or using non-volatile solvents, membrane technology can even be the only alternative to "de-bottleneck" the given application. In some cases, the membrane separation process as such will not be capable of completely isolating the product or catalyst from the organic solution, but only of concentrating it. Further purification or isolation will then take place using other techniques whose efficiency is increased thanks to the membrane pre-treatment. The different pressure-driven membrane applications in organic media reviewed below have been divided into four main categories: applications in catalysis, the food industry, petrochemistry and fine chemical synthesis. Most of these applications refer to solvent-resistant NF (SRNF), which is a young field of research rapidly gaining more attention. Some of the NF-membranes used are so dense that they can be considered in fact as RO-membranes. The main UF-applications in the chemical and refining industry on the other hand [14], are limited to the treatment of oily wastewaters, for instance to recover electrocoat paint from rinse water. Other industrial applications are also water based, such as protein separation from milk whey and juice clarification, and will thus not be discussed.

3.6.1.5.2 **Catalysis**

Membrane-assisted catalysis has already been applied in a number of processes [15]. When UF membranes are combined with catalysis, enlargement (e.g. via dendrimers) or immobilization (e.g. on organic or inorganic supports) of the homogeneous catalyst are still required. Other UF possibilities are the use of catalytic species inside dispersed droplets or of catalytic nanoparticles. HF, on the other hand, plays an important role in bridging the gap between homogeneous and heterogeneous catalysis. Off-the-shelf available, highly active and selective homogeneous catalysts can be used with all the advantages of heterogeneous catalysis, such as easy catalyst separation, facilitated product isolation and continuous operation. The actual catalysts in this case, mostly transition metal complexes (TMCs), are very expensive and toxic. They thus have to be separated from the product and for preference re-used.

With the advent of several commercially available types of SRNF membranes, this field seems to be finally re-activated since the decade-old pioneering work left untouched for years. As early as 1971, Westaway and Walker patented a separation process with CA membranes to recycle catalysts, mostly Rh-complexes, in the hydroformylation or dimerization of olefins [16].

De Smet et al. introduced a continuous process combining SRNF with homogeneous catalysis. The concept was proven with two enantioselective hydrogenation catalysts, Rh-EtDUPHOS and Ru-BINAP, separated from a methanol reaction mixture by the commercial MPF-60 membrane with rejection values of 97–98%. Hydrogen played the role of reagent for the catalysis and its pressure provided the driving force for the separation [17].

Depending on the specificity of the catalytic process, a semi-continuous operation mode with reaction and filtration in sequence can be more efficient. It allows reaction and separation to take place under different – and thus more optimized – conditions of temperature and pressure. The concept was illustrated with Pd-catalyzed Heck couplings, reactions typically taking place at temperatures that are not compatible with the stability of current SRNF membranes [18]. The Starmem 122 SRNF membrane, a PI membrane, rejected 88 to 98% of the TMC and the co-ligands, depending on the solvent system. Similar PI membranes (Starmem 142A and 142C) were used in the separation of the phase-transfer catalyst tetraoctylammonium bromide (TOABr) [19]. Aerts et al. followed a similar semi-continuous approach to recycle the Co-Jacobson catalyst in the hydrolytic kinetic resolution of epoxides using laboratory-made PDMS membranes with a rejection of 98% in diethyl ether [20].

Without reporting actual catalysis data, Scarpello et al. screened several commercial polymeric membranes for the separation of the Jacobsen catalyst, Pd-BINAP and Wilkinson catalyst out of relevant reaction solvents [21]. Turlan et al. even used zeolite membranes with a pore radius of 0.55 nm (silicalite) to separate Pd complexes from a DMAc-dissolved Heck coupling product [22].

3.6.1.5.3 **Food Industry**

Most processes in the food industry are aqueous, but in some cases, mainly in the processing of vegetable oils, organic solvents are required. Solvent extraction of vegetable oils is the best option when the seeds have oil contents below 25%. After extraction, "oil miscella" are obtained containing the oil, the solvent, phospholipids, free fatty acids, carbohydrates, proteins and pigments. Compared to the classical processes applied for further purification, membrane technology avoids the use of additional chemicals or high-temperature treatments, thus minimizing waste streams, energy consumption and thermal damage to the products.

The removal of phospholipids has been studied most. The use of UF in this "degumming" process is based on the fact that amphiphilic phospholipids form reversed micelles in apolar solvents. With a MW of 20 000, these micelles can easily be isolated. Gupta was the first to describe such an application. Phospholipids were completely removed with a 300-μm thick silicone membrane, and their concentration could be reduced to 16 ppm with a PSf membrane and to 23 ppm with a PAN membrane. Also, the color of the crude soybean oil was reduced and its metal content lowered [23]. Other work of Gupta mentions the addition of a solute, such as ammonia, to neutralize the crude oil resulting in an increased free fatty acid rejection [24]. Iwama indicated the problem of high transport resistance when passing viscous miscella through spiral modules. Owing to increased viscosity, the miscella could be concentrated only 20 times. With tubular membranes, the miscella could be concentrated 50 to 200 times [25]. For similar processes, Köseoglu et al. described the performance of DS-7, a thin PDMS membrane, showing 99.4% rejection for phospholipids in crude cotton-seed oil [26]. An important problem in degumming is the fouling of the membrane by the retained micelles on the membrane surface. Ochoa concluded that PVDF had a considerably better fouling resistance than polyethersulfone and PSf membranes [27].

Membranes have also been integrated at a second stage in edible-oil processing to recover the extraction solvents. Köseoglu reported the use of RO and UF in the separation of cotton-seed oil from isopropanol, ethanol and hexane. These membranes, originally developed for aqueous applications, showed acceptable performances in isopropanol and ethanol, but were often damaged or hardly permeable, as in the case of hexane [28]. Schmidt tested a PDMS membrane in the separation of corn-seed oil from hexane, with high permeability and rejection values of 90% [29]. Stafie reported similar rejection values for laboratory-made PDMS membranes [30].

The last step in edible-oil processing with integrated membrane technology, is the removal of free fatty acid (FFA) from a crude oil fraction. One way to do this deacidification is via extraction with an organic solvent, like methanol, with good selectivity for the FFA. The extractant (solvent with FFA) can then be processed with membranes to remove the FFA from the solvent. RO membranes, such as Desal 5 and NTR-759, performed well in the separation of oleic acid from methanol with rejections above 90%. An important problem is the significant drop in flux when solutions with higher concentration are processed [31]. Another approach is the treatment of the triglyceride extract with NF membranes, where the triglyceride is retained and the solvent and the FFA permeate through. Zwijnenberg investigated this principle with PEBAX and CA membranes, giving permeates with almost 100% pure FFA [32].

As well as in edible-oil processing, the use of organic solvents is also relevant in the synthesis of amino acids and their derivatives. In producing dipeptides such as L-aspartyl-L-phenylalanine methyl ester, better known as aspartame, from amino acids and derivatives via enzymatic processes in organic solvents such as butyl acetate and 2-methyl-2-butanol, unreacted amino acids should be recycled after the synthesis to make the process more efficient. Reddy et al. tested several commercial membranes and found a polyamide-polyphenylene sulfone composite to be promising for this application. However, further research is still needed here to apply other membrane materials, which are more resistant towards solvents like butyl acetate [33].

3.6.1.5.4 **Petrochemical Industry**
The largest scale on which membrane technology has been introduced in organic processes so far is in petrochemistry, more specifically in the refining of lubri-

cants. To remove waxy components from lubricants, catalytic dewaxing and solvent dewaxing are two possible routes. Membrane technology has been introduced to de-bottleneck problems in the existing solvent dewaxing plants, which have high energy and cooling-water demands and significant exhaust of organic solvents.

The solvent dewaxing process involves addition of solvent during the chilling of the waxy feed. The chilling process induces crystallization of the wax and the solvent is added to maintain the fluidity. The wax crystals are filtered by rotating drum filters, resulting in a lube oil filtrate and a slack wax with entrained solvent and oil. In both fractions, the solvent is recovered by successive vaporization and distillation.

In the late 1980s, several groups started research to separate the dewaxing solvents from the lube oil with membrane technology in order to support or even substitute the conventional solvent-recovery plants. Exxon filed several patents concerned with membrane developments, basically describing the use of different membrane materials [34–37]. Finally, a cooperation with W.R. Grace led to the development of a PI-membrane-based process to recover these solvents [38–40]. The lube oil filtrate is treated in an SRNF installation, with lube oil rejections of more than 95%. Cold solvent with a +99% purity is thus permeating and, at reduced temperature, directly recycled to the chilling feed stream. Because of the possibility of recycling cold solvent, this new technology enables the processing of more lube oil feed stocks with the same refrigeration capacity and less energy input.

The Shell Oil Company investigated silicone-based membranes. Recognizing the swelling problem of these rubbery polymers in the common dewaxing solvents, halogenated siloxanes were used to separate hydrocarbon oil from solvents such as toluene and methyl ethyl ketone. In a process design with three membrane units, concentration polarization was prevented by recirculating part of the concentrated lube oil fraction in the more diluted feed streams [41]. Similar separations were done by Pasternak using thermally cross-linked poly(aliphatic terpene) membranes [42].

Other membrane applications have also been investigated in petrochemistry, namely for the production of high-quality aromatics from the middle distillate range, containing a low fraction of aromatics. Low-boiling polar solvents, such as acetonitrile and nitroethane, extract these aromatic compounds, resulting in an extract with 75 to 90% aromatics. When the extract is then filtered over RO membranes prepared from PI or CA, a permeate stream with up to 99% aromatics can be obtained. With membrane technology, high-boiling polar solvents such as NMP, DMF, DMSO and DMAc can also be used to extract the aromatics. Black et al. succeeded in preparing a stable PA membrane for such a process [43].

Furfural is another extraction solvent of interest. Hardly permeating through common silicones, the lube oil was concentrated in the permeate stream to such an extent that demixing occurred owing to the limited solubility of the oil in furfural, thus allowing easy solvent recovery [44].

In the refining of gasoline, the removal of elemental sulfur from gasoline has been patented. The formed insoluble polysulfides were easily removed by UF or even MF [45].

3.6.1.5.5 Fine Chemical Synthesis

The synthesis of pharmaceuticals often involves multi-step reactions and each reaction may have a different optimal solvent. The final isolation of the product often takes preferentially place in yet another solvent. Solvent exchange is thus necessary in most syntheses. Because most intermediates are thermolabile, conventional distillation can only be done under high vacuum, but is excluded when high-boiling solvents have to be exchanged with low-boiling ones. This kind of process has been described with SRNF membranes capable of retaining compounds with molecular weights between 250 and 1000 Da. A certain volume of the starting solution containing the solute is filtered to a volume reduction of 80%. The second solvent is added to restore the original volume and the resulting is concentrated again. These steps can be repeated until the concentration of the first solvent is reduced to the desired level. The process was proven for erythromycin, exchanged from ethyl acetate to methanol [46]. Livingston tested the concept with solutions containing tetraoctylammonium bromide and tetrabutylammonium bromide. Toluene was exchanged for methanol and subsequently for ethyl acetate [47].

3.6.1.6
Current Market

The commercial market for solvent-resistant membranes is still very young and even though excellent membranes are available at present for certain applications, they may still be completely absent for other applications. The number of solvent-resistant membranes, marketed as such, that are commercially available for the moment is clearly still too limited. However, many membranes sold for aqueous applications could be well suited for applications in another solvent as long as the membrane materials and the membrane module as a whole are compatible with the applied solvent. Koch Membrane Systems (USA) was the first company to enter the market in the late 1990s with three different membranes designed for solvent applications. The MPF-60 membrane was taken off the market after a few years, but MPF-50 and MPF-44 are still available. The W.R. Grace Company (USA) first turned the dewaxing process with Mobil into a large-scale commercial success before bringing the PI-membranes to the market under the trade name STAR-MEM. The membranes, available with four different MWCOs (200, 220, 280 and 400 Da), are currently commercialized by M.E.T. (UK). Since about 2002, the Dutch company Solsep has entered the market for solvent-resistant membranes, commercializing two UF and two NF membranes with either polar or apolar character.

Important progress has also been made in the development of ceramic membranes, for which the German company HITK has brought a silylated TiO_2-membrane to the market, able, for example, to separate homogeneous catalysts from toluene at high fluxes.

3.6.1.7
Perspectives

New developments are hampered by the reluctance of process engineers to switch from conventional, well-established separation methods to membrane separation, together with the current focus of membrane companies on the more lucrative aqueous applications with annual growth potentials of more than 15%. However, the membrane research initiated by the urgent need to de-bottleneck certain running production facilities, has already been applied to large-scale plants giving year-round trouble-free operation. Such an industrial development will undoubtedly play an important role as an inspiring example to convince engineers and managers to implement solvent-resistant membrane separations in other related processes.

Parallel to these initiatives from process engineers, there is still a clear task for material scientists to develop new membranes with improved performances. Membranes with sharp MWCO curves would mean a huge step forward, enabling the rejection of a particular compound from compounds with rather similar MW or size. With really sharp MWCO curves in the range below 300 Da, the enormous potential of solvent separations could be realized, with aromatic/aliphatic and linear/branched separations as ultimate goals, taking into account the high osmotic pressures that must be overcome [14]. Apart from one remarkable 1964 *Nature* publication, which even reported high selectivities for different sets of solvent mixtures [48], such solvent separations have not yet been realized.

Aprotic solvents, such as DMA, DMF, NMP and dioxane, currently still form a problematic class of solvents, since they are excellent solvents for polymers, and thus require extremely high chemical stability from both top layers and support layers. Finally, membranes for high-temperature applications are not yet available, with current reported upper limits in the range of 40–80°C only.

Ceramic membranes could provide an answer to some of these questions, even though no single ceramic has even the theoretical potential to be suitable for all different solvents. Moreover, ceramic membranes will remain more expensive and suffer from a low surface/volume ratio when constructed in a module. Their use will thus basically remain limited to small-scale, high-value separations.

With the recent implementation of high-throughput experimental techniques in this type of membrane separation, the development of new membranes and novel applications will surely be accelerated [49]. Probably, tailor-made membranes will be needed to solve specific separation problems in industrial chemical processes.

In the field of UF, the fouling problem needs to be further addressed, but this is not expected to be straightforward. Better module design, automated cleaning procedures and the development of inherently less fouling membranes with modern nanotechnology may be possible ways forward.

Acknowledgements

L.E.M.G. acknowledges the I.W.T. for a grant as doctoral research fellow. The Belgian Federal Government for an I.A.P.-P.A.I. grant on Supramolecular Catalysis and the Flemish Government for a grant from the Concerted Research Action (G.O.A.) are gratefully acknowledged.

References

1 M. Mulder, *Basic Principles of Membrane Technology*, Kluwer Academic, Dordrecht, **1991**.

2 W. Eykamp, in *Membrane Separations Technology, Principles and Applications*, R.D. Nobel, S.A. Stern, eds., Elsevier, Amsterdam, **1995**, Ch. 2.

3 C.J.D. Fell, in *Membrane Separations Technology, Principles and Applications*, R.D. Nobel, S.A. Stern, eds., Elsevier, Amsterdam, **1995**, Ch. 4.

4 S. Loeb, S. Sourirajan, *Adv. Chem. Ser.* **1963**, 38, 117.

5 R.J. Petersen, *J. Membrane Sci.* **1993**, 83, 81–150.

6 Ivo F.J. Vankelecom, Koen De Smet, Lieven E.M. Gevers, Sven Aerts, Peggy Van De Velde, Filip Du Prez, Pierre A. Jacobs, *J. Membrane Sci.* **2004**, 231, 1–2, 99–108.

7 E. Matthiasson, Björn Sivik, *Desalination*, **1980**, 35, 59–103

8 J.C. Chen, Q. Li, M. Elimelech, *Adv. Colloid Interface Sci.* **2004**, 107, 83–108.

9 S.S. Sablani, M.F.A. Goosen, R. Al-Belushi, M. Wilf, *Desalination*, **2001**, 141, 269–289.

10 I.-S. Chang, P. Le Clech, B. Jefferson, S. Judd, *J. Environ. Eng.* November **2002**, 1018–1029.

11 M. Goosen, S.S. Sablani, H. Al-Hinai, S. Al-Obeidani, R. Al-Belushi, D. Jackson, *Sep. Sci. Technol.* **2004**, 39, 2261–2297,

12 Y.-J. Zhao, K.-F. Wu, Z.-J. Wang, L. Zhao, S.-S. Li, *J. Environ. Sci.* **2000**, 12, 241–251.

13 I.F.J. Vankelecom, K. De Smet, L.E.M. Gevers, P.A. Jacobs, in *Nanofiltration – Principles and Applications*, A.I. Schäfer, A.G. Fane, T.D. Waite, eds., Elsevier, Amsterdam, **2004**, Ch. 2.

14 R. Baker, in *Membrane Technology in the Chemical Industry*, S.P. Nunes, K.-V. Peinemann, eds., Wiley-VCH, Weinheim, **2001**, pp. 268–297.

15 I.F.J. Vankelecom, *Chem. Rev.* **2002**, 102, 3779–3810.

16 M.T. Westaway, G. Walker, US Patent 3,617,553, **1969**.

17 K. De Smet, S. Aerts, E. Ceulemans, I.F.J. Vankelecom, P.A Jacobs, *Chem. Commun.* **2001**, 5.

18 D. Nair, J.T. Scarpello, L.S. White, L.M. Freitos dos Santos, I.F.J. Vankelecom, A.G. Livingston, *Tetrahedron Lett.* **2001**, 42, 8219–8222.

19 S.S. Luthra, X. Yang, L.M. Freitas dos Santos, L.S. White, A.G. Livingston, *Chem. Commun.* **2001**, 1468.

20 S. Aerts, H. Weyten, A. Buekenhoudt, L.E.M. Gevers, I.F.J. Vankelecom, P.A. Jacobs, *Chem. Commun.* **2004**, 710–711.

21 J.T. Scarpello, D. Nair, L.M. Freitos dos Santos, L.S. White, A.G. Livingston, *J. Membrane Sci.* **2002**, 203, 71–85.

22 D. Turlan, E.P. Urriolabeitia, R. Navarro, C. Royo, M. Menéndez, J. Santamaria, *Chem. Commun.* **2001**, 2608–2609.

23 S. Gupta, K. Achintya, US Patent 4,533,501, **1985**.

24 S. Gupta, K. Achintya, US Patent 4,062,882, **1977**.

25 A.K. Iwama and N.K. Yoshiyasu, Purification of crude glyceride oil compositions, *EP302766*, **1983**.

26 L. Lin, K.C. Rhee, S.S. Köseoglu, *J. Membrane Sci.* **1997**, 134, 101–108.

27 N. Ochoa, C. Pagliero, J. Marchese, M. Mattea, *Se. Purific. Technol*, **2001**, 22–23, 417–422.

28 S.S Köseoglu, J.T. Lawhon, E.W. Lusas, *J. Am. Oil Chem. Soc.* **1990**, 67, 315–322.

29 M. Schmidt, K.-V. Peinemann, N. Scharnagl, K. Friese, R. Schubert, Radiation-modified siloxane composite membranes

for ultrafiltration of solutes from organic solvents, *DE19507584*, **1996**.

30 N. Stafie, D.F. Stamatialis, M. Wessling, *J. Membrane Sci.* **2004**, 228, 103–106.

31 L.P. Raman, M. Cheryan, N. Rajagopalan, *J. Am. Oil Chem. Soc.* **1996**, 73, 219–224.

32 H.J. Zwijnenberg, A.M. Krosse, K. Ebert, K.-V. Peinemann, F.P. Cuperus, *J. Am. Oil Chem. Soc.* **1999**, 76, 83–87.

33 K.K. Reddy, T. Kawakatsu, J.B. Snape, M. Nakajima, *Sep. Sci. Technol.* **1996**, 31, 1161–1178.

34 D.L. Wernick, US Patent 4,678,555, **1987**.

35 L.E. Black, Interfacially polymerized membranes and reverse-osmosis of organic solvent solutions using them, *EP 421676*, **1991**.

36 H.F. Shuey, W. Wan, US Patent 4,532,041, **1985**.

37 B.P. Anderson, US Patent 4,963,303, **1990**.

38 L.S White, A.R. Nitsch, *J. Membrane Sci.* **2000**, 179, 267–274.

39 R. M. Gould, L.S. White, C.R. Wildemuth, *Environ. Prog.* **2001**, 20, 12–16.

40 L.S. White, I.-F. Fan, S. Minhas, US Patent 5,429,748, **1995**.

41 J.G.A. Bitter, J.P. Haan and H.C. Rijkens, US Patent 4,748,288, **1988**.

42 M. Pasternak, US Patent 5,234,579, **1993**.

43 L.E. Black, P.G. Miasek, G. Adriaens, US Patent 4,532,029, **1985**.

44 J.G.A. Bitter, J.P. Haan, US Patent 4,810,366, **1989**.

45 J.L. Feimer, D.W. Kraemer, J. Mann, G.L. Wagner, Membrane process to remove elemental sulphur from gasoline, *CA2111176*, **1994**.

46 J.P. Sheth, Y. Qin, K.K. Sirkar, B.C. Saltzis, *J. Membrane Sci.* **2003**, 211, 251–261.

47 A.G. Livingston, Method, *WO02076588*, **2002**.

48 S. Sourirajan, *Nature* **1964**, 1348.

49 P. Vandezande, L.E.M. Gevers, J.S. Paul, I.F.J. Vankelecom, P.A. Jacobs, *J. Membrane Sci.* accepted for publication.

3.6.2
Vapor Permeation and Pervaporation

Thomas Schäfer and João G. Crespo

3.6.2.1
Introduction

Vapor permeation and pervaporation are membrane separation processes that employ dense, non-porous membranes for the selective separation of dilute solutes from a vapor or liquid bulk, respectively, into a solute-enriched vapor phase. The separation concept of vapor permeation and pervaporation is based on the molecular interaction between the feed components and the dense membrane, unlike some pressure-driven membrane processes such as microfiltration, whose general separation mechanism is primarily based on size-exclusion. Hence, the membrane serves as a selective transport barrier during the permeation of solutes from the feed (upstream) phase to the downstream phase and, in this way, possesses an additional selectivity (permselectivity) compared to evaporative techniques, such as distillation (see Chapter 3.1). This is an advantage when, for example, a feed stream consists of an azeotrope that, by definition, cannot be further separated by distillation. Introducing a permselective membrane barrier through which separation is controlled by solute–membrane interactions rather than those dominating the vapor–liquid equilibrium, such an evaporative separation problem can be overcome without the need for external aids such as entrainers. The most common example for such an application is the dehydration of ethanol.

In the following, the basic principles of vapor permeation and pervaporation will be presented, along with the most important technical aspects of operation. It will be shown that understanding the process fundamentals can lead to more efficient downstream processing, in a wide range of applications during chemical processing and with a large degree of versatility.

3.6.2.2
Process Fundamentals

3.6.2.2.1   **Principal Mass-Transport Phenomena**
Vapor permeation (VP) and pervaporation (PV) are membrane separation processes whose only difference lies in the feed fluid being a vapor (VP) or a liquid (PV), respectively. This difference has implications for feed fluid handling as well as the nature of the transport phenomena occurring in the feed stream, as in VP the feed fluid is compressible whilst in PV it is effectively not; however, this does not in any way affect the transport phenomena across and after the membrane barrier. For this reason, vapor permeation and pervaporation will be discussed simultaneously, with differences being explicitly emphasized where necessary.

The principle of VP/PV is illustrated in Fig. 3.6-9. A non-porous membrane separates a vapor or liquid feed from a downstream compartment to which a vacuum is applied. The difference in total pressure between the two sides of the non-porous membrane, P^{feed} and P^{perm}, creates a chemical potential gradient of the compounds, $\Delta\mu = \mu^{feed} - \mu^{permeate}$, which is the driving force of the process. Contrary to pressure-driven processes (see Chapter 3.6.1), the driving force is hence the gradient in the chemical potential μ_i (or partial pressure p_i) of a solute i, rather than the gradient in total transmembrane pressure P. Owing to the vacuum at the downstream (permeate) side of the membrane, the permeate is in the vapor state. In particular with regard to PV, this tempts us to interpret the process as an "evaporation across a membrane", as a phase change occurs from liquid on the membrane feed side to vapor on the permeate side. Whilst this in fact gave rise to the term "pervaporation" (*permeation + evaporation*), it is a wrong and misleading description of the process. No direct liquid–vapor phase change occurs during pervaporation, which is why it is not an evaporative technique, as we shall indicate.

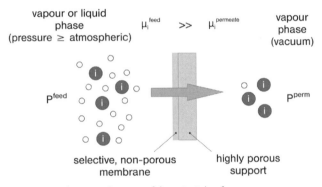

Fig. 3.6-9 Schematic diagram of the principle of vapor permeation/pervaporation.

Solution-diffusion transport mechanism When the feed contacts the membrane, the solutes (denoted i in Fig. 3.6-10) adsorb on and subsequently absorb in the membrane surface by solute–polymer interactions (Fig. 3.6-10A). Preferential solute–polymer interactions imply that the solvating power of the polymer is higher for the solutes than for the bulk solvent.

Under ideal conditions, a thermodynamic equilibrium will be reached when the chemical potential of the solute i is equal at the membrane surface and the feed phase adjacent to it. The sorption of these solutes at the membrane surface creates a solute concentration gradient across the membrane, resulting in a diffusive net flux of solute across the membrane polymer (Fig. 3.6-10B). In vapor permeation/pervaporation, any solute that has diffused toward the membrane downstream surface is ideally instantaneously desorbed and subsequently removed from the downstream side of the membrane (Fig. 3.6-10C). This can be achieved either by applying a vacuum (vacuum vapor permeation/pervaporation), or by passing an inert gas over the membrane downstream surface (sweeping-gas vapor permea-

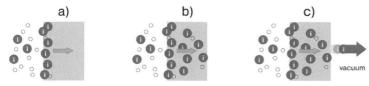

Fig. 3.6-10 Sorption (a), diffusion (b) and desorption (c) of permeating solutes during vapor permeation/pervaporation.

tion/pervaporation). As a consequence, the solute concentration on the membrane downstream surface remains practically zero, and a maximum concentration gradient is maintained between the two membrane surfaces. Also as a result, the diffusive net flux across the membrane is maximal. If the vacuum is not sufficiently low for desorption of all the solutes reaching the membrane downstream surface, the concentration of solute i at that surface will be unequal to zero. As a consequence, the concentration gradient will decrease and so will the diffusive net flux across the membrane. This illustrates clearly that the driving force in VP/PV is thus the gradient of the concentration, or more precisely of the chemical potential, of a solute between the bulk phases on each side of the membrane.

It should be noted that the role of the vacuum in VP/PV is nothing more than the efficient desorption and removal of solutes from the membrane downstream surface and hence the maintenance of the driving force of the process. The pressure difference between the two sides of the membrane does not *directly* affect the transport of an individual component i within the membrane polymer, as is the case in pressure-driven processes involving porous membranes (see Chapter 3.6.1). Ideally, the molecular motion of a component i within the membrane polymer is purely diffusive and thus independent of any operating conditions beyond the membrane surfaces. Of course, these operating conditions contribute to determining the concentration of a component i on the membrane surfaces, which in turn produces a certain diffusive net flux of this component across the membrane (see below in the section on "Membrane swelling"). In the case of vapor permeation the feed stream may be previously compressed in order to achieve a higher (volumetric) concentration of the feed compounds and, hence, increase the driving force.

3.6.2.2.2 Vapor Permeation/Pervaporation Separation Characterization

Permeability According to the aforementioned, both the sorption and the diffusion of a solute determine the mass transfer across, and hence the separation characteristics of the membrane. The product of the diffusion coefficient $D_{i,m}$ and the sorption coefficient S_i is denoted the "permeability" of the membrane for component i and is commonly designated as P_i. In order to avoid confusing permeability and pressure, however, here it will be denoted L_i (used commonly for the phenomenological constant)

$$L_i = D_{i,m} \cdot S_i \tag{4}$$

The solubility (sorption) and diffusivity of a solute in a membrane polymer are often inversely related. Solubility in the membrane polymer commonly increases with condensability of the solute. The latter, however, is *often higher the larger the molecule and thus the lower the diffusivity.*

Flux   The flux J_i of a solute across the membrane of a defined thickness z_m is given in its simplest form by the mass-transfer coefficient times the driving force, with the latter being the chemical potential gradient, which in practice is more conveniently replaced by the mole fraction, or concentration, or partial pressure of solute i in the feed and the permeate, respectively:

$$J_i = \frac{D_{i,m}\,S_i}{z_m}\,(p_{i,f} - p_{i,p}) = \frac{L_i}{z_m}\,(p_{i,f} - p_{i,p})$$

$$= \frac{L_i}{z_m}\,(x_{i,f}\,\gamma_{i,f}\,P_i^S - y_{i,p}\,P^P) = \frac{L_i}{z_m}\,(x_{i,f}\,H_i - y_{i,p}\,P^P) \qquad (5)$$

$$\text{with } p_{i,f} = x_{i,f}$$

$\gamma_{i,f}\,P_i^S$ being valid for solutions of i. It must be noticed that the activity coefficient $\gamma_{i,f}$ varies with the concentration of i in the feed solution. For dilute solutions of i the product $\gamma_{i,f}\,P_i^S$ is constant and may be described by the Henry coefficient, H_i. In vapor permeation the value of $p_{i,f}$ may be increased by compressing the feed vapor before processing it through the membrane unit. If the partial permeate pressure $p_{i,p}$ of solute i is negligible in view of its partial feed pressure $p_{i,f}$ (but not necessarily when the total permeate pressure P^P is close to zero, as is often stated in the literature), then it follows that

$$J_i = \frac{D_{i,m}\,S_i}{z_m}\,H_i\,x_{i,f} \qquad (6)$$

In practice, the value of J_i depends on how the driving force is expressed. Care needs to be taken when determining values for $D_{i,m}$ and S_i for VP/PV modeling. Different values are obtained if these two parameters are derived from thermodynamic principles or calculated from mass concentrations.

Enrichment and selectivity   The enrichment factor β_i is the ratio of the concentration of i in the permeate divided by its concentration in the feed. In practice it is often calculated as the ratio of the mass concentrations of i in the permeate, c_i^P (kg m^{-3}), and the mass concentration of i in the feed, c_i^f (kg m^{-3}):

$$\frac{c_i^P}{c_i^f} = \beta_i \qquad (7)$$

The ratio of the enrichment factors of a compound i and a compound j indicates the selectivity α_{ij} a membrane has for compound i in comparison with compound j:

$$\frac{\beta_i}{\beta_j} = \frac{x_{i,P}/x_{i,f}}{x_{j,P}/x_{j,f}} = \alpha_{ij} \qquad (8)$$

The selectivity (or separation factor, see Chapter 3.1) can then also be expressed in terms of the respective partial fluxes as follows (Cussler, 1997):

$$\alpha_{ij} = \frac{\beta_i}{\beta_j} = \frac{J_i / x_{i,f}}{J_j / x_{j,f}} = \frac{D_{i,m} S_i}{D_{j,m} S_j} \frac{H_i}{H_j} = \frac{L_i}{L_j} \frac{H_i}{H_j} \tag{9}$$

Equation (9) illustrates that in comparison to the selectivity of a simple liquid–vapor equilibrium based on the ratio of the respective volatilities, here expressed in terms of the Henry coefficients of solutes i and j, respectively, the membrane introduces a further selectivity given by the ratio of the permeabilities of the respective compounds i and j (assuming all the above assumptions to be valid). In other words, whilst the *driving force* for solute transport is *identical* in processes based on the vapor–liquid equilibrium and in vapor permeation or pervaporation, the latter can exceed the vapor–liquid equilibrium selectivity whenever

$$\frac{L_i}{L_j} = \frac{D_{i,m} S_i}{D_{j,m} S_j} > 1 \tag{10}$$

For a given separation problem, differences in selectivity of the two processes result therefore from both the diffusivity in, and the affinity of solutes i and j for, the membrane polymer. This is the principal difference and advantage that pervaporation has in comparison with separation processes based on the vapor–liquid equilibrium, such as distillation. It must certainly be noted that while selectivities can be higher in pervaporation, the flux (a "kinetic" parameter) per unit mass transfer area will in general be lower, owing to the additional mass-transport resistance presented by the membrane. However, this can be compensated for by increasing the membrane area, which contribution to the overall process costs has became less relevant during recent years.

For illustration, rubbery polymeric membranes, whose polymeric network is sufficiently elastic and mobile to allow comparatively large organic compounds to diffuse through it (Table 3.6-2), are in general used for the recovery of organic compounds from aqueous solutions. Because of its small size, the bulk solvent, water, unfortunately diffuses through the membrane even better. This is why in organophilic pervaporation the selectivity is mainly achieved and determined by the ratio of the solubility coefficients (sorption selectivity, Table 3.6-2). Membrane selectivity, as defined in Eq. (7), is an intrinsic parameter and can differ from the overall *process* selectivity, as will be shown later.

Table 3.6-2 Selectivity of a pervaporation membrane for ethyl hexanoate and isobutyl alcohol with respect to water; sorption coeffi- cients obtained on the basis of weight fractions and diffusion coefficients based on Fick's Law.

Compound	Sorption coefficient S_i	Diffusion coefficient D_i [m^2 s^{-1}]	Sorption selectivity S_i/S_{water}	Diffusion selectivity D_i/D_{water}	Permeability selectivity L_i/L_{water}
Water[a]	0.0005	2.2×10^{-10}	1	1	1
Ethyl Hexanoate[a]	241.3	2.1×10^{-12}	482 600	0.0096	4607
Isobutyl Alcohol[b]	1.0	5.4×10^{-12}	2000	0.025	49

[a] data for PDMS (Lamer et al., 1994); [b] data for POMS (Schäfer, 2002).

3.6.2.3
Non-ideal Phenomena

The solution-diffusion model is valid only in strictly ideal systems, namely when dealing with solutions of infinite dilution. As soon as one departs from such ideal solutions, it becomes to some extent subjective what can still be considered as "almost ideal" and "highly dilute". For the pervaporation of isobutyl alcohol, for example, a feed concentration of 50 mg kg^{-1} would lead to a membrane surface concentration of 50 mg kg^{-1} (according to the sorption coefficient listed in Table 3.6-2). For the same feed concentration, ethyl hexanoate would yield a membrane surface concentration about 240 times higher, namely 12 g kg^{-1}, which may not be considered "ideal" anymore. The stronger the (desired) solute–polymer affinity, the more pronounced can be the non-ideal phenomena, with the most relevant being discussed below.

It should at this point be emphasized that the term *"non-ideal"* refers to phenomena that cannot be described by or are not expected to occur on the basis of the strongly simplified models that are commonly employed in membrane technology for modeling VP/PV performance. Nevertheless, the interested reader will find a large amount of literature trying to adjust empirical models to observed "non-ideal" phenomena, sometimes leading to intimidating constructs, rather than understanding the underlying molecular interactions.

3.6.2.3.1   Membrane Swelling and Flux Coupling

Swelling   Membrane swelling is the increase of the volume of a pure polymer due to the dissolution of penetrants (solutes). Just as in liquids, the dissolution of solutes in the membrane polymer is a mixing process, the difference being that the high molecular weight polymer chains are much less mobile than a low molecular weight solvent. The extent to which a solute at a given concentration in the feed liquid dissolves in the membrane is defined by the so-called sorption (partition) coefficient introduced earlier. When sorbing in the membrane polymer, the solutes cause a membrane swelling which is all the more pronounced the higher the solute concentration in the membrane and the greater the degree of rearrangement of the polymer chains induced by the solute. At high feed concentrations of high-affinity solutes, the membrane polymer can swell to such an extent that its intrinsic properties are significantly altered, even to losing a degree of its intrinsic selectivity. For this reason, VP/PV are most commonly, but not necessarily, employed for the recovery of reasonably dilute target compounds from solutions.

Because a vacuum is applied for the removal of the solutes on the membrane downstream face, this side of the membrane is ideally "dry" in comparison to the more swollen (if polymeric membranes are employed) and hence more flexible membrane upstream face resulting from the solute uptake. This anisotropy of the membrane in the direction of the diffusion of the solute always exists for polymeric membranes and results in a non-uniform diffusivity of solute within the membrane. In other words, the diffusion coefficient of solute i in the membrane, $D_{i,m}$, will be position-dependent and not constant across the membrane.

According to the solution-diffusion model, mass transport of a solute across the non-porous membrane is purely diffusive during VP/PV. It should be noted that under the circumstance of excessive swelling, this assumption might no longer be valid. Convective transport can occur as clusters of solvent or solute molecules permeate across the membrane, and mass-transport phenomena might approximate those observed in pressure-driven membrane separations (see Chapter 3.6.1). Generally known assumptions on the principal transport phenomena occurring during VP and PV need therefore to be validated for individual applications, rather than taken as paradigms.

For the description of such interactions as well as of polymer swelling, models based on the Flory–Huggins Theory (Flory, 1953; Mulder, 1991) and UNIQUAC are often applied for mixtures in general and, for binary mixtures, also the Solubility Parameter Theory if the feed components are hydrophobic (Hildebrand and Scott, 1949; Mulder, 1991). The application of both Solubility Parameter Theory and the Flory–Huggins Theory can be found explained in more detail in Mulder (1991) and Huang and Rhim (1991). The UNIQUAC (UNIversal-QUAsi Chemical) model was originally developed for vapor–liquid equilibrium calculations (see Chapter 3.1) and later extended to polymer systems. It has been applied modified for individual applications to account for free volume effects (Free Volume Theory), or for specifics such as hydrogen bonding, which can cause a strong non-ideality in a polymer–penetrant system. It is in principle capable of describing multi-component mixtures whilst requiring only binary parameters. Examples for applications of these models can be found in the literature (Heintz and Stephan, 1994a), as well as an overview over the background and the limits of UNIQUAC-related models and the modifications of the Flory-Huggins Theory (Jonquières et al., 1998). The Free Volume Theory is a phenomenological approach, which considers the mobility of the permeating solute as a function of the free space available within the membrane polymer network (Huang and Rhim, 1991). In general, however, it may be stated that the solute–polymer interactions are so complex that up to now no model is universally applicable. In the future, molecular modeling approaches can therefore be expected to replace empirical models, as has already been observed in fluid dynamics.

Flux coupling Flux coupling can be imagined as a permeant of low diffusivity ("slower" permeant) being dragged through the membrane polymer by a permeant of higher diffusivity ("faster" permeant) resulting in higher fluxes than expected of the slower permeant. The opposite might also happen, namely the slowing down of the diffusion of the faster permeant by a slower one. The so-called Maxwell–Stefan approach is capable of describing such non-ideal phenomena (Wesselingh and Krishna, 1990; Heintz and Stephan, 1994b). This theory assumes that the driving force for a solute i within a multicomponent mixture equals the sum of the frictional resistances between solutes resulting from their relative motion while diffusing through the membrane polymer,

3.6.2.3.2 Concentration Polarization

Feed-side concentration polarization When feed components sorb in the membrane, a local concentration gradient develops in the feed phase adjacent to the membrane upstream face. Owing to this gradient, transport of components from the bulk into the bulk–membrane interface occurs, thus replenishing the components that were absorbed by the membrane. The transport of a solute across the phase adjacent to the membrane can be either convective or diffusive, depending on the solute concentration as well as the fluid dynamic conditions over the membrane surface.

In pervaporation, as the feed fluid is a liquid, a thin, stagnant boundary layer always exists over the membrane surface in which the solute transport is diffusive (Fig. 3.6-11). The thickness of this boundary layer (stagnant liquid film) can be calculated from well-established boundary layer equations (for critical reviews on the use of the most common correlations see, for example, Gekas and Hallström, 1987 and Cussler, 1997). If the flux of a solute i across the concentration boundary layer toward the membrane is lower than the maximum (for the respective solute bulk feed concentration) attainable solute flux across the membrane, then solute i will be depleted in the boundary layer over the membrane upstream surface. As a consequence, the concentration of i in the membrane upstream surface will also be lower (assuming a constant sorption coefficient), the concentration gradient over the membrane will decrease and hence so will the trans-membrane flux.

This phenomenon is denoted "feed-side concentration polarization" and, in practice, affects mainly the fluxes of compounds of high sorption coefficient, even under turbulent hydrodynamic conditions over the membrane, as their permeability (and hence flux across the membrane) is high. It should at this point be emphasized that contrary to the non-ideal transport phenomena discussed earlier, feed-side concentration polarization is not a membrane-intrinsic phenomenon, but stems from poor design of the upstream flow conditions; in practice it may in fact not be overcome owing to module design limitations (Baker et al., 1997).

a)

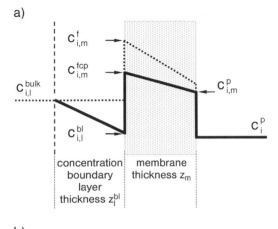

b)

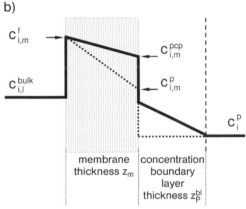

Fig. 3.6-11 The principle of concentration polarization (a) on the membrane upstream side in contact with the feed liquid and (b) on the membrane downstream side facing the vacuum. The straight lines indicating the concentration profiles are strongly oversimplified.

In vapor permeation, feed-side concentration polarization is much less prone to occur than in pervaporation, owing to the high mass-transfer rate of the solute in the vapor feed phase. In fact, this feature is one of the main factors that distinguish *the two processes*.

Permeate-side concentration polarization   Because one aims at employing selective membranes as thin as possible for obtaining high fluxes, most membranes are composites consisting of a thin selective membrane and a macro-porous support for mechanical stability and easier handling. Asymmetric supports are often used that are sufficiently tight on top to allow casting the selective dense film, and highly porous on the downstream side in order to minimize mass-transport resistance during desorption of the solutes permeated. Even so, the narrow pores of the support constitute in practice a resistance to the mass transport (Huang and Feng, 1993; Rautenbach and Helmus, 1994). This resistance results from hindered transport of solutes away from the membrane downstream surface, causing an accumulation of solute in the pores of the support, a local increase of the solute partial pressure and hence a decrease in driving force (see Eq. (5) and Fig. 3.6-12). If the pressure drop in the macroporous support causes the solute partial pressure p_i^P to rise to its saturation vapor pressure, the solute will even condense in the pores.

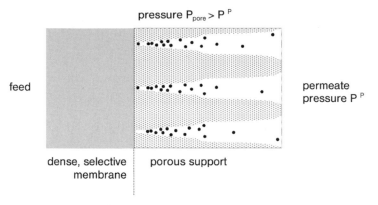

Fig. 3.6-12 Increase of downstream pressure due to accumulation of solute.

An illustrative example for permeate-side concentration polarization is the pervaporation of vanillin, a high-boiling aroma compound. The boiling point of vanillin is about 558 K and its saturation vapor pressure correspondingly low (0.29 Pa at 298.15 K). Using composite membranes, the pressure can quickly rise inside the macroporous support structure during operation until the saturation vapor pressure of vanillin is reached, resulting in crystallization of vanillin within the support. If, however, the feed temperature is maintained slightly above the ambient (as is the case during bioconversions by which vanillin can be produced) and a homogeneous membrane without a porous support is used, vanillin will just desorb from the membrane downstream face and will subsequently cool to ambient temperature within the vacuum duct and crystallize (Böddeker et al., 1997). The elegance of this process is the fact that lower-boiling compounds such as water will pass the vacuum duct uncondensed. The selectivity in this case may be infinite.

3.6.2.4
Technical Aspects of Vapor Permeation/Pervaporation

A typical pervaporation set-up is depicted in Fig. 3.6-13. The liquid feed is re-circulated continuously between the feed reservoir (1) and the pervaporation module (2).

The solutes permeating through the membrane leave the membrane downstream face as a vapor owing to the vacuum which is established initially by the vacuum pump (4). The phase transition of the permeating solutes from the liquid to the vapor state involves heat consumption corresponding to the heat of vaporization of the solutes. This heat is taken up from the environment, namely the bulk feed, which consequently cools. In modules with a large membrane area this causes a temperature drop between the feed and the retentate and has to be compensated.

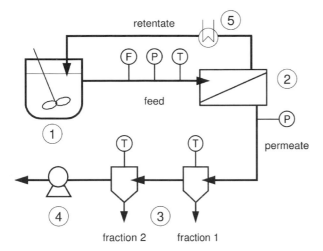

Fig. 3.6-13 A schematic standard pervapora- pump; 5: heat exchanger; F: flow control;
tion set-up. 1: feed tank; 2: pervaporation P: pressure indicator/control; T: temperature
module; 3: condensation unit; 4: vacuum control.

On the industrial scale, heat exchangers (5) are therefore included in the feed circuit. In laboratory units this temperature drop may be neglected, because the membrane area is commonly small in comparison to the recirculation rate of the feed, which is maintained at a controlled temperature. The permeate vapor is recovered in condensers at a temperature that allows a quantitative condensation of the vapor. The condensation can optionally be carried out in a series of condensation stages at different temperatures in order to achieve a permeate fractionation and increased enrichment of target compounds (3).

Similarly to other traditional equipment used in separation processes, the main objectives when designing a vapor permeation or a pervaporation unit are the attainment of the highest possible mass-transfer surface to volume ratio, while maintaining adequate conditions to avoid detrimental mass-transport phenomena. These criteria, together with the need for simple operation and easy maintenance procedures, determine to a great extent the principles for module design.

Industrial-size plate-and-frame modules, for example, consist of a stack of tightly packed membranes over which the feed solution is recirculated (Mulder, 1997). The membranes are separated by spacers and the permeate is withdrawn by a central permeate pipe (Stürken, 1994). Pressure losses occur on both the feed and the permeate side of the packed membranes and need to be accounted for in the module design. On the feed side, the fluid dynamic conditions over the membrane may be less uniform than on the laboratory scale, resulting in more pronounced concentration polarization. On the permeate side, the packed configuration of the membranes may lead to considerable pressure losses, rendering the instantaneous removal of solutes from the membrane downstream surface more difficult. Both aspects may cause solute fluxes lower than expected (Chapter 3.2) and a possible shift in selectivities, resulting from the way each solute is affected by concentration polarization effects at both sides of the membrane.

In principle, nevertheless, scaling-up is quite straightforward in vapor permeation and pervaporation, as happens in other modular processes; but care must be taken when using laboratory data for plant design because it is quite common to perform laboratory-scale studies in conditions close to ideal – optimized feed-fluid dynamic conditions, extremely low downstream pressure, oversized condensers and extremely low condensation temperature – which are not economic for industrial operation.

3.6.2.4.1 Feed-Fluid Dynamic Conditions

As previously discussed, operation under optimal feed-fluid dynamic conditions – meaning operation under conditions that overcome polarization of concentration of a given solute (or solutes) – may not be attainable, even under extremely turbulent mass-transfer conditions. This is particularly true for the recovery/removal of hydrophobic organics from aqueous streams; these compounds exhibit a highly favorable partitioning towards the membrane and therefore concentration polarization is severe even in highly turbulent membrane modules. For these compounds, measured separation factors can be 10% to 20% of the intrinsic separation factors in the absence of concentration polarization (Baker et al., 1997). As a consequence of concentration polarization, the selectivity of the overall process towards a given solute will be due not only to the intrinsic membrane selectivity but also to the selectivity imposed by the external mass-transfer conditions. Under these circumstances it is possible to operate from conditions where the separation process is membrane-controlled to conditions where the control is due to mass transfer at the membrane upstream surface.

The external mass-transfer conditions have to be selected taking into consideration the potential gain in solute flux (and selectivity) and the restrictions imposed by module design and the energy costs involved. Additionally, when processing streams sensitive to shear stress, as may happen when coupling pervaporation with active biological reactors or with coordination complex catalytic reactors, care has to be taken to not affect the (bio)catalysts.

3.6.2.4.2 Downstream Pressure and Condensation Strategy

On the industrial scale, the vacuum pump is shut off once vacuum is established in the downstream circuit and ideally the condensation unit alone maintains the vacuum. This procedure allows operating in a more economical way allowing minimization of energy costs.

Instead of using a vacuum for maintaining the driving force, an inert gas can be swept over the membrane downstream surface to remove the permeate solutes. This mode of operation is gaining renewed interest, especially when hybrid processes such as combined pervaporation–air stripping are used for complementary removal of less-volatile (pervaporation) and more-volatile (air stripping) organic compounds from water streams (Shah et al., 2004). Otherwise, sweep-gas pervaporation is less efficient for maintaining the driving force, and it renders the condensation of the permeate solutes from the sweeping gas stream more difficult.

If non-condensable gases permeate the membrane and enter the downstream compartment, continuous operation of the vacuum pump is required to remove these non-condensable gases, which otherwise would lead to an increased downstream pressure. The flux of non-condensable gases through the membrane requires a higher energy input in order to guarantee a desired downstream pressure and efficient transport of the permeating solutes from the downstream surface of the membrane. This problem is particularly important in cases where a pervaporation unit is coupled to an active fermentation so that off-gases, such as carbon dioxide, can permeate through the membrane in large quantities. On the industrial scale, capture of these gases may prove difficult, unless extremely low temperatures are employed in the condensation unit, raising the overall operating costs. Therefore, these gases may pass the condensation unit unaffected; the consequence can be a strongly decreased condenser efficiency resulting in a considerable loss of the more volatile permeate vapor compounds.

The permeate vapor is typically recovered in condensers at a temperature that allows a quantitative condensation of the vapor. Condensation can optionally be carried out in a series of condensation stages, at different temperatures, in order to achieve a permeate fractionation and increased enrichment of target compounds. Because the condensation efficiency is determined by both the condenser temperature and the residence time of the permeate vapor in the condenser, careful design of the condensation unit becomes crucial. New ways of capturing the permeating vapors have been suggested in the literature but no successful application has been reported. The concept is attractive but the low molecular density of the vapor permeate does not favor its capture by adsorptive and/or absorptive techniques.

The technical aspects outlined above illustrate that for a process optimization *all* process parameters must be considered, since they strongly affect each other. It should also be pointed out that although vapor permeation and pervaporation are membrane separation process, their selectivity and effectiveness may be significantly determined by mass and heat transport phenomena occurring beyond the membrane surfaces.

3.6.2.5
Implementation of Vapor Permeation/Pervaporation in Chemical Processes

Vapor permeation and pervaporation are emerging membrane technologies that can be used for various industrial applications for a predefined task; however, the optimal process design is unlikely to consist solely of them. Often, the optimized solution is a hybrid process combining these membrane techniques with one or more other separation technologies (Lipnizki et al., 1999). This section will focus on vapor permeation/pervaporation-based hybrid processes that have been realized on a commercial scale or that represent a relevant breakthrough with a high potential impact. Particular attention will be devoted to hybrid processes involving vapor permeation/pervaporation combined with evaporation/distillation and with (bio)catalytic reactors.

3.6.2.5.1 Hybrid Processes Involving Evaporation/Distillation

The most successful application of pervaporation on the industrial scale has been the dehydration of ethanol (see also Chapter 3.1). Distillation of ethanol/water mixtures reach an azeotropic point at about 95.6 wt.% of ethanol. The azeotrope can only be broken and the remaining water removed by adding an external agent, able to alter the liquid–vapor equilibrium of the original ethanol–water system. The use of these external agents, usually known as entrainers, involve adding, for example, benzene to the ethanol/water mixture. Entrainers may constitute an additional environmental problem as well as requiring further downstream processing for their subsequent removal from the final product. Using hydrophilic polyvinylalcohol (PVA) membranes, pervaporation was successfully coupled to distillation for water removal, yielding a final purity of ethanol of about 99.95 wt.% without any need to employ entrainers. PVA is a so-called "glassy" polymer, with the polymer chains constituting a rigid network of little flexibility (Mulder, 1997). Both ethanol and water sorb in PVA, although, because of its larger molecular size, the diffusivity of ethanol through the rigid PVA-network is strongly hindered in comparison to water. Because the use of glassy polymers has mainly been restricted so far to the selective removal of water from organic solvents, such as dichloroethylene, isopropyl alcohol and tetrahydrofuran, this separation is widely denoted "hydrophilic" pervaporation. The industrial success of this hybrid process has been quite significant in the past and has led to the use of the technique for the dehydration of several other solvents, as can be seen in Table 3.6-3.

Table 3.6-3 Solvents routinely dehydrated in vapor permeation/pervaporation units

Isopropanol, ethanol: Standard applications for pervaporation, typically dehydrated from their azeotropes to fractions of a percent of water. De-bottlenecking of entrainer plants.

Ethyl acetate, butyl acetate: Form azeotropes in the miscibility gap. Pervaporation or vapor permeation is easily the best technique for dehydration.

Acetone: Does not form an azeotrope with water but requires a large reflux when distilled. Pervaporation is ideal for final dehydration or for de-bottlenecking existing distillation systems.

Acetonitrile: Forms an azeotrope with water; fully miscible with water. Can easily be dehydrated to low water concentrations. Avoids messing with contaminated salt solution and re-distillation of salt-contaminated organic phase.

Pyridine: Forms fully miscible, water-rich azeotrope, easily split by pervaporation or vapor permeation. Final dehydration feasible. Avoids entrainers and messy salt/alkali solutions.

Tetrahydrofuran: Easily dehydrated by pervaporation down to a few hundred ppm of water. No messy chemicals.

Methylethyl ketone: Distillation is only possible with an entrainer because the azeotropic composition is nearly identical to the miscibility limit. Pervaporation is far superior.

n-*Butanol*, n-*propanol:* Form azeotropes with high water content so the distillation/phase separation process involves massive recycle streams. Pervaporation plants are less costly to build and easier to operate.

Information from http://www.sulzerchemtech.com

Vapor permeation was found to be a good option for solvent dehydration in cases of impure feed streams containing non-volatile compounds aggressive to the membranes, or when the feedstock contains dissolved solids that can precipitate at the membrane surface, affecting its performance and lifetime. In those cases the liquid feed is evaporated prior to processing by vapor permeation so that the membranes are not exposed to compounds that may affect their performance. The vapor permeation unit also employs hydrophilic polyvinylalcohol (PVA) membranes. Vapor permeation has been also successfully applied for direct processing of vapor streams from distillation columns. In this way it is not necessary to condense the vapor stream prior to membrane processing.

3.6.2.5.2 Hybrid Processes Involving (Bio)catalytic Reactors

Vapor permeation and pervaporation may also be used in integrated hybrid processes involving active (bio)catalytic reactors. Again, the strategy involves the removal of water during the time-course of the reaction in order to shift the chemical equilibrium towards the formation of target end-products. This approach has been followed during condensation, elimination and esterification reactions, where the continuous removal of one of the end-products (water) permits extension of the conversion of the substrates to the desired product. This approach has been successfully used by different authors using chemical catalysts (Li et al., 2001) as well as biocatalysts (Gubicza et al., 2003; Iz k et al., 2004). In this case the membranes are also hydrophilic.

This methodology can be extended to the recovery/removal of organics from aqueous streams during the time-course of the reaction. The pervaporative removal of organic compounds from aqueous solutions, also called "organophilic" pervaporation, requires the use of a membrane with a more flexible polymer structure. Almost all organic compounds are of a larger molecular size than water and their diffusivity is hindered in glassy structures. Membranes used for the separation of organic compounds from water are therefore of a "rubbery" nature with a more flexible polymeric network. Evidently, an easier diffusivity for the organic compounds goes along with that of water, resulting in a considerable loss of membrane selectivity (Section 2.2.3). Therefore, special attention is given to modifying rubbery polymers so that strong solute–polymer interactions are promoted in order to greatly enhance the sorption of the organic solute in the membrane polymer in comparison to the sorption of water. Elastomeric polymers used in organophilic pervaporation include polydimethylsiloxane (PDMS) and its derivatives, such as polyoctylmethylsiloxane (POMS), as well as membranes using polyether block amides (PEBA).

Glassy and rubbery membranes differ strongly with regard to the flexibility of their polymeric structures. Within the more rigid polymeric network of a glassy membrane the diffusivity of components will be strongly related to their molecular volume, whereas in the more flexible structure of an elastomeric membrane it is less the diffusivity that differs strongly for different components but far more the solute–polymer interaction (sorption). This is why the selectivity of glassy membranes is more diffusion controlled, whilst the selectivity of rubbery membranes is more sorption controlled (Section 2.2.3). Since the selectivity in rubbery membranes is essentially sorption controlled, a significant effort has been made to produce them as thin rubbery films supported in porous supports in order to guarantee the necessary mechanical strength. Thinner films would then reduce the antagonist role played by diffusion through the membrane, which favours transport of the smaller solvent molecules (water).

A significant number of applications have been suggested for the recovery of organics during the time-course of (bio)catalytic reactions or for product recovery from post-reaction media. Common examples include the recovery of end-products from condensation, elimination and esterification post-reaction media and recovery of biological products from active bio-reactors. As pervaporation can be operated continuously at low temperature, does not require any extraction aid and does not exert high stress on the active biocatalyst (Schäfer and Crespo, 2002), it has been extensively studied on the laboratory-scale for the recovery of biological products, namely flavor compounds from the culture media (Baudot and Marin, 1997; Karlsson and Trägård, 1993). In contrast to pressure-driven membrane processes, membrane fouling is a minor problem in pervaporation because the membranes used are non-porous. Organophilic pervaporation linked to a bioconversion process has until now solely been studied for the recovery of individual aroma compounds or inhibiting metabolic products (Böddeker, 1994; Lamer et al., 1996; Rajagopalan et al., 1994). Most research on aroma recovery by organophilic pervaporation has been conducted using aqueous aroma model solutions (Börjesson et al., 1996; Baudot and Marin, 1997) with the emphasis being on the engineering aspects of pervaporation. An evaluation of the actual organoleptic value of an aroma concentrate obtained by pervaporation has been reported (Schäfer et al., 1999).

3.6.2.5.3 Other Relevant Separation Applications

A large variety of applications using either vapor permeation or pervaporation has been reported. These include the use of pervaporation for the removal of toxic organics from water (Schnabel et al., 1998) and wastewater streams (Moulin et al., 2002), sometimes using hybrid approaches with adsorptive techniques; the use of pervaporation membranes in direct methanol fuel cells (Pivovar et al., 1999); and, more recently, the resolution of isomeric mixtures (Kusumocahyo et al., 2004) and membrane-assisted enantiomer enrichment (Paris et al., 2004), in both cases using membranes containing specific complexation agents such as cyclodextrins.

One of the most active areas of research involves the use of pervaporation for organic/organic separations. The industrial interest is extremely high and several authors have reported the use of pervaporation for separation of aromatic/aliphatic hydrocarbon mixtures at high temperature (Matsui and Paul, 2003), as well as the separation of binary organic mixtures involving methyl *tert*-butyl ether (MTBE) and methanol (Yoshida and Cohen, 2003). As an example, Sulzerchemtech has already reported the possibility for industrial recovery of methanol and ethanol from organic mixtures.

3.6.2.5.4 Analytical Applications

Vapor permeation and pervaporation have been also proposed as pre-enrichment steps in the preparation of samples for analytical processing (Mishima and Nakagawa, 2004). This area of application is attracting significant interest because, owing to the high selectivity of vapor permeation/pervaporation for organic volatile compounds when using hydrophobic membranes, it is possible to selectively enrich a given sample of the target compounds prior to analysis. This approach may be applied to samples collected from the headspace of reactors or other vapor streams, and also to samples derived from liquid media. If correctly designed (see previous sections) it is possible to adjust the enrichment degree according to the analytical requirements. This technique may be applied together with gas chromatography, mass spectrometry and electronic sensing systems such as the so-called "electronic noses" (Pinheiro et al., 2002).

3.6.2.6
Perspectives

The principles of vapor permeation/pervaporation and their advantages for the recovery of volatile compounds have been presented and discussed. It has been shown that these versatile processes should be understood as a whole in order to be adapted to individual separation problems.

While vapor permeation and hydrophilic pervaporation have readily found well-established areas for industrial application, organophilic pervaporation has been struggling for the past years to surmount the difficult step from a "highly interesting laboratory technique" to an "industrially adopted technology". The reason for these difficulties can be found in the intrinsic character of these processes: (1) – in vapor permeation the feed stream is a vapor that can be compressed, which allows an increase in the volumetric concentration of a target permeant solute and, hence, the driving force for transport; additionally, the external mass-transfer conditions can be easily optimized in the feed vapor phase, overcoming possible concentration polarization problems; (2) – in hydrophilic pervaporation, concentration polarization of the target permeant (water) is not a problem because transport is mainly regulated by diffusion, and not by selective sorption to the membrane; (3) – organophilic pervaporation suffers from different problems, concentration polarization being one of the most important, as previously discussed.

In spite of the improvement in module design and the development of new approaches for better mass-transfer conditions – for example the recent work on the use of Dean vortices or the assessment of full-scale vibrating pervaporation membrane units – organophilic pervaporation has not succeed in finding a relevant industrial application. This situation may change dramatically if new hybrid processes are able to bring clear added value, allowing for better design of traditional processes. In particular, integrated catalytic reactors with in situ product recovery by pervaporation is an exciting area of research, where new opportunities may arrive from the development of novel tailor-made membranes incorporating complexation/catalytic agents such as zeolites and enantioselectors (e.g. enantioselective ionic liquids).

Symbols

c Mass concentration (kg m^{-3})
D Diffusion coefficient or Diffusivity (m^2 s^{-1})
H Henry coefficient (Pa)
J Flux (m s^{-1})
L permeability (m^2 s^{-1})
p Partial pressure (Pa)
P Pressure (Pa)
S Sorption coefficient (–)
x Mole fraction (mol mol^{-1})
x Liquid phase mole fraction (–)
y Vapor phase mole fraction (–)
z thickness (m)

Greek letters

α Selectivity (separation factor) (–)
β Enrichment (–)
γ Activity coefficient (–)

Indices

bl   Boundary layer
cp   Concentration polarization
f   Feed
i Component i
j Component j
m   Membrane
mol   molar
p   Permeate
S   Saturation

References

Baker, R.W., Wijmans, J.G., Athayde, A.L., Daniels, R., Ly, J.H., Le, M. *J. Membrane Sci.* **1997**, 137, 159.

Baudot, A., Marin, M. *Trans IChemE* **1997**, 75C, 117.

Böddeker, K.W. in: *Membrane Processes In Separation And Purification*, J.G. Crespo, K.W. Böddeker, eds., Kluwer, Dordrecht, **1994**.

Böddeker, K.W., Gatfield, I.L., Jähnig, J., Schorm, C. *J. Membrane Sci.* **1997**, 137, 155.

Börjesson, J., Karlsson, H.O.E., Trägårdh, G. *J. Membrane Sci.* **1996**, 119, 229.

Cussler, E.L. *DiffusionMass Transfer in Fluid Systems*, Cambridge University Press, Cambridge, 2nd Edition, **1997**.

Flory, P.J. *Principles of Polymer Chemistry*, Cornell University Press, Ithaca, NJ, **1953**.

Gekas, V., Hallström, B. *J. Membrane Sci.*, **1987**, 30, 153.

Gubicza, L., Nemestothy, N., Frater, T., Belafi-Bako, K. *Green Chem.* **2003**, 5(2), 236.

Heintz, A., Stephan, W. *J. Membrane Sci.*, **1994**, 89, 143.

Heintz, A., Stephan, W. *J. Membrane Sci.*, **1994**, 89, 153.

Hildebrand, J.H., Scott, R.L. *The Solubility of Non-Electrolytes*, Plenum Press, New York, **1949**.

Huang, R.Y.M., Feng, X. *J. Membrane Sci.*, **1993**, 84, 15.

Huang, R.Y.M., Rhim, J.W. in: *Pervaporation Membrane Separation Processes*, R.Y.M. Huang, ed., Elsevier, Amsterdam, **1991**.

Izák, P., Mateus, N.M.M., Afonso, C. A. M., Crespo, J. G. *Sep. Purif. Technol.* **2005**, 41 (2), 141.

Jonquières, A., Perrin, L. Arnold, S. and Lochon, P. *J. Membrane Sci.*, **1998**, 150 (1), 125.

Karlsson, H.O.E., Träghård, G. *J. Membrane Sci.* **1993**, 76, 121.

Kusumocahyo, S.P., Kanamori, T., Sumaru, K., Iwatsubo, T., Shinbo, T. *J. Membrane Sci.* **2004**, 231(1–2), 127.

Lamer, T., Rohart, M.S., Voilley, A., Baussart, H. *J. Membrane Sci.* **1994**, 90(3), 251.

Lamer, T., Spinnler, H.E., Souchon, I., Voilley, A. *Process Biochem.* **1996**, 31(6), 533.

Li, X.H., Wang, L.F. *J. Membrane Sci.* **2001**, 186(1), 19.

Lipnizki, F., Field, R.W., Ten, P.K. *J. Membrane Sci.* **1999**, 153 (2), 183.

Matsui, S., Paul, D.R. *J. Membrane Sci.* **2003**, 213(1–2), 67.

Mishima, S., Nakagawa, T. *J. Membrane Sci.* **2004**, 228(1), 1.

Moulin, P., Allouane, T., Latapie, L., Raufast, C., Charbit, F. *J. Membrane Sci.* **2002**, 197(1–2), 103.

Mulder, M. *Basic Principles of Membrane Technology*, Kluwer Academic, Dordrecht, **1997**.

Mulder, M.H.V. in: *Pervaporation Membrane Separation Processes*, R.Y.M. Huang, ed., Elsevier, Amsterdam, **1991**.

Paris, J., Molina-Jouve, C., Nuel, D., Moulin, P., Charbit, F. *J. Membrane Sci.* **2004**, 237(1–2), 9.

Pinheiro, C., Rodrigues, C.M., Schäfer, T., Crespo, J.G. *Biotechnol. Bioeng.* **2002**, 77(6), 632.

Pivovar, B.S., Wang, Y.X., Cussler, E.L. *J. Membrane Sci.* **1999**, 154(2), 155.

Rajagopalan, N., Cheryan, M., Matsuura, T. *Biotechnol. Techniques* **1994**, 8(12), 869.

Rautenbach, R., Helmus, F.P. *J. Membrane Sci.* **1994**, 87, 171.

Schäfer, T., Bengtson, G., Pingel, H., Böddeker, K.W., Crespo, J.P.S.G. *Biotechnol. Bioeng.* **1999**, 62(4), 412.

Schäfer, T., Crespo, J.G. in: *Transport Phenomena in Food Processing*, G. Barbosa-C novas, J. Vélez-Ruíz, J. Welti-Chanes, eds., CRC Press, Boca Raton, FL, **2002**.

Schäfer, T. PhD Thesis, Universidade Nova de Lisboa, Portugal, **2002**.

Schnabel, S., Moulin, P., Nguyen, Q.T., Roizard, D., Aptel P. *J. Membrane Sci.* **1998**, 142(1), 129.

Shah, M.R., Noble, R.D. *J. Membrane Sci.* **2004**, 241(2), 257.

Stürken, K. PhD Thesis, GKSS Forschungszentrum Geesthacht GmbH, Germany, **1994**.

Wesselingh, J.A., Krishna, R. *Mass Transfer*, Ellis Horwood, New York, **1990**.

Yoshida, W., Cohen, Y. *J. Membrane Sci.* **2003**, 213(1–2), 145.

3.7
Nanostructures in Separation

3.7.1
Functionalized Magnetic Particles

Costas Tsouris, Jeremy Noonan, Tung-yu Ying, Ching-Ju Chin, and Sotira Yiacoumi

3.7.1.1
Introduction

What makes a separation process "green," or environmentally friendly? The first characteristics that come to mind are low energy requirements and minimal waste production. Imagine a separation process that relies on a force generated naturally from the materials involved in the separation itself and that is so selective for a particular target molecule that it requires only small amounts of adsorbent materials, which are relatively simple to recover and reuse. The potential to realize such "green" ideals is found in functionalized magnetic particles.

An ideal functionalized magnetic particle is a nano- or micron-sized solid that consists of a magnetic core embedded in a coat or matrix onto which surface-reactive functional groups are attached (see Fig. 3.7-1). Separations using such particles consist of three main stages. The first stage is the separation of the target substance from the fluid (liquid or gas) medium by selective adsorption onto the surface of the particle. The second stage is the separation of the target-containing particle by deposition onto the surface of a magnetic filter. The third stage is the regeneration of the particles. Indeed, the novelty and advantage of the concept of a functionalized magnetic particle lie in its combination of these steps into a single separation technology. Particles that are functionalized for a specific compound but are not magnetic may succeed in the first step. However, the particles themselves are difficult to remove. Particles that are magnetic but not functionalized are easily recovered, but such particles have limited ability to remove selectively another target compound. The third step is equally important. If the particles cannot be regenerated, they cannot be reused and the separated compound cannot be recovered. Thus, all three dimensions must be considered when evaluating the performance of a magnetic particle: its ability to separate, to be separated, and to be regenerated.

Green Separation Processes. Edited by C. A. M. Afonso and J. G. Crespo
Copyright © 2005 WILEY-VCH Verlag GmbH & Co. KGaA, Weinheim
ISBN 3-527-30985-3

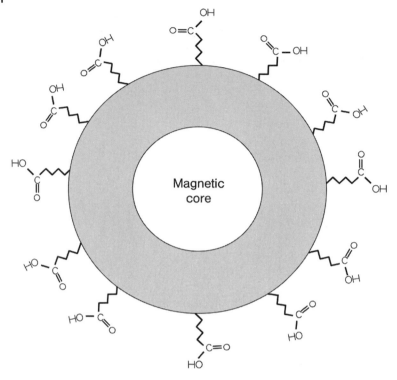

Fig. 3.7-1 General schematic of a functionalized magnetic particle with carboxylic acid functional groups.

The ability of magnetic particles to be separated depends on how they respond to
magnetic filtration. The ability of magnetic particles to remove target compounds
depends on at least two basic factors: the particle must have sufficient adsorption
capacity (large interfacial area) and a specific affinity for the target molecule (selectivity). The ability of the particle to be regenerated depends on the strength of the
chemical or physical bond between the surface chemical group and the target molecule. When evaluating the usefulness of a magnetic particle for achieving a separation goal, one must consider these criteria. Together, these factors provide the ra-
tionale for the structure and important properties of functionalized magnetic particles.

Can the particle be separated by magnetic filtration?

The properties that determine the answer to this question are the magnetic susceptibility and size of the particle. Magnetic susceptibility, as the name implies,
describes the impact of a magnetic field on the magnetization of a material. This
property is the primary factor that governs the ease of separation and is reported
in dimensionless units (as the volume magnetic susceptibility) or in SI units of
$m^3 \, kg^{-1}$ (as the specific magnetic susceptibility). The basic principle of removal is
this: the more highly magnetic the material, the weaker is the induced external
magnetic field necessary for removal.

The material composing the core of the particle gives the particle the magnetic properties that make it susceptible to removal by magnetic filtration. In general, materials can be classified as strongly magnetic, weakly magnetic, or nonmagnetic, depending on the degree of their magnetic susceptibility. A class of materials called ferrites (often classified in the literature as ferrimagnetic) contains the most common substances used for particle cores. [1]Strongly magnetic, ferrites are distinguished from other magnetic materials by the order of their atomic moments, which are antiparallel but unbalanced, that is, the magnitude of the moments in one direction is greater than that in the opposite direction. These materials can be recovered easily by magnetic separation using a weak magnetic field strength of up to 0.15 T (Svoboda, 1987). Among the ferrites, magnetite is the material of choice because it is chemically stable, inexpensive, relatively easy to produce in the laboratory, and has a very high magnetic susceptibility; a small fraction of magnetite (less than 1% iron by weight; Leun and Sengupta, 2000) in a functionalized particle is enough to make the particle easily recoverable. Weakly magnetic materials, which are typically paramagnetic, include some elemental metals (e.g. aluminum, chromium, and titanium) as well as some iron oxides (e.g. hematite). The removal of these materials requires a stronger magnetic field of up to 0.8 T (Svoboda, 1987). Nonmagnetic particles, which cannot be removed by magnetic separation because they have a negative magnetic susceptibility, are called diamagnetic. These include some of the heavy metals (e.g. copper, lead, zinc, and arsenic), which are common targets for separation by functionalized particles.

The requirement of such a weak magnetic field for the separation of magnetite may be satisfied by permanent magnets, such as rare earth magnets, which are naturally capable of generating a magnetic field of up to 0.5 T. Because these magnets do not depend on any external energy source, the energy demand for this process is very low, and is determined mainly by the energy needed for pump operations. When the field is removed, the particles may be recovered easily from the filter and regenerated, creating a minimal amount of waste. Another advantage of magnetic filtration as compared with other filtration processes is that magnetic filters, in general, have a higher capacity for colloidal particles before breakthrough can occur.

Particle size is another factor affecting the recoverability of a particle by magnetic filtration. The ability of a magnetic separator to capture a particle is governed by the interaction of various forces on the particle. For a particle to be captured by a wire in a magnetic filter, the attractive magnetic force between the wire and particle must exceed the forces driving the particle away from the wire – that is, the hydrodynamic, gravitational, inertial, and diffusional forces. The relative impact of these forces varies with particle size. As particle size decreases from microscale to nanoscale, the effect of hydrodynamic and diffusional forces on the movement of particles through a magnetic filter may become great enough to compete with the magnetic force, thus decreasing the removal efficiency of the filter (Ying et al., 2000). Thus, while the removal of micron-size magnetic particles by magnetic filtration has proven consistently successful, the removal of true nanoparticles is more challenging and represents a frontier for future research.

Magnetic seeding

In many magnetic filtration applications, bare magnetite particles are used as magnetic seeds, aggregating with nonmagnetic or weakly magnetic materials that could not otherwise be removed by magnetic filtration. This magnetic seeding process has been demonstrated to remove contaminants from wastewater efficiently. Selectivity in the magnetic seeding process can be achieved if the nonmagnetic particles that aggregate with magnetite are engineered functionalized particles designed to selectively adsorb dissolved molecules in the solution. Particles are designated "functionalized" only when they are specifically tailored to select target substances for separations. This functionalization is conferred by the coat and surface functional groups added to the particle. In summary, selectivity in a magnetic separation process can be achieved directly by functionalized magnetic particles or indirectly by functionalized nonmagnetic particles used in a magnetic seeding process. The drawback of the latter approach is that regeneration will require an additional step; therefore, functionalizing magnetic particles is a better approach.

Does the particle have sufficient loading capacity for the target substance?

This performance factor is related to the particle coat. Polymers such as polysaccharides, acrylics, and copolymers are most commonly used as the material for the coat (Leun and Sengupta, 2000). In biomolecular applications, phospholipid coats are also used (Bucak et al., 2003). Table 3.7-1 lists the kinds of polymers that are employed, along with some of their pertinent properties. The particle coat or matrix enhances the separation capabilities of a magnetic particle by providing a surface to support the reactive functional groups that give the particle its specific sorption affinity and by stabilizing the particles so that they do not aggregate, which would limit the available surface area for adsorption, thus reducing the removal capacity of the particles. Polymers and surfactants stabilize the particles by creating repulsive surface forces, such as steric or electrostatic forces.

Table 3.7-1 Some common surface functional groups with their targets.

Name	Molecular Structure	Affinity
Carboxylic Acid		Heavy metals
Bis(trimethylpentyl)- phosphinic Acid		Heavy metals
Polypropylene Oxide (PPO)		Synthetic organics
Copper Phthalocyanine		Polycyclic organic dyes

The central property relevant to the loading capacity of a particle is its exposed surface area, which is a function of its size, colloidal stability, and porosity. In general, the smaller the particle radius, the greater is its surface area relative to its volume (assuming that the shape is essentially spherical). For example, in one study a three-orders-of-magnitude increase in the ratio of surface area to volume corresponded to a reduction in particle size from 10 µm to 10 nm (Moeser et al., 2002). Therefore, size is an important property affecting the loading capacity of a particle. In addition, smaller particles can be easily fluidized, thus increasing mass-transfer rates. The challenge is to have particles that are sufficiently small to take advantage of the high surface area and mass-transfer rates but large enough to be easily separated by magnetic filtration.

The effect of particle size on interfacial surface area explains why colloidal stability matters. Unstable particles are prone to aggregation. While particle aggregation is necessary to achieve nonselective separation of contaminants in conventional water treatment, it is undesirable for selective separation by magnetic particles. When a particle aggregates with another particle, its effective size increases and the portion of its surface area in contact with the second particle is no longer available for adsorption. Therefore, in general, particle aggregation limits the loading capacity by decreasing the surface area. As described above, however, polymeric coats may be crafted to ensure particle stability.

Porosity is a third property affecting particle surface area. A porous particle provides more locations for adsorption because it exposes additional surface area inside the particle. A particle can be made more or less porous by altering the structure of its polymer matrix. Though porous materials may increase sorption capacity, they do so at the expense of sorption kinetics. For target compounds to adsorb inside these pores, they must diffuse along the surface of the particle. This diffusion step represents another layer of mass-transfer resistance, which slows the rate of adsorption. Therefore, a trade-off exists between surface area and the rate of removal. However, porosity is a significant factor only for larger magnetic particles (microscale). Using a nonporous nanoparticle, one can obtain a similar ratio of surface area to volume and at the same time achieve rapid sorption kinetics (Moeser et al., 2002).

The large surface area-to-mass ratio for adsorption provided by nanoparticles means that, overall, less material is necessary to achieve adequate separation. Handling reduced quantities of materials leads to less energy and waste. Thus, the small size of magnetic particles is one factor that makes this a green technology.

Will the particle selectively remove the target substance?

The surface-reactive groups confer the quality of specificity to the particle, which makes it truly functionalized. Highly selective separation is achieved by the adsorption of target molecules onto these groups or by ion exchange. A wide variety of functional groups exists, categorized by their affinity for a particular target substance. Examples of functionalized groups reported in the literature are shown in Table 3.7-1. Among the functional groups shown to be selective for heavy metals are carboxylic groups, which work by ion exchange (Phanapavudhikul et al., 2003), and chelating materials such as bis(trimethylpentyl) phosphinic acid (Kaminski and

Nunez, 1999). Surfactants and lipids are effective for the separation of biomolecules such as proteins (Bucak et al., 2003), while hydrophobic polypropylene oxide side chains have a high affinity for synthetic organic compounds (Moeser et al., 2002). The selectivity of copper phthalocyanine dye for organic compounds, specifically polycyclic organic dyes, has also been demonstrated (Safarik, 1994).

Whether these surface groups can achieve effective separation depends on their affinity for the target. The sorption properties of a functional group and the solution characteristics regulate a particle's ability to take up a specific substance selectively. Sorption properties relevant to specific affinity include complexation reactions and surface-charge interactions. For adsorption to take place by attractive electrostatic forces, a functional group must have a charge opposite to that of the target solute. Otherwise, a repulsive electrostatic force will form between the particle surface and the solute. A functional group and the solute must also have similar affinities with water. For example, hydrophobic functional groups are necessary for the adsorption of synthetic organic compounds (Moeser et al., 2002). Ionic strength is an important solution characteristic affecting the selective removal of a solute (Bucak et al., 2003; Phanapavudhikul et al., 2003). If a functional surface group has a negative charge, raising the ionic strength of a solution will increase the concentration of positively charged ions surrounding the particle. These cations may shield the particle from the electrostatic attractive force between it and the positively charged solute, thereby limiting adsorption (Bucak et al. 2003).

The high degree of selectivity is another factor that gives rise to the environmental soundness of this process. To achieve the same quality of removal of a specific compound with nonselective adsorbents requires much more material because nontarget compounds will compete with target compounds for adsorbent sites. The result is that much of the adsorbent material is not used to achieve the intended separation. These multiloaded adsorbents are difficult to regenerate because they might require multiple steps to remove the different adsorbed materials. In contrast, highly selective functionalized particles will effectively exclude molecules that are not of interest. Therefore, the process calls for less-adsorbent material, which is then simpler to regenerate because it is loaded with only one kind of molecule.

3.7.1.2
Examples of Functionalized Magnetic Particles in Separations

Additional examples of functionalized magnetic materials described in the literature for specific applications are summarized in this section. Magnetic microparticles with embedded silicotitanate have shown excellent adsorption ability for radioactive cesium isotope (^{137}Cs) from HEDPA (1-hydroxyethane-1,1-diphosphonic acid), although dissolution of magnetite in acidic solutions in the regeneration step leads to a decrease in magnetic susceptibility (Kaminski and Nunez, 2002).

Mesoporous silica has become important owing to its large surface area for adsorption and its uniform nanopores for nanosynthesis. It can also be functionalized to enhance its selectivity and reactivity in adsorption or separation applications.

Mesoporous silica functionalized with thiol layers, for instance, showed specific adsorption for heavy metal ions (Feng et al., 1997). Wu et al. (2004a) incorporated magnetite into mesoporous silica by coating silica on magnetite nanoparticles using sol–gel synthesis. A full coverage of magnetite by a silica shell of only 100 nm was achieved and confirmed by two findings. First, the zeta potential of the composite particles was very similar to that of natural silica. Second, no leakage of iron occurred when the composite particles were placed in an acid solution. The surface area of the composite particles was also higher than that of pure magnetite particles (Wu et al., 2004a; Oliveira et al., 2002, 2003). Composite magnetic particles with activated carbon and clay were also synthesized by co-precipitation of ferrous and ferrite salts in an alkaline suspension of activated carbon or clay (Oliveira et al., 2002, 2003).

It is not necessary that composite magnetic particles be coated with or embedded in large-surface-area materials. Such particles may be complex metal oxides, which themselves are magnetic. Wu and coworkers (2004b) used $CuFe_2O_4$ to remove azo-dye Acid Red B (ARB). They determined that $CuFe_2O_4$ particles not only adsorbed ARB but also served as a catalyst when ARB was completely decomposed to SO_2, CO_2, H_2O, and nitrate by combustion. Materials such as Ni/polystyrene/TiO_2 magnetic particles also possess both magnetic and catalytic characteristics.

Besides chemical separations, functionalized magnetic particles have also been developed for biomedical applications such as cell separations, drug delivery, and immunomagnetic array. Composite particles for biomedical applications usually consist of a magnetite core with a polymer coat or nanoscale iron oxide particles distributed in a polymer matrix such as polyvinyl alcohol, dextran, or even silica. Noble metal coatings are also being considered (Carpenter, 2001). Many functionalized magnetic particles are composite particles because the surface functional groups such as carboxyl groups, protein, biotin, DNA, avidin, and other molecules on the coat or matrix can be easily modified for specific adsorption (Koneracka et al., 1999, 2002). Commercially available microscale composite particles are coated with hydrophobic styrene–divinylbenzene; however, adsorption of DNA may lead to false results for polymerase chain reactions (PCR) (Španová et al., 2003). Magnetic hydrophilic polymer microspheres of 1.7-µm diameter were thus designed to have low specific adsorption of biologically active compounds (Španová et al., 2004). These hydrophilic magnetic particles, consisting of 200-nm magnetite particles encapsulated in methacrylate-based polymers, were prepared by coating oleic acid on magnetite nanoparticles to prevent aggregation of poly(2-hydroxyethyl methacrylate-*co*-ethylene dimethacrylate) and successfully showed no interference for PCR.

The magnetic characteristics are influenced by the size and composition of composite particles. For collagen-coated magnetic particles, the magnetic susceptibility was found to decrease with increasing particle size (Ali-Zade, 2004). Composite magnetic particles with activated carbon and clay from Oliveira's work (Oliveira et al., 2002, 2003) showed that as the ratio of the nonmagnetic component increases, the specific magnetic polarization decreases. The polarization dropped from 62 to

$21\ J\ T^{-1}\ kg^{-1}$ when pure iron oxide particles were mixed with the same amount of activated carbon.

Kohler et al. (2004) modified magnetite nanoparticles with a bifunctional self-assembled trifluoroethylester (TFEE) polyethylene glycol (PEG) saline, which can conjugate with amine and carboxylic groups present in many targets. Magnetoliposomes, which are magnetite nanoparticles coated with phospholipids, easily interact with target chemicals without complicated chemical manipulations (Cuyper et al., 2003). Besides being coated onto the surfaces of magnetite nanoparticles, organic molecules can also directly link with magnetite nanoparticles during synthesis. Mornet et al. (2000) synthesized magnetite nanoparticles in solutions in the presence of DNA. DNA and the Fe^{2+} and Fe^{3+} cations formed a complex salt, which formed DNA–magnetite composite nanoparticles once an alkaline solution was introduced.

3.7.1.3
Theory of Magnetic Separations

Besides particle regeneration, selective adsorption of a target molecule by functionalized magnetic particles and magnetic filtration of the particles are two aspects of separation that must be taken into consideration when designing a separation process using magnetic particles. Having the means to predict with reasonable accuracy the ability of a particle to separate and be separated would enable one to functionalize the particles with the right coat and surface groups for a particular target and to construct the best magnetic filtration system for removing the particles. Some useful models have been developed to make these predictions and aid in the design process.

3.7.1.4
High-gradient Magnetic Separation Modeling

The goal of magnetic filtration design is to optimize filter performance. Filter performance is evaluated by such factors as removal efficiency, operating costs, and throughput. Expressed in terms of a simple ratio of feed-to-effluent concentrations, removal efficiency measures how well the filter separates the target substances from the feed. A filter is performing well if the removal efficiency is high enough that the effluent concentration remains below an acceptable limit. Although it is the most important performance factor, removal efficiency is not the only consideration. A filter might have a superb removal efficiency but be very costly to operate. Over time, the filter will exceed its removal capacity, and the particles will begin to break through into the filtrate at concentrations that exceed the established threshold. At this point, the filter must be regenerated in order to be useful again, which requires time and energy. A filter that requires frequent regeneration may prove too expensive. Certain kinds of metals may be excellent magnets but be prone to rust over time. Thus, replacement costs must be considered. In the case of high-gradient magnetic separation (HGMS), one must also consider the energy requirements to run electromagnets. With a sufficiently strong magnetic field, a magnetic filter can efficiently remove almost anything. However, the energy needed to generate this field may be too expensive. Industrial applications require filters that can handle large quantities of wastewater. In such cases, throughput – the volume of feed that can pass through the filter per unit time – is another important aspect of filter performance.

Engineers must consider these factors when choosing the design parameters for a magnetic filtration system. Generally speaking, there are two kinds of parameters: filter and feed. Filter parameters include its dimensions, packing material and density (kind and amount), and magnetic field strength. Fluid parameters include the size and magnetic susceptibility of the suspended particles and the flow rate. These are parameters that can be easily manipulated or measured.

The mechanisms of particle capture in a magnetic filter and the conditions under which the capture may be achieved have been studied by many researchers. In general, existing magnetic separation models can be divided into two categories: trajectory analysis and build-up models. Trajectory analysis is used to study the interaction between a single particle and a single matrix element (e.g. a single cylindrical wire). By solving a force-balance equation including external forces, such as magnetic and drag forces, and interparticle forces, such as van der Waals and inertial forces, the trajectory of a paramagnetic particle as it approaches a matrix element can be obtained (Fig. 3.7-2). The derivations of the forces used in the force balance equation are given by Watson (1973), Lawson et al. (1977), and Schewe et al. (1980).

One important objective of trajectory analysis is to determine the limiting trajectory and thus the critical radius of particle capture. As shown in Fig. 3.7-2, the limiting trajectory is defined as the exact path that divides the particle trajectories into those leading to capture by the collector and those passing by the collector. The distance between the limiting trajectory and the axis is then defined as the critical radius (R_c), which can be used to determine the removal efficiency of magnetic filtration (Watson, 1973):

$$\text{Removal efficiency (RE)} = 1 - C_{\text{out}}/C_{\text{in}} = 1 - exp\left[\frac{-4(1-\varepsilon)LR_c}{3\pi a}\right] \tag{1}$$

where L is the length of a filter matrix that is packed randomly with ferromagnetic wire at a packing fraction of ε; C_{out} and C_{in} are the effluent and influent concentrations, respectively; and a is the wire radius. Substitution of the term $4(1-\varepsilon)R_c/3\pi a$ by the parameter λ, leads to the following macroscopic equation:

$$C_{out} = C_{in} exp(-\lambda L), \tag{2}$$

where λ is the filter coefficient that has been used extensively to describe filtration processes.

The trajectory model is limited to a clean wire. Once particles are deposited on the surface of matrix elements or onto other particles already deposited, changes in the surface characteristics of the collector will affect subsequent capture of par-

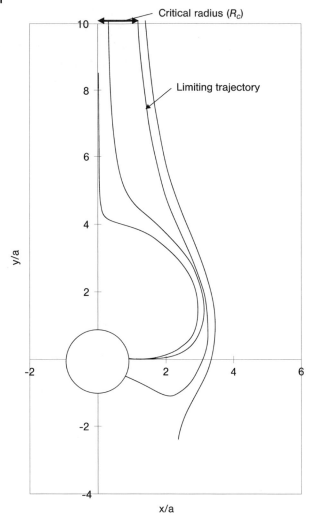

Fig. 3.7-2 Limiting trajectory for small paramagnetic particles captured by a wire; *a* is the radius of the collector.

ticles. The build-up model is therefore used to investigate the dynamic behavior of particle loading on the surface of a matrix element and thus determine the loading capacity of the ferromagnetic matrix. An example of the calculation of the build-up model is shown in Fig. 3.7-3 (Ying et al., 2000). For a particle to remain attached on the surface of a collector, the net force in the radial direction should point toward the center of the wire and the net force in the tangential direction should be opposite to the direction of the drag force. The area satisfying both conditions (the area bordered by the bold line in Fig. 3.7-3) is defined as the particle build-up or loading area.

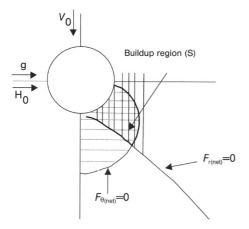

V_0

Buildup region (S)

g

H_0

$F_{r(net)}=0$

$F_{\theta(net)}=0$

Fig. 3.7-3 Loci of zero net radial and tangential forces around one quadrant of a wire: wire diameter = 62.5 μm; H_0 = 0.2 T; b = 0.5 μm; Q = 22.5 ml min^{-1}; χ_p = 0.006.

Luborsky and Drummond (1976) developed a build-up model to describe the build-up configuration and fully loaded conditions of a collector. They formulated the hydrodynamic force by considering the boundary-layer thickness of a wire and calculated the critical angle for each layer of particles. They also obtained the loading volume of particles retained by each wire. Following the work by Luborsky and Drummond (1976), Nesset and Finch (1978, 1981) calculated the boundary-layer thickness by employing the Blasius solution. A good prediction was found at high Reynolds number under upstream capture.

In general, the approaches of the trajectory analysis and the build-up models are based on the force-balance equation. Among the forces (external and interparticle) involved in the system, magnetic and hydrodynamic forces, the most significant ones, also compete with one another. The performance of magnetic separation is, therefore, examined in terms of magnetic velocity (V_m) and superficial velocity (V_0) (Watson, 1973),

$$Separation\ performance \propto \frac{V_m}{V_0} \tag{3}$$

where

$$V_m = \frac{2b^2\mu_0\Delta\chi MH_0}{9\eta a}, \tag{4}$$

where b is particle radius, μ_0 (= $4\pi \times 10^{-7}$ H m^{-1}) is the permeability of free space, $\Delta\chi$ is the difference of the volume magnetic susceptibilities between the particle and background carrier (e.g. water), H_0 is the applied magnetic field strength, M is the magnetization value of the matrix material, and η is the fluid dynamic viscosity. The size and magnetic susceptibility of target particles are system parameters, while the applied magnetic field, properties of matrix material, and superficial velocity are operating parameters.

In most studies in which particles are in submicron and micron size range, a high separation performance could be achieved by controlling the strength of the applied magnetic field and the superficial velocity. However, in the separation of magnetic nanoparticles, this approach may not be effective. For example, under similar operating conditions, the magnetic velocity of a 10-nm particle is 100 times less than that of a 0.1-μm particle. In order to achieve the same performance, a very strong magnetic field and a slow superficial velocity are required, which may, however, not be practical. The performance of magnetic separation is, therefore, limited by the particle size. To overcome the limitation of particle size, magnetic-seeding flocculation has been investigated by many researchers including Tsouris et al. (1995) and Yiacoumi et al. (1996) and used to enhance the size and magnetic susceptibility of target particles. In this process, strongly magnetic particles are introduced to flocculate with target particles in the suspension and subsequently form paramagnetic flocs of larger size and higher magnetic susceptibility (Ying, 2001). The enhancement of particle size and magnetic susceptibility could greatly improve the magnetic separation efficiency of nanoparticles. As discussed in the previous section, magnetite particles have a relatively high magnetic susceptibility and thus are good candidates for magnetic seeding of nanoparticles.

References

Ali-Zade, R. A., **2004**, *Inorg. Mater.* 40(5):509–515.

Bucak, S., D. A. Jones, et al., **2003**, *Biotechnol. Prog.* 19(2): 477–484.

Carpenter, E. E., **2001**, *J. Magn. Magn. Mater.* 225:17–20.

Cuyper, M. D., P. Müller, et al., **2003**, *J. Phys.-Condens. Mater.* 15: S1425–S1436.

Feng, X., G. E. Fryxell, et al., **1997**, *Science* 276(5314): 923.

Kaminski, M. D. and L. Nunez, **1999**, *J. Magn. Magn. Mater.* 194(1–3): 31–36.

Kaminski, M. D., and L. Nunez, **2002**, *Sep. Sci. Technol.* 37(16): 3703–3714.

Kohler, N., E. Glen, et al., **2004**, *J. Am. Chem. Soc.* 126: 7206–7211.

Koneracka, M., P. Kopcansky, et al., **2002**, *J. Mol. Catal. B-Enzym.* 18: 13–18.

Koneracka, M., P. Kopcansky, et al., **1999**, *J. Magn. Magn. Mater.* 201: 427–430.

Lawson, W. F., Jr., W. H. Simons, and R. P. Treat, **1977**, *J. Appl. Phys.* 48: 3213.

Leun, D. and A. K. Sengupta, **2000**, *Environ. Sci. Technol.* 34(15): 3276–3282.

Luborsky, F. E. and B. J. Drummond, **1976**, *IEEE Trans. Magn.* 12: 463.

Moeser, G. D., K. A. Roach, et al., **2002**, *Ind. Eng. Chem. Res.* 41(19): 4739–4749.

Mornet, S., A. Vekris, et al., **2000**, *Mater. Lett.* 42: 183–188.

Nesset, J. E. and J. A. Finch, **1978**, in *Proc. Int. Conf. On Ind. Appl. Of Magn. Separation,* M.H. Ridge, publ. *IEEE,* 169–173.

Nesset, J. E. and J. A. Finch, **1981**, *IEEE Trans. Magn.* 17: 1506.

Oliveira, Luiz C. A., Rachel V. R. A. Rios, et al., **2002**, *Carbon* 40: 2177–2183.

Oliveira, Luiz C. A., Rachel V. R. A. Rios, et al., **2003**, *Appl. Clay Sci.* 22: 169–177.

Phanapavudhikul, P., J. A. Waters, et al., **2003**, *J. Environ. Sci. Health* A 38(10): 2277–2285.

Safarik, I., **1994**, *Water Res.* 29(1): 101–105.

Schewe, H., M. Takayasu, and F. J. Friedlaender, **1980**, *IEEE Trans. Magn.* 16: 149.

Španová, A., B. Rittich, et al., **2003**, *J. Chromatogr. A* 1009(1–2): 215–221.

Španová, A., D. Hor k, et al., **2004**, *J. Chromatogr. B* 800(1–2): 24–32.

Svoboda, J., *Magnetic Methods for the Treatment of Minerals,* Elsevier, New York, **1987**.

Tsouris, C., S. Yiacoumi, and T. C. Scott, **1995**, *Chem. Eng. Commun.* 137: 147.

Watson, J. H. P., **1973**, *J. Appl. Phys.* 44: 4209.

Wu, P., J. Zhu, et al., **2004a**, *Adv. Funct. Mater.* 14(4): 345–351.

Wu, R., J. Qu, et al., **2004b**, *Appl. Catal., B*:48: 49–56.

Yiacoumi, S., D. A. Rountree, and C. Tsouris, **1996**, *J. Colloid Interface Sci.* 184: 477.

Ying, T.-Y., S. Yiacoumi, and C. Tsouris, **2000**, *Chem. Eng. Sci.* 55:1101.

Ying, T.-Y., **2001**, Ph.D. Dissertation, Georgia Institute of Technology, Atlanta.

**3.7.2
Dendrimers**

Karsten Gloe, Bianca Antonioli, Kerstin Gloe, and Holger Stephan

3.7.2.1
Introduction

An aim of both green chemistry and engineering is to minimize or, if possible, eliminate pollution of the environment with hazardous substances. Two topics of investigation to achieve this goal have been the improvement of separation processes that allow waste reuse and recovery by removal of toxic components, and the development of relevant new technologies, including solvent-free processes or the use of benign solvents to replace volatile organic compounds.

Solvent extraction has evolved to be a unit operation widely used in chemical, mining, pharmaceutical and food industries for the separation, concentration and recycling of different materials, especially metals and organic chemicals. A continuing challenge facing such applications is centered on how to enhance the selectivity and efficiency of these processes and how to control solvent losses through entrainment or dissolution in associated aqueous discharge streams. As a consequence manifold efforts are being made to find environmentally friendly and more sustainable solutions to these problems. The design of alternative tailor-made extractants and benign solvents, especially on the basis of supramolecular chemistry, is one promising approach; another is directed at the development of new or modified technological processes [1].

Dendrimers constitute a unique class of three-dimensional molecules with a defined highly-branched symmetric architecture. These molecules are playing a continuously increasing role in different areas of science and technology. Their unusual chemical and physical properties result in them having very considerable potential for use in green separation processes. The typical behavior of dendrimers is a reflection of their compact, tree-like molecular structure, providing an arrangement of inner and outer molecular functionalities which results in them being useful receptor and carrier molecules at the nanometer level [2].

It is the intention of this chapter to give an overview of the possible uses of dendrimers in separation science and also to describe recent progress in testing application-specific dendrimers for the separation of cations, anions and neutral molecules.

Several excellent and comprehensive reviews of dendrimers and their applications have been published, and the reader is referred to these for further details of their synthesis, properties and potential uses [3].

3.7.2.2
Dendrimers – Promising Reagents for Separation Processes

As illustrated schematically in Fig. 3.7-4, dendrimers are nearly perfect highly-branched and monodispersed macromolecules. Their three-dimensional architecture is characterized by the presence of a central core unit, several internal cavities and a large number of functional end groups (Fig. 3.7-4a), all of which can bind guest molecules using physical or chemical interactions (Fig. 3.7-4b). These symmetrically built molecules, first synthesized by F. Vögtle et al. in 1978 [4], show unique physical, chemical and biological properties that provide a basis for promising applications in a number of areas, including chemistry, physics, engineering, biology and medicine. Their well-defined geometrical and chemical structure can be controlled and varied by a bottom-up strategy that includes varying the number of generations (see Fig. 3.7-5).

As shown in Fig. 3.7-6 there are three different strategies (divergent, convergent and supramolecular) for stepwise dendrimer preparation, each of which permit clear control over critical molecular parameters such as size, shape and surface/interior structure. The divergent strategy (Fig. 3.7-6a) starts at the central core and adds layers of repeating units to branching points leading to an additional number of generations, with corresponding increases in size, surface functions and molecular weight. In contrast, the convergent strategy (Fig. 3.7-6b) builds the dendrimer

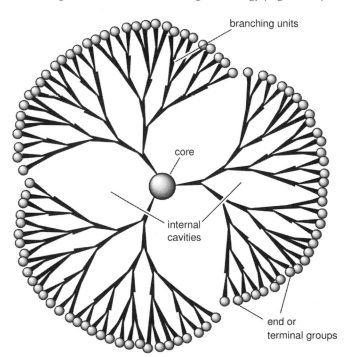

Fig. 3.7-4 Schematic representation of the dendrimer structure:
(a) main structural compartments

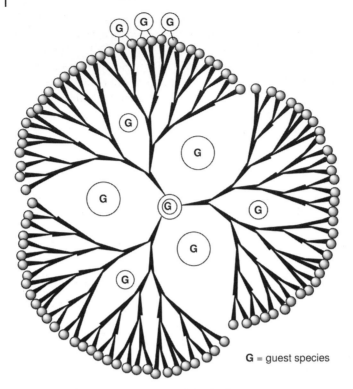

Fig. 3.7-4 Schematic representation of the dendrimer structure:
(b) binding possibilities for guests molecules.

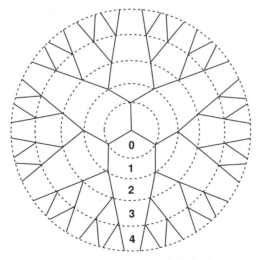

Fig. 3.7-5 Repetitive organization of the dendrimer structure.

from its periphery and is directed inwards, with the dendrimer segments being finally covalently linked to a multifunctional core molecule. The supramolecular strategy (Fig. 3.7-6c) is generally based on a convergent procedure. In this third approach the dendrimer is assembled by a self-organization process involving functional dendritic branches (so called dendrons) using non-covalent interactions with defined core components or through coordinative bonding to a metal ion center via incorporated donor functions. Because there is a wide range of available core units, building blocks and surface functions, an almost unlimited variety of dendrimers exhibiting designed properties are in principle possible.

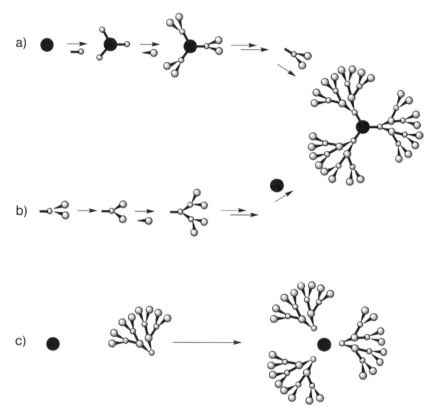

Fig. 3.7-6 Synthetic approaches for dendrimers: (a) divergent method; (b) convergent method; (c) self-assembly.

Three representative dendrimers, with different numbers of generations, are shown in Fig. 3.7-7. They are a poly(amidoamine) dendrimer (PAMAM) synthesized by D.A. Tomalia, which incorporates an ethylenediamine core and terminal amino groups (Fig. 3.7-7a), a poly(propyleneamine) product (POPAM) reported by E.W. Meijer with a butylenediamine core as well as terminal amino groups (Fig. 3.7-7b) and a poly(arylether)-containing dendrimer from J.M.J. Fréchet with a biphenylether core and terminal carboxylic acid moieties (Fig. 3.7-7c).

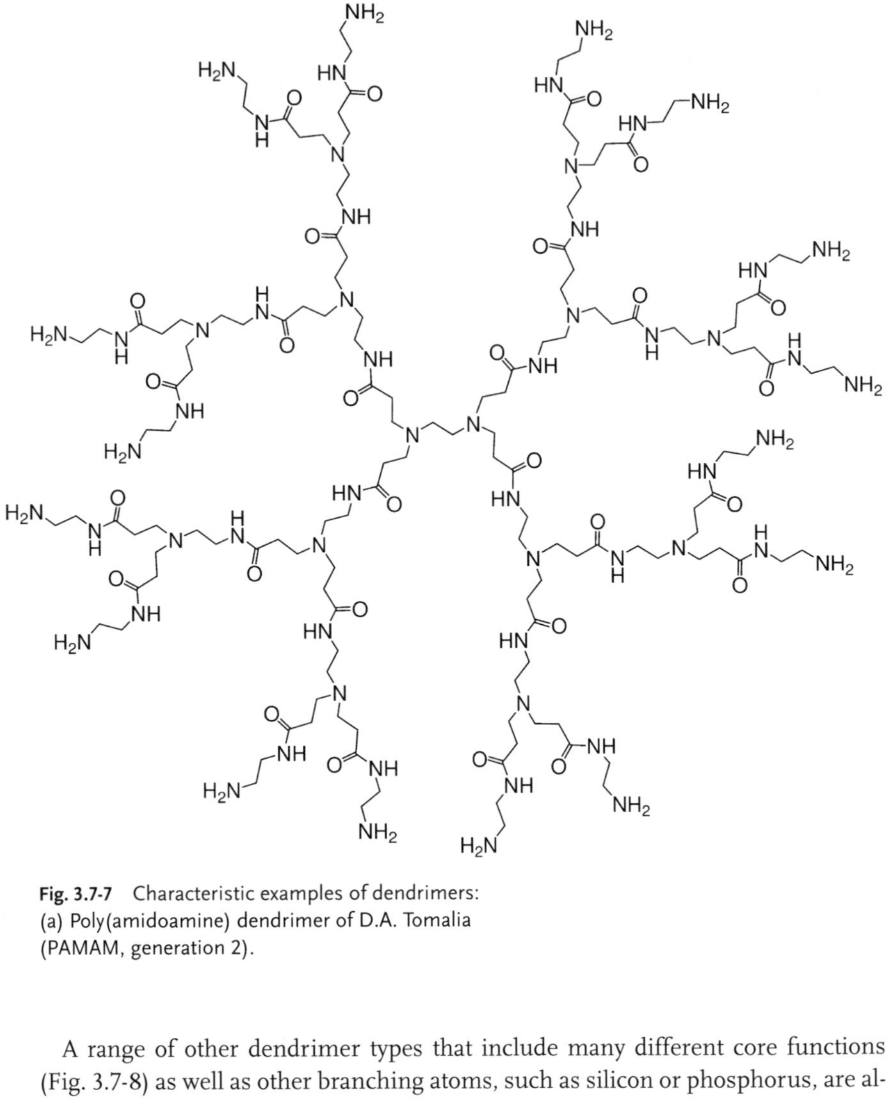

Fig. 3.7-7 Characteristic examples of dendrimers:
(a) Poly(amidoamine) dendrimer of D.A. Tomalia
(PAMAM, generation 2).

A range of other dendrimer types that include many different core functions
(Fig. 3.7-8) as well as other branching atoms, such as silicon or phosphorus, are also known. At present, derivatives of the first two dendrimer types are also commercially available [5].

Because of their specific globular structure, dendrimers behave quite differently from traditional linear polymers and their properties show some analogy to biological macromolecules such as proteins or enzymes. Experimental and theoretical
studies indicate that dendrimers possess a fluctuating molecular structure in solution, which is affected by their basic structure and inherent charge distribution
as well as being influenced by the polarity of the solvent, the ionic strength and the
pH of the solution. The fluctuation process involves the backfolding of the end
groups towards the center of the molecule [6]. This has wide consequences for the
availability of the internal cavities or peripheral functions for guest binding. Figu-

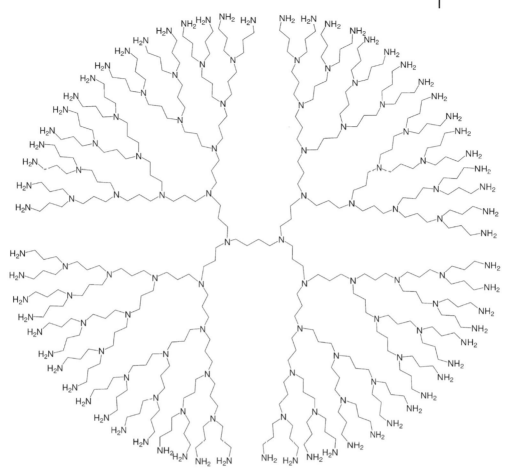

Fig. 3.7-7 Characteristic examples of dendrimers:
(b) Poly(propyleneamine) dendrimer of E.W. Meijer (POPAM,
generation 4).

re 3.7-9 illustrates the structure change that occurs for a neutral POPAM dendrimer on changing from generation 1 with four amine end groups to generation 5 with sixty-four amine end groups. The respective molecular sizes increasing along this sequence from about 1 nm to 5 nm. This rises to about 10 nm for generation 8. On increasing the number of generations, the macroscopic properties of the corresponding dendrimers are increasingly dominated by the nature of the peripheral groups. In particular, the lipophilicity and solubility can be readily tuned by this means to yield systems ranging from typically hydrophilic (based on end groups such as $-COOH$, $-OH$, $-NH_2$ or polyether) to strongly hydrophobic (based on alkyl or aryl moieties). The above possibilities are of very considerable importance for tailoring the physical properties of dendrimers for use in a range of binding and separation processes in solution.

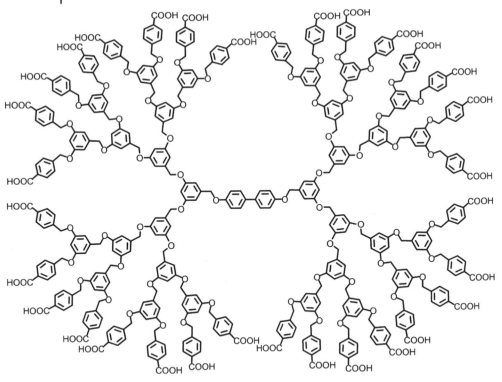

Fig. 3.7-7 Characteristic examples of dendrimers:
(c) Poly(arylether) dendrimer of J.M.J. Fréchet.

As discussed earlier, dendrimers may act as supramolecular endo- or exorecep-
tors for the molecular recognition of cations, anions and neutral molecules. In this
manner their behavior resembles that of many biomolecules. The molecular recognition process will be strongly influenced by the particular framework employed as well as by the nature of the end groups and the number of generations present. Generally, dendrimers are associated with a high uptake of guest species,
which may bind using a range of reversible weak supramolecular interactions, ranging from electrostatic, π- and hydrophobic interactions to hydrogen and coordinative bonding. As a consequence of the multiple recognition processes possible,
both the selectivity and the efficiency of binding can often be controlled. Furthermore, multiple guest binding may lead to cooperative effects such as enhanced
host–guest interaction. The exploitation of internal dendritic cavities for guest binding may be aided by a structure incorporating electrostatic repulsion between
charged branches or by the presence of solvent molecules in the structure before
guest uptake.

In analogy with similar behavior in supramolecular systems, the dendrimer
functionalities can be tailored towards the binding of selected species using the
now well documented principles of molecular recognition. Selected examples will
now be discussed that illustrate some representative behavior of this type.

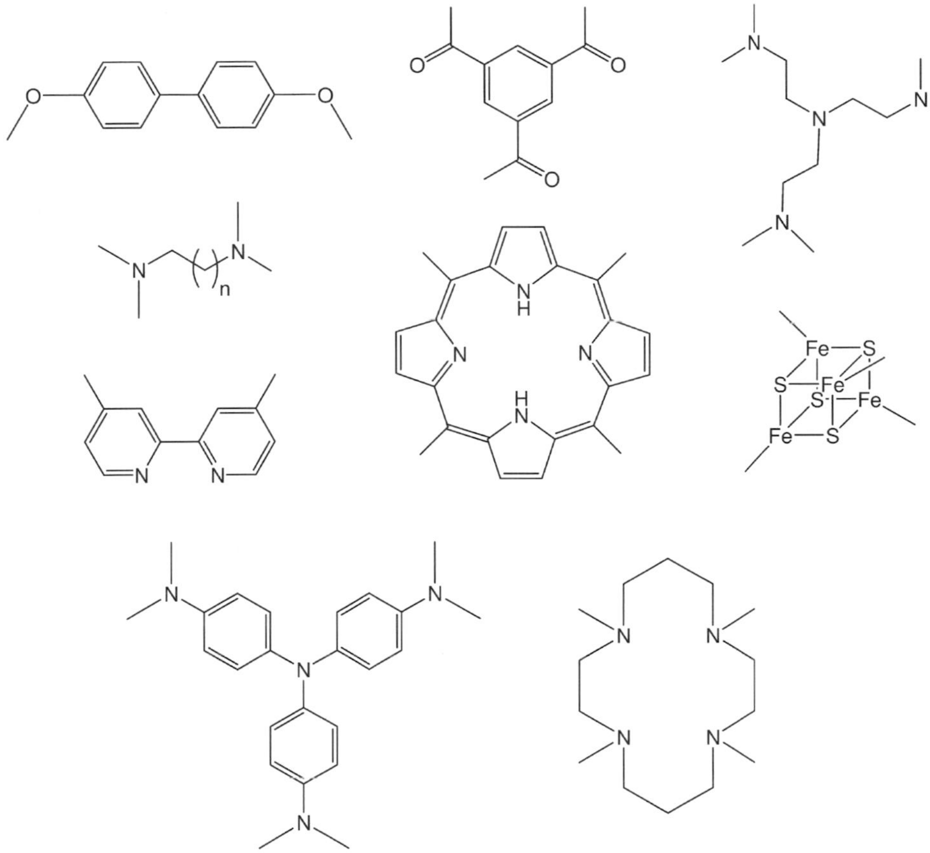

Fig. 3.7-8 Typical core units of dendrimers.

The exceptional potential of dendrimers to act as host molecules was first demonstrated by E.W. Meijer, using the so-called "dendritic box" to bind dye molecules reversibly. In this case the binding involves an acid–base reaction that results in electrostatic interaction between positively charged amino branches of the POPAM dendrimer employed and dye anions. The guest molecules may then be mechanically locked in the dendrimer by grafting bulky amino acid residues onto its periphery. The subsequent stepwise chemical cleavage of the peripheral dendrimer functions allows the controlled release of encapsulated guests. The PAMAM and POPAM dendrimers (see Fig. 3.7-7) containing tertiary amine functions have been shown to bind both cations and anions, with the binding controlled by the pH of the aqueous solution. Whereas such dendrimers readily interact with anions in their protonated forms, neutral molecules incorporating donor atoms are capable of forming strong coordinative bonds with transition metal ions bonded to the tertiary amine nitrogen. In the case of anion binding, when amide or urea moieties are present significant synergistic effects can occur through additional favorable hydrogen bond formation [7].

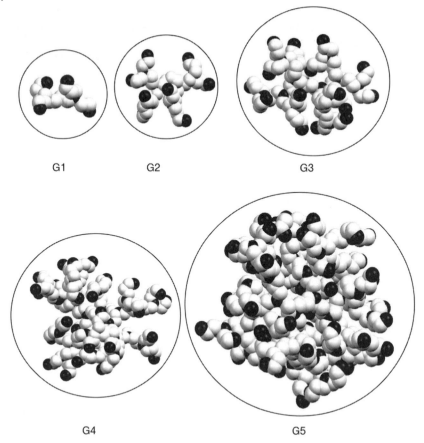

Fig. 3.7-9 MM$^+$ calculated structures of POPAM dendrimers for generations 1 to 5 (C white; N black; H omitted for clarity).

Studies of metal ion binding [Cu(II), Ni(II), Fe(III), Mn(II), Ag(I), Au(III), Pd(II), Pt(II), Ru(III)] by different OH- and NH$_2$-terminated PAMAM derivatives in aqueous solution show that these metal ions are bound to the branching tertiary amine functions with high capacity, with the observed uptake clearly dependent on the number of generations and on the pH. In the case of amine-terminated derivatives, the metal ions can also bind to these end groups, resulting in a further increase in uptake [8].

Other examples of strong metal ion binding involve bound catecholamide, a typical metal chelator, attached to the dendritic scaffold of POPAM and PAMAM, each of the latter with different numbers of generations. In all cases very efficient Fe(III) uptake by these chelating end groups was observed and such behavior was found to be independent of the size or structure of the dendrimer framework [9]. Other proposed systems involve chelating branching units for binding metal ions via incorporated amide functions for binding lanthanide ions and phenylazomethine groups for Fe(III) and Sn(II) [10]. Crown ether moieties have also been of interest for metal binding and have possible applications as the core, branching and peripheral functions [7,11]. In particular, the incorporation of such rings in the dendritic structure gives the prospect of controlling the location and number of incorporated metal ions more precisely. Metal ion incorporation in such a structure could them be combined with a subsequent chemical transformation to generate, for example, uniform nano-sized particles of zero-valent metals, sulfides, selenides and so forth [8].

As mentioned already, in an early study the binding of organic dye anions to poly-ammonium dendrimers of the POPAM family was investigated. Further studies showed that the combination of positively charged ammonium or ferricinium cations with additional hydrogen-bonding functions derived from urea or amide also resulted in the binding and phase transfer of inorganic anions such as phosphate, sulfate, halides, nitrate, pertechnetate and perrhenate. Even though in these cases the binding affinity was observed to be controlled in part by the dimensions of the internal cavities, the anion selectivity was found to be rather limited [12].

The above discussion demonstrates the many possibilities for employing dendritic derivatives for binding both charged and uncharged hosts. Nevertheless, further research is clearly required in order to gain deeper insight into the factors influencing selective host binding and its control.

3.7.2.3
Examples of Dendrimers in Separation Processes

The first attempt to prepare essentially water-insoluble, amine-containing dendrimers was realized by conversion of the primary amino groups on the periphery of POPAM dendrimers into amide analogues with long-chain alkyl acid chlorides [13]. In particular, the palmitoyl amides can be employed as efficient host molecules for the extraction of anionic xanthene dyes. These "inverted unimolecular dendritic micellar" structures possess a polar interior of tertiary amines capable of binding anionic guests. The apolar periphery shields the interior from the aqueous environment resulting in improved distribution behavior of dendritic hosts compared to hydrophobic tertiary amines. In the acidic range, where the tertiary amine groups are fully protonated, Fluorescein, Erythrosin B and Rose Bengal are quantitatively extracted by palmitoyl-POPAM dendrimers into the organic phase. The efficiency of xanthene dye extraction correlates with the acid–base behavior of the dendrimers and the lipophilicity of the dye molecules [14]. Back-extraction can easily be performed by increasing the pH, which leads to deprotonation of the tertiary amine groups. The dendritic hosts show a high loading capacity for the dyes investigated. Thus, a generation 4 dendrimer with thirty tertiary amine groups is able to bind thirty dye molecules [15]. Further, a complex with fifty Rose Bengal molecules inside a generation 5 dendrimer was observed in the organic phase. Interestingly, hydrophobic poly(propylene) fibers form stable blends with POPAM dendrimers that have fatty acid amides on the periphery. These blends can be employed with hollow fibers for separation processes.

Hydrophobic urea-functionalized dendrimers are readily available by reaction of POPAM dendrimers with isocyanates [16]. Such polyurea dendrimers with hexyl, octyl and dodecyl alkyl chains (see Fig. 3.7-10) have proved to be efficient carriers for oxoanions. Pertechnetate, perrhenate, and even the highly hydrophilic nucleotide ATP can be extracted from weak acid solution with remarkable efficacy (Fig. 3.7-11). The extraction of these oxoanions with urea-functionalized dendrimers follows the same behavior as that observed for hydrophobic dendritic amides discussed above.

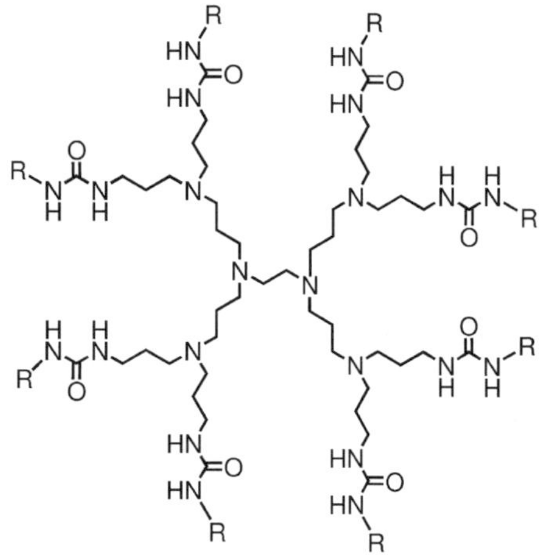

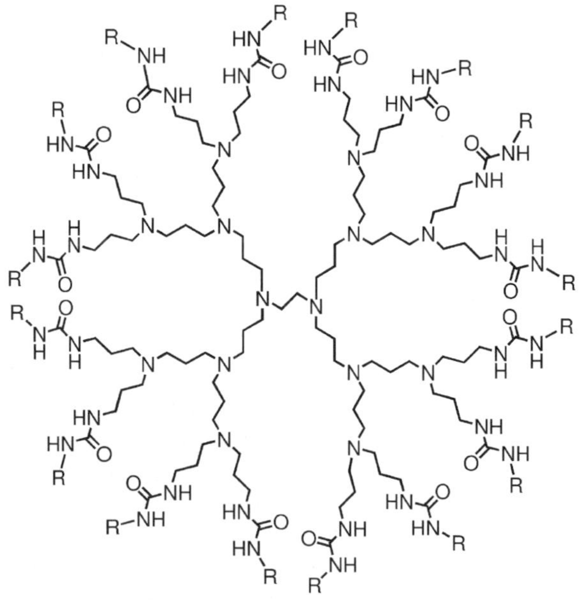

Fig. 3.7-10 Lipophilic urea-terminated POPAM dendrimers **Ia, b** (generation 2) and **IIa, b** (generation 3) with R = hexyl (a) and R = dodecyl (b) [16]

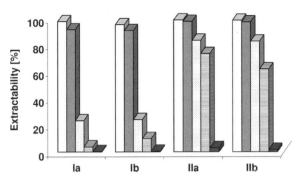

Fig. 3.7-11 Extractabilities of pertechnetate, perrhenate and nucleotides with POPAM dendrimers **Ia, b** and **IIa, b** [16].

$[KTcO_4] = [NH_4ReO_4] = [nucleotide] =$ 1×10^{-4} M; pH = 5.4; [dendrimer] = 1×10^{-3} M in $CHCl_3$. □ = TcO_4^-, ▨ = ReO_4^-, ▨ = ATP, ▤ = ADP, ■ = AMP

The guest-binding behavior can be tailored by variation of the dendritic periphery. For example, grafting pH-sensitive moieties such as methyl orange onto the periphery of POPAM dendrimers gives the possibility of pH-controlled inclusion and release of anionic guests [17]. In acidic media, the resulting positively charged surface prevents the penetration of anions into the interior of the dendrimer. It is only when the terminated methyl orange groups are completely deprotonated that the anions can be bound by the "internal" tertiary amine groups of POPAM dendrimers. Thus, host–guest binding behavior can be controlled by tailor-made surface modification of dendrimers.

Multi-crown dendrimers (Fig. 3.7-12) were found to exhibit good solubility in solvents of low polarity, making them attractive as extractants for separation processes. The combination of protonated tertiary amine groups as anion binding sites together with cation-active crown ether moieties represents an approach for the simultaneous binding of cations and anions. Extraction studies performed with sodium pertechnetate and mercury(II) chloride have shown that the guest molecules are mainly bound in the interior of the protonated polyamine skeleton [18].

Complete shielding of the metal ions extracted can be achieved by arrangement of the binding moiety as part of the core element. In this context, the development of PAMAM dendrons incorporating 8-hydroxyquinoline chelating units at the focal point have been reported (Fig. 3.7-13). By varying the chain length on the terminating ester groups the solubility of the dendrons, and consequently of the complexes formed, can be adjusted [19].

The dendritic host molecules discussed above represent promising extractants possessing particularly high separation efficiency and loading capacity. The apolar periphery is expected to have a favorable effect on the coalescence behavior and organic phase solubility in comparison to simple solvating and chelating agents. However, the problem of diluent entrainment in the aqueous phase may continue to be present. Thus, diluents may need to be separated from the raffinate in an additional process step, for example treatment with active carbon, macroporous resins

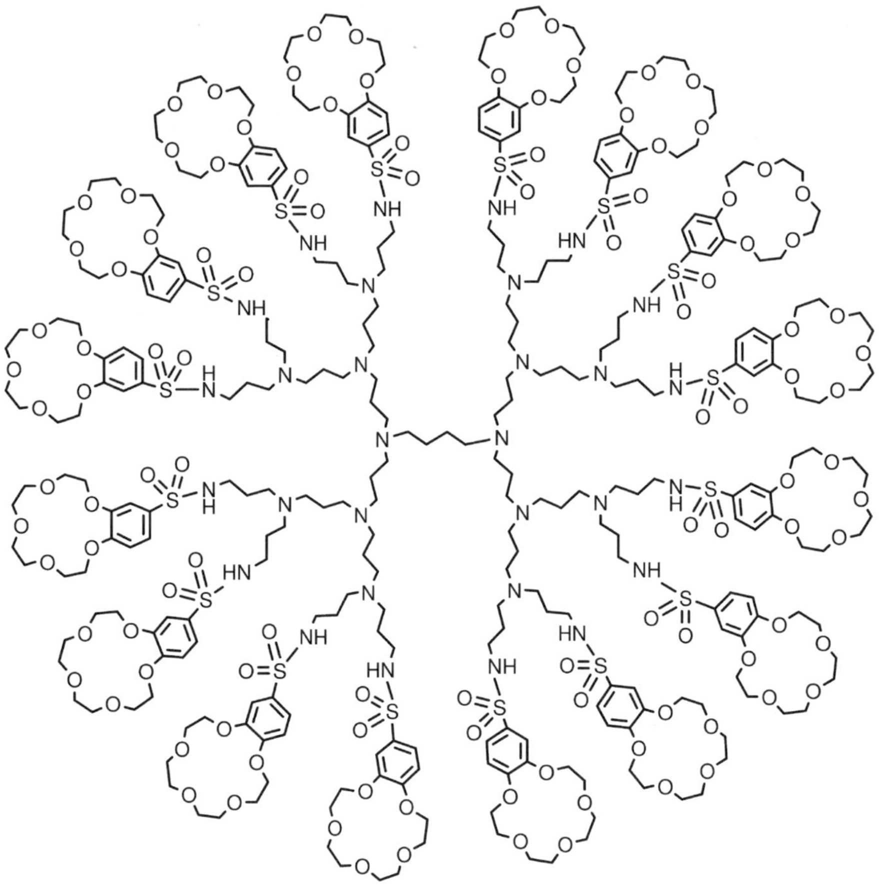

Fig. 3.7-12 Generation 3 multi-crown POPAM dendrimer [18].

or steam stripping. Recent developments show that selected dendrimers can themselves be employed for the removal of diluents from water. Thus, polyurea dendrimers are capable of separating aromatic hydrocarbons from water down to the level of a few parts per billion [20]. Alternatively, water-insoluble solvents can be used. In this context, hyperbranched polyester and polyesteramide polymers appear highly attractive [21]. On applying both hyperbranched polymers and dendritic molecules simultaneously, high separation efficiency and selectivity combined with non-contaminated raffinates may be achieved. It should be mentioned that fluorinated dendrimers dissolved in liquid carbon dioxide can transport polar dyes such as methyl orange, as well as the inorganic anions permanganate and dichromate, from water into supercritical CO_2 [22, 23]. However, some modification of dendritic hosts will normally be necessary in order to achieve appropriate mass-transfer and loading capacity for such supercritical extraction processes.

Owing to some advantages, in particular low energy requirements and fast reaction kinetics, polymer-supported ultrafiltration is gaining in importance as a pro-

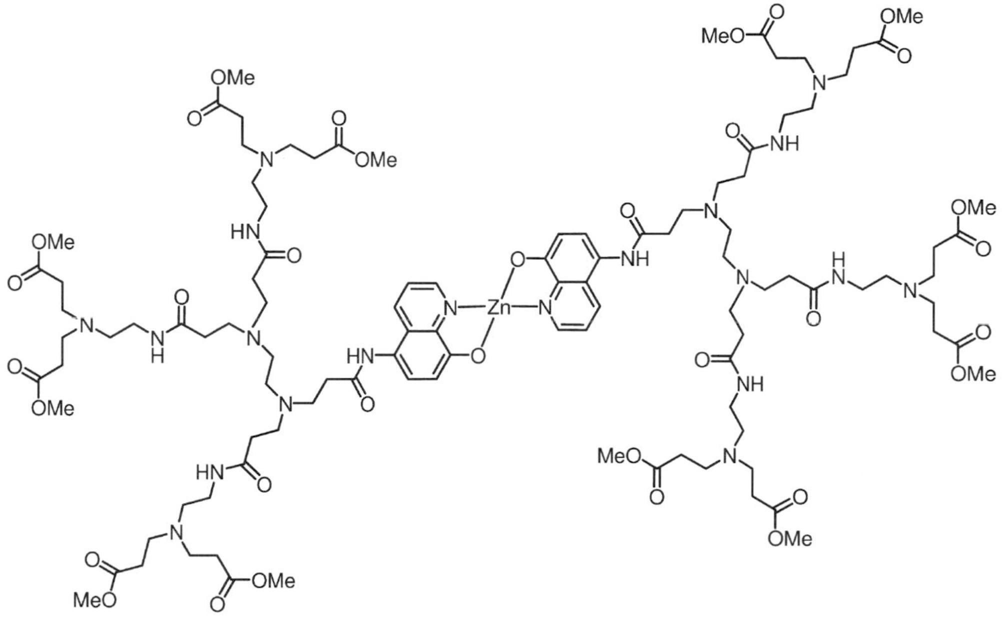

Fig. 3.7-13 Zinc(II) complex of an 8-hydroxyquinoline-containing PAMAM dendrimer [19].

cess for the removal of toxic metal ions from aqueous solution. In this connection, water-soluble dendrimers show very promising properties for acting as highly efficient polymeric binding systems. Even commercially available PAMAM dendrimers with terminal primary amine groups possess extremely high loading capacities. Thus, a generation 8 dendrimer can bind about 150 Cu(II) ions [24]. Both the primary and tertiary amine groups of PAMAM dendrimers are involved in complex formation, with the highest binding efficacy being achieved in weak acid solution. Regeneration of the dendrimers can be easily carried out by decreasing the pH. Weak sorption tendencies of dendrimers onto ultrafiltration membranes qualify these complexing ligands as well suited for use in separation processes. The separation selectivity can be improved by the addition of selective chelators to the dendrimer surface. Treatment of PAMAM dendrimers incorporating terminal piperazine groups with benzoylisothiocyanates gave *N*-benzoylthiourea-modified, water-soluble polymers (see Fig. 3.7-14) [25]. The resulting dendritic ligands show remarkable efficiency and selectivity for the separation of heavy metal ions. The highest binding capacity was found for Cu(II) and Hg(II). For most of the heavy metals investigated, regeneration of the loaded ligand occurred on decreasing the pH. Only in the case of Hg(II) was the use of thiourea solution necessary.

New applications can be expected in the area of dendrimer-immobilized liquid membranes. By using polyvinylidene fluoride membrane films containing a generation zero PAMAM dendrimer, highly selective CO_2 separation from gas mixtures is possible [26]. Dendrimers consisting of a cationic core shell and an apolar

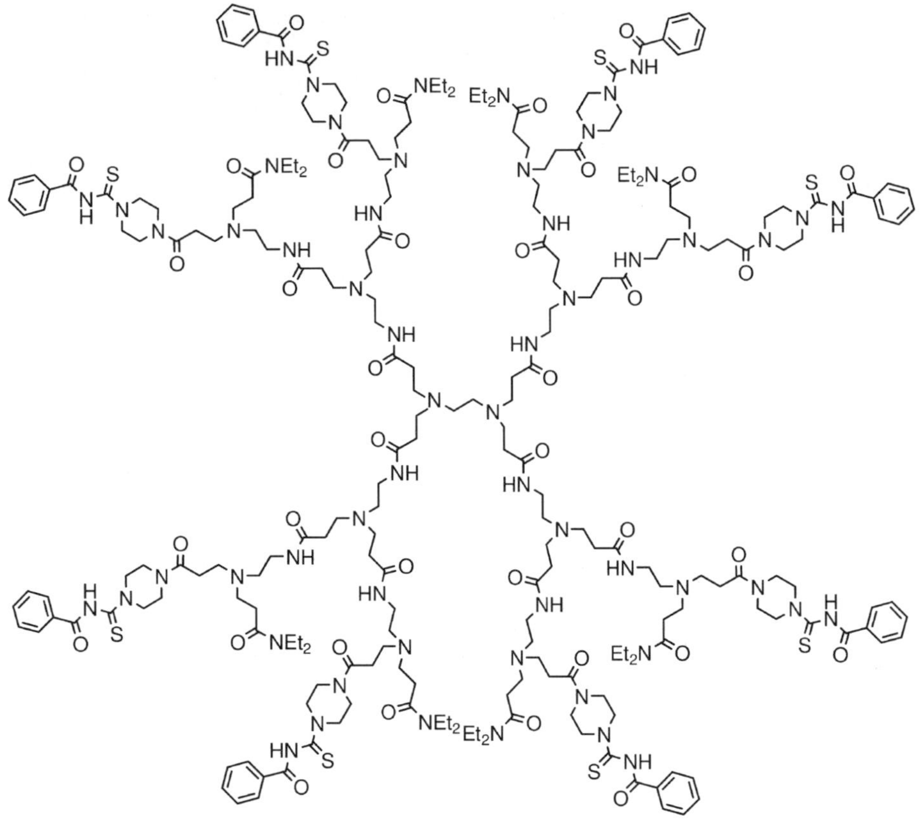

Fig. 3.7-14 Benzoylthiourea-containing PAMAM dendrimer [25].

outer shell have been demonstrated to show interesting anion exchange properties in liquid-membrane systems [27]. The anionic species can bind to the interior of the dendrimer via ammonium groups attached to an arylsilyl core element through electrostatic interaction. Such polycationic dendrimers have a high potential as shape-selective carriers for anionic substrates for application in nanofiltration, nanoreactors and membrane processes.

Dendrimers bonded to solid substrates such as silica gel or organic copolymers open the way to achieving improved stationary phases for chromatographic applications. Silica gel bound with Fréchet dendrons has been employed for capillary electrochromatography, leading to both the suppression of electro-osmotic flow and improved separation resolution for aromatic hydrocarbons and proteins compared to reversed-phase chromatographic material [28]. A highly selective chiral stationary phase was developed employing polymer beads with linked proline-5-indananilide monodendrons, and this product enabled the separation of racemic amino acid alkylamides [29]. PAMAM immobilized on silica gel has been employed for the pre-concentration and separation of Pd(II) [30]. Clearly, such PAMAM-coated silica particles offer good prospects for developing tailor-made size-exclusion chromatography columns for use in many separation problems [31].

3.7.2.4
Conclusions and Future Prospects

Dendrimers represent a promising class of monodispersed hyperbranched macromolecules with tailored chemical functionalities in their core and branching units and at their periphery. Their topological properties at the nanometer level open up manifold application possibilities in different fields of science and technology including separation science. This potential is reflected by the growing number of research groups engaged in dendrimer research and the extraordinary increase in publishing and patenting dendrimer research during the last decade.

The unusual properties of dendrimers may be attributed to their defined molecular architecture. An impressive example, as mentioned previously, is the so-called "dendritic box", which allows the selective reversible binding and transport of organic guest molecules. A number of potential applications are also based on the high loading capacity of functionalized dendrimers for guest ions or neutral molecules. This feature may, for example, find use for the separation of environmentally unfriendly species, such as heavy metal ions, inorganic anions or aromatic hydrocarbons, from aqueous solution via solvent extraction, sorption and membrane processes or via ultrafiltration. Examples of the efficient and selective binding of gas molecules from mixtures by dendrimer-immobilized liquid membranes are also known. In the majority of cases the commercially available poly(amidoamine) and poly(propyleneamine) dendrimers have been used directly or in modified form for separation processes. Nevertheless, the synthetic art in dendrimer chemistry continues to flourish, with the tailoring of systems for selected separation problems in part providing a motivation this activity.

As a rule, dendrimers fulfill the essential requirements of a benign extractant. They generally display high stability, solubility, low toxicity, good biocompatibility, high loading capacity and reversible binding. It is noted that a common environmental problem with solvent extraction arises from the need to employ an organic solvent as the basis of the organic phase. The solvents normally employed in traditional extraction systems are toxic, flammable and volatile organic compounds (VOCs). Recent studies with dendrimers show that such VOCs can be successfully substituted by environmentally more friendly fluorinated solvents, ionic liquids [32] or by supercritical CO_2. Furthermore, the use of water-soluble dendrimers also allows the binding of toxic species directly in aqueous solution. Their separation by nanofiltration leads to pure water-based processing without any additional organic compounds.

A further possible advantage of the use of dendrimers as sequestering reagents in comparison with more conventional complexing agents results from the inherently different design of dendrimers, whose internal and external functionalities allow, as mentioned already, the creation of an inner "binding" microenvironment with defined coordination sites and a solvent-compatible, hydrophilic or hydrophobic periphery. With respect to this, it is expected that new tailor-made dendrimer systems will continue to be created making wide use of the principles of molecular recognition and self-organization.

At present, the introduction of dendrimers for industrial processes is restricted by the expensive, multi-step synthesis of the former; their introduction is unfavorable on cost–benefit criteria. As a consequence, the application of dendrimers can be expected, at least initially, only in those applications where the unique structure–property relationship of a particular dendritic reagent is crucial. It follows that future work will almost certainly focus on optimizing synthetic procedures for dendrimers in order to simplify their preparation and hence reduce the corresponding costs [33]. This goal may likely be achieved more readily by using supramolecular self-assembly processes. A further possibility exists in the use of micro-reaction devices for the process, reducing drastically the amount of the expensive complexing agent.

It has been an aim of this chapter to indicate that dendrimers have very significant potential for application in separation technologies. Nevertheless, at present it is difficult to predict whether their use will be limited to specific applications or whether they will be introduced more widely. However, dendritic hyperbranched polymers possessing a polydisperse and irregular structure can be synthesized more easily via a one-step procedure and these appear to be an economical alternative to dendrimers [34]; it may be noted that their use in particularly interesting applications, with significant benefits for the separation of azeotropic mixtures by extractive distillation or solvent extraction, are at present under consideration [35].

Environmental issues are of overwhelming societal importance and dendrimer chemistry clearly provides a range of interesting prospects for solving relevant problems. Thus the further development of dendrimer research will undoubtedly increase the impact of this field on environmental chemistry. Consequently, future progress in this area will clearly benefit from an interdisciplinary approach involving chemists, physicists and engineers in order to adapt new dendritic systems to both established and new technologies.

Acknowledgements

The authors are especially grateful to Prof. Len Lindoy, Sydney University, for stimulating discussions on the topic of this review and to Prof. Fritz Vögtle, University of Bonn, for continuous fruitful cooperation within the field of dendrimer chemistry.

Financial support of the authors' dendrimer research by the Saxon State Ministry of Science and Art, Dresden, is acknowledged with thanks.

References

1 J. Rydberg, M. Cox, C. Musikas, G. R. Choppin, eds., *Solvent Extraction Principles and Practice*, Marcel Dekker, New York, **2004**.

2 (a) G. R. Newkome, C. N. Moorefield, F. Vögtle, *Dendrimers and Dendrons*, Wiley-VCH, Weinheim, **2001**; (b) J. M. J. Fréchet, D. A. Tomalia, eds., *Dendrimers and Other Dendritic Polymers*, J. Wiley & Sons, Chichester, **2001**.

3 Special Issues on Recent Developments in Dendrimer Chemistry: (a) *Tetrahedron* **2003**, *59*, 3787–4024 (D. K. Smith, Ed.); (b) *Compt. Rend. Chim.* **2003**, *6*, 709–1212 (D. Astruc, Ed.); (c) K. Inoue, *Prog. Polym. Sci.* **2000**, *25*, 453–571; (d) G. M. Dykes, *J. Chem. Technol. Biotechnol.* **2001**, *76*, 903–918; (e) P. J. Gittins, L. J. Twyman, *Supramol. Chem.* **2003**, *15*, 15–23; (f) U. Boas, P.M. H. Heegard, *Chem. Soc. Rev.* **2004**, *33*, 43–63.

4 E. Buhleier, W. Wehner, F. Vögtle, *Synthesis* **1978**, 155–158.

5 (a) DSM Hybrane BV, POB 18, Geleen 6160 DM, The Netherlands, (www.dsm.com) for POPAMs; (b) Dendritech Inc., 3110 Schuette Drive, Midland, MI 48642, USA, (dendritech.com) for PAMAMs.

6 M. Ballauff, C. N. Likos, *Angew. Chem. Int. Ed.* **2004**, *43*, 2998–3020.

7 M. W. P. L. Baars, E. W. Meijer, *Top. Curr. Chem.* **2000**, *210*, 131–182.

8 R. M. Crooks, M. Zhao, L. Sun, V. Chechik, L. K. Yeung, *Acc. Chem. Res.* **2001**, *34*, 181–190.

9 S. M. Cohen, S. Petoud, K. N. Raymond, *Chem. Eur. J.* **2001**, *7*, 272–279.

10 (a) V. Balzani, F. Vögtle, *Compt. Rend. Chim.* **2003**, *6*, 867–872; (b) R. Nakajima, M. Tsuruta, , M. Higuchi, K. Yamamoto, *J. Am. Chem. Soc.* **2004**, *126*, 1630–1631.

11 I. M. Atkinson, J. D. Chartres, A. M. Groth, L. F. Lindoy, M. P. Lowe, G. V. Meehan, *Chem. Commun.* **2002**, 2428–2429.

12 (a) D. L. Stone, D. K. Smith, *Polyhedron* **2003**, 763–768; (b) K. Gloe, H. Stephan, M. Grotjahn, *Chem. Eng. Technol.* **2003**, *26*, 1107–1117.

13 S. Stevelmans, J. C. M. van Hest, J. F. G. A. Jansen, D. A. F. J. van Boxtel, E. M. M.

de Brabander-van den Berg, E. W. Meijer, *J. Am. Chem. Soc.* **1996**, *118*, 7398–7399.

14 M. W. P. L. Baars, P. E. Froehling, E. W. Meijer, *Chem. Commun.* **1997**, 1959–1960.

15 P. E. Froehling, *Dyes Pigments* **2001**, *48*, 187–195.

16 H. Stephan, H. Spies, B. Johannsen, L. Klein, F. Vögtle, *Chem. Commun.* **1999**, 1875–1876.

17 H. Stephan, H. Spies, B. Johannsen, C. Kauffmann, F. Vögtle, *Org. Lett.* **2000**, *2*, 2343–2346.

18 H. Stephan, H. Spies, B. Johannsen, K. Gloe, M. Gorka, F. Vögtle, *Eur. J. Inorg. Chem.* **2001**, 2957–2963.

19 L. Shen, F. Li, Y. Sha, X. Hong, C. Huang, *Tetrahedron Lett.* **2004**, *45*, 3961–3964.

20 M. Arkas, D. Tsiourvas, C. M. Paleos, *Chem. Mater.* **2003**, *15*, 2844–2847.

21 M. Seiler, D. Köhler, W. Arlt, *Sep. Purif. Technol.* **2003**, *30*, 179–197.

22 A. I. Cooper, J. D. Londono, G. Wignall, J. B. McClain, E. T. Samulski, J. S. Lin, A. Dobrynin, M. Rubinstein, A. L. C. Burke, J. M. J. Fréchet, J. M. DeSimone, *Nature* **1997**, *389*, 368–371.

23 E. L. V. Goetheer, M. W. P. L. Baars, L. J. P. van den Broeke, E. W. Meijer, J. T. F. Keurentjes, *Ind. Eng. Chem. Res.* **2000**, *39*, 4634–4640.

24 M. S. Diallo, L. Balogh, A.Shafagati, J. H. Johnson, W. A. Goddard III, D. A. Tomalia, *Environ. Sci. Technol.* **1999**, *33*, 820–824.

25 A. Rether, M. Schuster, *React. Funct. Polym.* **2003**, *57*, 13–21.

26 A. S. Kovvali, H. Chen, K. K. Sirkar, *J. Am. Chem. Soc.* **2000**, *122*, 7594–7595.

27 A. W. Kleij, R. van de Coevering, R. J. M. Klein Gebbink, A.-M. Noordman, A. L. Spek, G. van Koten, *Chem. Eur. J.* **2001**, *7*, 181–192.

28 H. C. Chao, J. E. Hanson, *J. Sep. Sci.* **2002**, *25*, 345–350.

29 M. Xu, E. Brahmachary, M. Janco, F. H. Ling, F. Svec, J. M. J. Fréchet, *J. Chromatogr. A* **2001**, *928*, 25–40.

30 X. Z. Wu, P. Liu, Q. S. Pu, Q. Y. Sun, Z. X. Su, *Talanta* **2004**, *62*, 918–923.

31 K. Sakai, T. C. Teng, A. Katada, T. Harada, K. Yoshida, K. Yamanaka, Y. Asami, M. Sa-

kata, C. Hirayama, M. Kunitake, *Chem. Mater.* **2003**, *15*, 4091–4097.

32 A. E. Visser, R. P. Swatlowski, W. M. Reichert, S. T. Griffin, R. D. Rogers, *Ind. Eng. Chem. Res.* **2000**, *39*, 3596–3604.

33 S. Hecht, *J. Polym. Sci. A* **2003**, *41*, 1047–1058.

34 (a) B. I. Voit, *Compt. Rend. Chim.* **2003**, *6*, 821–832; (b) P. Froehling, *J. Polym. Sci. A* **2004**, *42*, 3110–3115.

35 M. Seiler, *Chem. Eng. Technol.* **2002**, *25*, 237–253.

3.8
Separations Using Superheated Water

Anthony A. Clifford

3.8.1
Introduction

The term superheated water strictly means water below its critical temperature of 374°C, but usually refers to liquid water under pressure between 100°C and 374°C. It is also sometimes referred to as subcritical water or pressurized hot water. Water is also used as a supercritical fluid above 374°C, most notably for the oxidation and destruction of waste and toxic materials [1], but supercritical water will not be discussed here. Superheated water has also been used for processes other than separation, for example waste treatment by the so-called wet-air oxidation process [2], recrystallization and chemical reactions in supercritical water, which have been thoroughly reviewed [3–5]. Destruction of waste molecules by both superheated and supercritical water can be used following extraction and this will be described below. This chapter will concentrate on work using superheated water as a replacement for organic solvents for extractions, chromatography and related processes. Much of this work has been restricted to the range 100–300°C. At these lower temperatures water is not highly compressible, and the pressure of the medium does not have much effect as long as it is high enough to maintain the water in the liquid phase.

The manipulation of water properties with temperature to achieve these ends has been made the subject of a patent [6], which is worth reading for those interested in the subject as it contains much useful information. It can be seen at the US patent website, http://patft.uspto.gov, and searching using its number, 6,352,644. Examples of separations described in this patent include: the removal of pesticides from contaminated soil including removal in situ; the removal of organic pollutants from wastewater; the extraction of organic compounds from solids; the extraction of compounds from solids coupled with degradation; the extraction and degradation of chemical warfare agents; the extraction of synthesis contaminants and organic compounds from polymers or plastics; the extraction of biologically active organic compounds from plant tissue; the extraction and reaction of

Green Separation Processes. Edited by C. A. M. Afonso and J. G. Crespo
Copyright © 2005 WILEY-VCH Verlag GmbH & Co. KGaA, Weinheim
ISBN 3-527-30985-3

compounds from plant tissue to produce flavors and fragrances; and the use of superheated water as a mobile phase for liquid chromatography.

Most of the work described below is on a laboratory scale and some of it is directed towards chemical analysis. However, some larger-scale processes are under consideration, mainly for environmental reasons. Chemical reactions, which bring about modification of the extracted compounds, can occur during extraction. In some cases this is an advantage, e.g. when pollutants are destroyed or flavors produced. The work related to these is described in so far as it is publicly available. Some important features of superheated water are first briefly discussed.

3.8.1.1
Polarity

Water changes dramatically when its temperature rises, because of the breakdown in its hydrogen-bonded structure with temperature. The high degree of association in the liquid causes its relative permittivity (more commonly called its dielectric constant) to be very high, around 80 under ambient conditions, but as the temperature rises hydrogen bonding breaks up [7] and the dielectric constant falls, as shown in Fig. 3.8-1 [8]. By 205°C its dielectric constant has fallen to equal that of methanol (i.e. 33) at ambient temperature. Thus, between 100 and 200°C superheated water is behaving like a water–methanol mixture. When the critical temperature is reached, water behaves as a collection of individual molecules, with the dielectric constant of a gas. This behavior forms the basis of the use of superheated water as a replacement for organic solvents. It means that it has a limited ability to dissolve organic molecules.

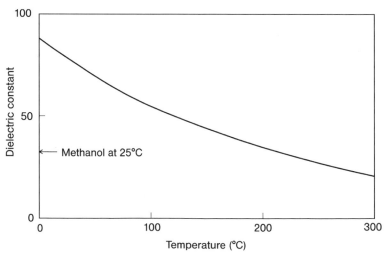

Fig. 3.8-1 The dielectric constant (relative permittivity) of liquid water as a function of temperature at its vapor pressure [8].

3.8.1.2
Solubilities of Organic Compounds

Partly because of its fall in polarity with increased temperature, superheated water can dissolve organic compounds to some extent, especially if they are slightly polar or polarizable, as are aromatic compounds. The solubility of an organic compound in superheated water is often many orders of magnitude higher than its solubility in water at ambient temperature for two reasons, one being the polarity change. The other is that a compound with low solubility at ambient temperature will have a high positive enthalpy of solution, and thus a large increase in solubility with temperature.

An early measurement [9] showed that naphthalene forms a 10 mass% solution in water at 270°C. It was later shown [10] that both benz[e]pyrene and nonadecyl-benzene reach the same concentration at 350°C. A number of other early measurements of solubility/phase behavior of high quality have been made, but are not comprehensively reviewed here. More-recent work has been carried out on a range of compounds of relevance to extraction and related processes, for example compounds obtained from the extraction of plant material [11] and fatty acids [12]. Work has also been carried out on the solubilities of polyaromatic compounds and pesticides, between 25 and 200°C [13–15]. Some of these data have been analyzed [16] to obtain a method of roughly estimating high-temperature solubilities. The analysis first takes into account the effect of temperature thermodynamically and then adds in the effect of the change in dielectric constant empirically. From the limited data available, there is no trend with molecular type, and so an equation to roughly estimate the solubilities of poorly soluble compounds is given as follows [16]:

$$\ln[x_2(T)] \approx (T_0/T)\ln[x_2(T_0)] + 15[(T/293) - 1]^3 \tag{1}$$

In Eq. 1, x_2 is the mole fraction of the solute in the solution, T_0 is a temperature near ambient at which the solubility is known and T the temperature of the required solubility. Comparisons of predictions from this equation for choranthonil, the worst fitting compound, are given in Table 3.8-1. Agreement within a factor of about 5 was obtained, although the solubilities rise by several orders of magnitude over the temperature range.

Table 3.8-1 The solubility of chloranthonil in water.

T (K)	$x_{2(exp.)}$	$x_{2(calc.)}$
323	5.41×10^{-8}	5.05×10^{-8}
373	1.89×10^{-6}	6.09×10^{-7}
423	6.43×10^{-5}	8.05×10^{-6}
473	1.58×10^{-3}	2.15×10^{-4}

Data from Ref [16].

3.8.1.3
Laboratory-scale Extraction

Because of the greater solubility of some organic compounds in superheated water, this medium can be considered for the replacement of conventional organic solvents in extraction and other processes. A simple example of an extraction apparatus for work on a small scale is shown schematically in Fig. 3.8-2, which was designed for plant-material extraction. The system is readily adapted for related work, the main consideration being the pressure required at the temperature to be studied [8]. Often water is deoxygenated, by sparging with helium for half an hour. It is then pumped at around 1 ml min⁻¹ into the extraction cell in an oven at the required temperature, through a long tube that acts as a heat exchanger to bring the water up to temperature. On exiting from the oven, the water stream is cooled using a coil of tubing in air and passes through a narrow tube into a collection vial. The narrow tube acts as a restrictor to maintain the pressure in the system above the vapor pressure, typically at 50 bar. A valve can be used instead of the restrictor. Minor refinements are not shown here. Other studies, referred to below, use very similar experimental systems. For kinetic experiments, to study extraction rates, the vial is replaced at known intervals. The extract is than back-extracted into a solvent, such as pentane, and then analyzed and quantified, for example by gas chromatography

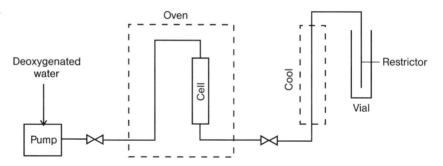

Fig. 3.8-2 Schematic diagram of a small laboratory system for superheated water extraction.

3.8.1.4
Pilot-plant Equipment

Figure 3.8-3 is a schematic diagram of a pilot plant located at Critical Processes Ltd in the UK. It will serve as an example of a possible configuration. It has an extraction cell of 32-l volume and will operate at up to 200°C. The maximum water flow rate is 60 l h⁻¹. The oil heater is 6 kW and the flow rate of oil is 120 l h⁻¹. The oil flow rate is higher because of the lower heat capacity of oil. In the system water is pumped from a container through two heat exchangers into the cell. The first heat exchanger is heated by the product emerging from the cell, and the second by oil from a heater. The flow rate is measured and the pressure is controlled by a valve at the exit. Temperatures are measured throughout the system and an example of temperatures is shown in the next section.

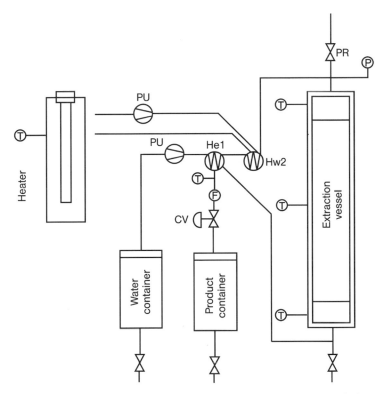

Fig. 3.8-3 Schematic diagram of a pilot plant for superheated water extraction: T, temperature measurement; P, pressure measurement; PR, pressure relief valve; CV, pressure control valve; PU, pump; and He1 and He2 the two heat exchangers.

3.8.1.5
Energy Considerations

Because of the high temperatures used, it might be imagined that superheated water processes were costly in energy terms. However, this is not the case because water stays as a liquid and the latent heat of evaporation is not involved. As an example, we compare superheated water extraction with steam distillation. A greater mass of superheated water may be needed for a given mass of material to be extracted. However, only 505 kJ kg^{-1} is required to heat liquid water from 30°C to 150°C, compared with 2550 kJ kg^{-1} required to convert water at 30°C to steam at 100°C [8]. Moreover, it is relatively easy to recycle the heat in a superheated water process by passing the water leaving the extraction cell through a heat exchanger to heat the water flowing to the cell. Realistically, 80% of the energy can be recovered in this way, depending on the characteristics of the heat exchangers. By contrast it is difficult to recover heat in steam distillation, because a relatively large amount of cooling water is needed to condense the steam and only part of this can pass into the boiler. The amount of heat that can be recovered in steam distillation is of the order of 5%. These advantages mean that in a well-designed process, in spite of the greater amount of water required, less energy is needed. In the pilot plant described in the previous section 75% of the heat is recovered when it is operated at 175°C. The temperatures required throughout the plant to achieve this are shown in Fig. 3.8-4.

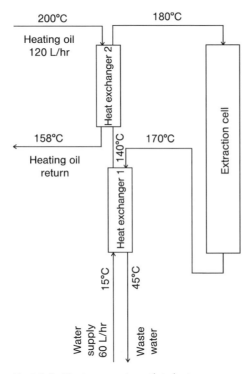

Fig. 3.8-4 Heat recovery in a pilot plant.

3.8.2
Extraction and Degradation Studies

Studies of the extraction of pollutants from soils, sediments, sludges and other matrices have been carried out extensively on a laboratory scale. In some cases degradation of the extracted compounds occurs during the process. In other studies, a degradation step, such as supercritical water oxidation, has been added in-line to destroy the extracted compounds. These laboratory studies have been aimed at developing both analytical procedures and clean-up processes and in some cases studies have been taken to pilot-plant scale.

3.8.2.1
Extraction From Solids and Semi-solids Other Than Biomass

The extraction of polyaromatic hydrocarbons from soil and urban particulates by superheated water was reported in 1994 [17]. Extraction of compounds up to benzo[*a*]pyrene was virtually complete in 15 min at 250°C, with a flow rate of 1 ml min⁻¹ and a sample of 0.5 g. Good but less complete results were obtained when extracting urban air particulates. The pressure did not influence the extraction behavior, provided it was sufficient to maintain water as a liquid. The extraction of polychlorinated biphenyls from soil and a river sediment was also found to be complete in 15 min at 250°C [18]. Work with a wider range of compounds showed that extraction was class selective [6, 19], with phenols and lighter aromatics being extracted at 50 to 150°C, polyaromatic hydrocarbons and lighter aliphatics at 250 to 300°C, but the heavier aliphatics only removed by steam at 250 to 300°C. This selectivity has been compared to other extraction methods [20]. The extraction of agrochemicals from soil has also been studied [6].

Since this earlier work, a large number of studies have been reported on the extraction of the same and other compounds from solid or semi-solid matrices. These studies include the extraction of saturated and unsaturated organic compounds, both polar and nonpolar [6, 17, 19–28], phenol derivatives including chlorinated phenols [6, 19, 29–31], surfactants [32], agrochemicals [6, 33–37], antibacterial agents [38], polychlorinated organics including PCBs [18, 39–40], polybrominated flame retardants [41] and metals [42]. The solid and semi-solid matrices from which these compounds were extracted include soils, river and coastal sediments [17, 18, 20–23, 25–28, 30, 31, 33–36, 40–42], airborne particulates [17, 20, 24], sewage sludge [32, 38], process sludge and dust [19, 29, 37] and inert matrices, such as sand, spiked with contaminants [31].

Some of the work carried out for analytical purposes has involved the coupling of superheated water extraction to other analytical methods. Coupling to superheated water chromatography has also been carried out, but this is discussed in Section 3.8.4. Hot phosphate-buffered water extraction has been coupled on line with liquid chromatography/mass spectrometry for analyzing contaminants in soil [43]. It has been coupled with enzyme immunoassay for efficient and fast polyaromatic hydrocarbon (PAH) screening in soil [21]. In a number of studies both static and dynamic superheated water extraction has been coupled to solid-phase microextraction [15, 25, 28, 30, 35, 38], sometimes with other analytical methods also coupled. It has been coupled with gas chromatography–mass spectrometry [31], capillary electrophoresis [31], liquid chromatography–mass spectrometry [32] and liquid chromatography–gas chromatography [41]. Sometimes other chemicals are added to the water used, such as acid [42] or phosphate buffer [43]. Different trapping methods for analytical extraction have been examined [44].

Work on the superheated water extraction of coal has been carried out, more directed towards possible processes. Sulfur can be extracted from Illinois Basin coal with acidified superheated water [45] to reduce its sulfur content from ~6% to ~1.5%. Some other elements can also be removed from coal with acidified superheated water, such as arsenic, selenium, mercury and other ash-forming elements [46, 47].

Treatment of polymers by superheated water at high temperatures can result in their decomposition/hydrolysis to monomers or lower oligomers [6]. At lower temperatures the polymer is stable and the small amounts of monomers, initiators, low oligomers and other small compounds can be extracted from polymers with superheated water [6]. For example, at 200°C styrene, alkylbenzene contaminants and styrene dimers were extracted from polystyrene without destroying the polymer. Although stable at 200°C, at 250°C polystyrene was decomposed into substituted benzenes [6].

3.8.2.2
Extraction with Simultaneous Degradation

Very often compounds being extracted by superheated water react in the medium by hydrolysis or otherwise. It is know from other studies involving pure contaminants that they will react, for example chlorinated hydrocarbons are often dechlorinated and converted into hydrocarbons. In other cases benign materials are obtained from pollutants. In the extraction of the explosives TNT, RDX and HMX from contaminated soil, decomposition occurs non-dramatically and completely to benign substances [48]. These compounds contain an oxidative reagent within the molecule. Soil obtained from a bomb disposal site contaminated with 120 000 ppm (12%) of TNT, after treatment in a static cell at 275°C for 1 h, contained only 2 ppm and the water remaining 4 ppm. Dioxins in contaminated soil treated for 4 h at were found to be reduced by 99.4%, 94.5% and 60% at temperatures of 350°C, 300°C and 150°C, respectively [49].

Extraction of contaminants has been combined with oxidation either within the extraction cell or as a second step. These processes avoid the need for clean-up of the water effluent from an extraction process. Wet-air oxidation [2] has been carried out within the cell during the extraction of PAHs from soil [50]. In excess of 99% of the hydrocarbons were removed, which was much higher than the percentage removal when extraction only was carried out. PAHs from solids have also been treated by extraction followed by on-line wet-air oxidation in a second cell [51, 52]. Supercritical water oxidation [1] has also been coupled as the second step [53].

3.8.2.3
Pilot-scale Studies of Decontamination

Soil decontamination has developed from the small-scale superheated water extraction of PAHs, pesticides and polychlorinated biphenyls (PCBs). This has been carried out on a pilot scale and at 250°C almost complete soil clean-up is achieved in less than 1 h [54]. The plant has an 8-l cell and operates batchwise. A device has also been developed which can be hammered into soil to carry out decontamination in situ [6, 55]. A pressure cell is formed when the device penetrates a clay layer. Alternatively the soil at the bottom of the device is frozen with liquid carbon dioxide. The natural organic (humic) content of the soil largely remains after these processes and can be used for successful cultivation. Hydrocarbons are re-precipitated as the water cools, whereas some compounds, such as pesticides, are degraded to less-toxic materials during the extraction process. Soil containing explosives has also been treated in the same pilot plant using static extraction [48]. The destruction of TNT, RDX and HMX was 99.99%, 99.9% and 97.9%, respectively.

3.8.3
Extraction of Biomass

3.8.3.1
Laboratory-scale Extractions

It should first perhaps be mentioned that pressurized cold water has been successful in plant extraction. Hypericin, protohypericin, pseudohypericin and proto-pseudohypericin have been extracted from St. John's wort (*Hypericum perforatum*) [56] and terpene trilactones from the leaves of *Ginkgo biloba* [57]. Much more of the compounds are extracted when the water is under pressure than with water at one atmosphere. This is thought to be due to greater penetration of the fluid, perhaps because air included in the plant leaves is compressed or dissolved. This effect may partly explain the success of superheated water extraction.

The first biomass studies with respect to superheated water reaction involved rosemary (*Rosmarinus officinalis*). This was extracted with water at 100 and 200°C [6] and it was found that the extracts contained the oxygenated compounds in the essential oil, such as 1,8-cineol, camphor, borneol and linalyl propanoate. The extract at 200°C contained small quantities of the monoterpene hydrocarbons α-pinene and camphene, but the extract at 100°C did not contain traceable amounts of the monoterpenes. In essential oils, the valuable components, which give the flavor and fragrance of the essential oil, are the oxygenated compounds, whereas the monoterpenes are the less-valuable components. Most conventional processes, such as steam distillation, give essential oils that consist mainly of terpenes (often in high proportion) and there are many processes operated worldwide for the "deterpenation" of essential oils. Thus the experiments showed that extraction with superheated water not only had environmental advantages but could produce a more valuable extract of higher quality.

A more detailed study of the extraction of rosemary [58] was subsequently carried out. Figure 3.8-5 shows results obtained at 150°C for the extraction of rosemary with liquid water, presented in the form of the percentage of each of eight compounds of interest obtained after a given time compared with the amount that can be obtained after a long extraction. The monoterpenes extracted are camphene and limonene, whose extraction profiles are indistinguishable, and α-pinene. These extract relatively slowly. The ester isobornyl acetate is extracted a little more quickly. A more rapidly extracted group is formed by the cyclic ether 1,8-cineole and the alcohol borneol. The ketones are removed most rapidly. Of the latter, verbenone, which has a double carbon–carbon bond conjugated with the ketone bond, is extracted more rapidly than camphor. Very small and unquantifiable amounts of heavier hydrocarbons were also extracted. Analysis of the rate curves obtained showed that if extraction was stopped after 70% of the total material had been extracted, 90% of the essential oil extract were oxygenated flavor and fragrance compounds.

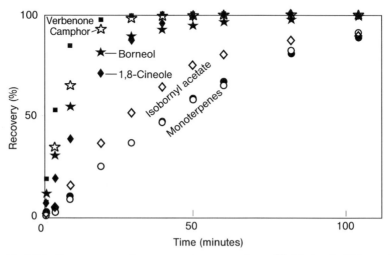

Fig. 3.8-5 Recovery curves for compounds extracted from rosemary with liquid water at 150 °C [58]: ●, camphene and limonene; ○, α-pinene; ▽, 1,8-cineole; ▼, borneol; ▲, verbenone; △, camphor.

Of course many other compounds are also extracted from plants by superheated water. This gives rise to problems in isolating desired compounds from the extract for a particular process, as discussed in Section 3.8.3.3. It has been shown, for example, that antioxidants are extracted from rosemary [59]. Experiments were carried out between 25 and 200°C and showed high selectivity of extraction towards compounds exhibiting high antioxidant activity.

It has been found that the total amount of fragrance compounds obtained by superheated water extraction was greater than that that could be obtained by steam distillation. This could be because penetration of the plant material is better with hot water under pressure, as proposed above. As an example of these comparisons, the work on wild marjoram (*Thymus mastichina*) [60] will be briefly outlined. Steam distillation was carried out over 3 h, whereas superheated water extraction at 150°C was completed in 15 min. The authors used about 10.5 times more water per gram of plant material in superheated water extraction than in steam distillation and said that the steam distillation seemed to have gone to completion. The yield from superheated water extraction was 5.1 times that obtained by steam distillation. This is in accord with previous results, although the ratio of yields is larger than obtained elsewhere. Table 3.8-2 shows the percentages of selected compounds in the oils obtained by both methods. The major component, eucalyptol, is present in roughly the same proportion in both oils, but the monoterpenes (the first three compounds listed) are much less prominent in the oil obtained by superheated water extraction. However, the superheated water extract contains a higher proportion of most other oxygenated compounds, except for geranyl acetate. These compounds give the oil its particular fragrance. The results are in general agreement with those reported in other studies.

Table 3.8-2   Percentages of components obtained from wild marjoram by steam distillation and water extraction at 150°C.

	Steam distillation	Water extraction
α-Pinene	2.5	0.1
β-Pinene	4.4	0.5
β-Myrcene	5.0	0.3
Eucalyptol	67	65
Linalool	2.1	4.7
α-terpineol	6.0	10.1
Geraniol	3.6	6.1
Geranyl acetate	3.6	2.6

Data from Ref. [60].

In general, the extraction curves are exponential in form. This can be seen in Fig. 3.8-6, which shows data for the extraction of eugenol and eugenyl acetate from clove buds [61]. The plots are of the quantity $\ln(m/m_0)$, where m is the mass of compound remaining in the cloves and m_0 the mass of the total compound originally in the cloves. The exponential form of the recovery curves indicates that the extraction rate is controlled by partitioning of the compounds between the water and the plant material. Release from the plant material and diffusion through it appear to be very rapid in this extraction process, in contrast to extraction by liquids and supercritical carbon dioxide.

Since the work of Clifford et al., several studies of the small-scale extraction of essential components have been made and the work and methods have been reviewed [62–64]. Work has been carried out on eucalyptus [65] peppermint [66, 67], savory [67], laurel [68] and *Thymbria spicata* [69]. These reach similar conclusions to those made above for rosemary, wild marjoram and clove and in some cases comparisons with steam distillation and other methods have been made [58, 60, 61, 66, 68]. In addition to essential oil components, work is also being carried out on the extraction of compounds for analytical purposes [70, 71], including the extraction of metals [72, 73]. Other extractions of medical and industrial interest are being car-

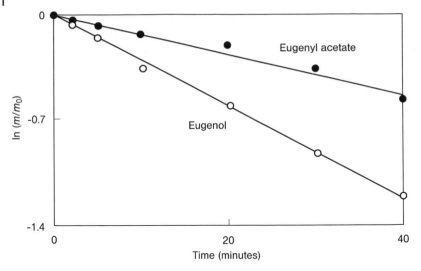

Fig. 3.8-6 Logarithmic plots of recovery of two compounds from clove buds [61] showing exponential recovery.

ried out. The author is aware of some of this work, which is subject to commercial confidentiality, although some has been published [74–79].

3.8.3.2
Extraction with Reaction

During extraction of flavor and fragrance compounds from plants, some of the components undergo reactions, such as oxidation and hydrolysis, to a minor extent, which may even improve the product. The monoterpenes α-terpene and terpinene are oxidized to verbenone and terpineol, respectively, and linaloyl acetate is hydrolysed to linalool, increasing the fragrance of the oils. In relation to plant materials, the hydrolysis of starch and cellulose to polysaccharides and sugars has been studied widely up to the present [80]. Esters are hydrolyzed in superheated water [81] and the medium has been used as a solvent and reagent for the hydrolysis of triglycerides, some containing unsaturated acids, in the temperature range from 260 to 280°C, although higher temperatures resulted in degradation [82, 83]. This opens up the possibility of the extraction of starches from natural products followed by the production of polysaccharides and sugars and the extraction of lipids followed by separation of fatty acids.

In an attempt to produce sugars directly from barley seeds, these were extracted with superheated water in our laboratory. Unfortunately, at the temperature required for hydrolysis, a dark brown extract was obtained, which smelled of burning. At lower temperature, however, a pleasant cooking aroma was obtained, due to the Maillard reaction. Similar results were obtained with other vegetable material, and superheated water could be used to manufacture food flavors [6]. The most interesting results were obtained with green Java coffee beans. Extraction at around 200°C produced a brown liquid, which had the aroma of coffee. Extractions were carried out using both deoxygenated water and water oxygenated by air at atmospheric pressure and by oxygen at 5 bar, although the product was not much affected by the presence of oxygen. Following extraction the cell contents were examined. After all experiments the beans were found to be whole. Water appeared to have permeated throughout the individual beans, and all beans were homogeneously dark brown. They had expanded to about double their initial size. The strong smell from the beans was bitter and burnt. The extracts were analyzed by headspace gas chromatography coupled to mass spectrometry. In some cases the effluent from the chromatograph was split so that the separated compounds could be sniffed to aid identification. The chromatograms obtained were compared with those from filter coffee obtained from the same beans after roasting and grinding (i.e. in the normal way). The "coffee" obtained from superheated water gave a much more concentrated extract, but the components and their proportions were very similar to conventional coffee. Thus a process similar to roasting is occurring for the green beans. These experiments may give rise to a process for producing coffee flavors for the food industry [6].

3.8.3.3
Process Development

The developments for extraction of a polymer are in an advanced stage and other polymer extractions are being researched. Unfortunately these are not being divulged at the present time because of commercial confidentiality. For essential oils, the advantages in yield and quality indicated, as well as the energy, environmental and clean-product benefits give rise to interest in developing processes from these small-scale experiments [64]. Process development is, however, at an early stage. Not much work is yet published, but the challenges are being tackled. The main problems to be solved are the separation of the desired components at the end of the process both from the extracting water and from highly water-soluble material also present in the plant material and therefore in the water extract. The extract is often coloured brown like tea and frequently contains a precipitate in which the oxygenated compounds are absorbed. On the small scale these are back extracted using an organic solvent such as pentane for analysis, but this would not be acceptable in an environmentally friendly process.

One suggestion is to back extract with supercritical carbon dioxide, which is also a clean solvent. Another alternative is to do the reverse, i.e. extract the plant material first with carbon dioxide to obtain a "concrete" or "oleoresin", which contains heavy materials, such as plant waxes. The concrete can then be treated with superheated water, as described below in the section on liquids. However, the use of supercritical carbon dioxide is probably not economically viable for most products.

In some cases there is no problem. For example, the extraction of flowers, such as rose petals, leads to a white precipitate containing the fragrance compounds. This may be a product that can be accepted as an additive to soaps and other cosmetics and could be described as a solid fragrance. Another situation is the production of flavor materials that are normally obtained by hot-water infusion, such as spices. The use of superheated water seems to give a more concentrated extract and thus may be more cost-effective.

3.8.4
Extraction of Liquids

Experiments have been carried out on the partitioning of aromatics between gasoline and diesel and superheated water [84]. The increase in the partition coefficient between ambient and 200°C was ~10 for benzene, toluene, ethyl benzene and xylenes and ~60 for naphthalene, for example. This behavior could be the basis of a process for the removal of benzene from gasoline in particular, and for the removal of aromatics from petroleum products in general.

Studies on liquid extraction may be carried out using an experimental system modified from that shown in Fig. 3.8-2. The cell is filled with packing material and the oil pumped through a tube so that it enters 20 mm above the bottom of the cell. It then trickles up over the packing material and collects at the top of the cell. Some experiments are reported [64] on the extraction of essential oils obtained by cold-pressing or steam distillation. These oils consist largely of terpenes, mainly limonene, with about 1% of oxygenated compounds. The superheated water extract, however, consists primarily of the oxygenated flavor and fragrance compounds, which are the valuable components of the oils. These experiments could form the basis of a method for deterpenating essential oils [6].

3.8.5
Chromatography

The first use of superheated water for chromatography was published in 1981 [85] and was designated thermal aqueous liquid chromatography or TALC. Chromatography was carried out up to 150°C using silica and polymer-based mobile phases. A flame ionization detector (FID) was also used, but not above 100°C. Following the interest in superheated water for extraction, its use as the mobile phase for reverse-phase liquid chromatography was revived [6, 86, 87]. This is of interest both for analytical chemistry and for the possibility of the use of superheated water chromatography for a process, either alone or following superheated water extraction. In the initial work, water up to 210°C was used with a polymer stationary phase to separate a wide variety of compounds, and gave chromatograms comparable to those obtained with solvent mixtures under ambient conditions. Since then there have been numerous studies, some of which have looked at the method in general [88–90].

There are two major challenges in the technique. The first is that if the effluent from the column needs to be cooled before entering the detector, the coiled tubing used for this can widen the chromatographic peaks. This is because its diameter needs to be much larger than typical spaces within the column, giving it a much greater plate height. A way of avoiding the latter problem is not to cool the effluent but to use an FID. Water has the advantage of not being detected in an FID, but, in order not to extinguish the FID flame, a low mobile phase flow velocity must be used, say 0.1 ml min^{-1} [87]. Alternatively the effluent may be split so that only a small proportion passes through the FID detector [91 ,92].

The second challenge is the stability of stationary phases at the higher temperatures used. Various studies of the stability of column packings have been made. Silica-based columns are particularly unstable and an ODS-bonded column was found to be unstable even below 100°C [93, 94]. Poly(styrene-divinylbenzene) columns have been much studied and are generally agreed to be stable up to 150°C [93–96], although they have been used at higher temperatures [87]. A graphitic carbon column was found to be stable up to 200°C. Work is being carried out on special stationary phases based on alumina, titania and zirconia.

Superheated water extraction has been coupled to superheated chromatography both on line [97] and off line with an intermediate sorbent concentration step [98]. Mixtures with other solvents have been used at high temperatures, such as dimethyl sulfoxide [99] and methanol [100]. Superheated deuterium chromatography has been used so that it could be coupled to NMR spectroscopy [101].

References

1 M. Modell, US Patent 4,338,199, **1982**.
2 F.J. Zimmernam, US Patent 2,665,249, **1954**.
3 M. Siskin and A.R. Katritzky, *Chem. Rev.*, **2001**, *101*, 825.
4 A.R. Katritzky, D.A. Nichols, M. Siskin, R. Murugan and M. Balasubramanian, *Chem. Rev.*, **2001**, *101*, 837.
5 S.M. Fields, C.Q. Ye, D.D. Zhang, B.R. Branch, X.J. Zhang, N. Okafo and P.E. Savage, *Catal. Today*, **2000**, *62*, 167.
6 S.B. Hawthorne, D.J. Miller, A.J-M, Lagadec, P.J. Hammond and A.A. Clifford, US Patent 6,352,644, **2002**.
7 A. Khan, *J. Phys. Chem. B*, **2000**, *104*, 11268.
8 L. Haar, J.S. Gallagher and G.S. Kell, *NBS/NRC Steam tables*, Hemisphere Publishing, Washington, DC, **1984**.
9 K. von Brollos, K. Peter and G.M. Schneider, *Ber. Bunsen-Ges. Phys. Chem.*, **1970**, *74*, 834.
10 N.D. Sanders, *Ind. Eng. Chem. Fundam.*, **1984**, *25*, 171.

11 D.J. Miller and S.B. Hawthorne, *J. Chem. Eng. Data*, **2000**, *45*, 315.
12 P. Khuwijitjaru, S. Adachi and R. Matsuno, *Biosci. Biotech. Biochem.*, **2002**, *66*, 1723.
13 D.J. Miller and S.B. Hawthorne, *Anal. Chem.*, **1999**, *70*, 1618.
14 D.J. Miller and S.B. Hawthorne, *J. Chem. Eng. Data*, **2000**, *45*, 78.
15 J. Rezo, J. Trejo and L.E. Vera-Avila, *Chemosphere*, **2002**, *47*, 933.
16 D.J. Miller, S.B. Hawthorne, A.M. Gizir and A.A. Clifford, *J. Chem. Eng. Data*, **1998**, *43*, 1043.
17 S.B. Hawthorne, Y. Yang and D.J. Miller, *Anal. Chem.*, **1994**, *66*, 2912.
18 Y. Yang, S. Bøwalt, S.B. Hawthorne and D.J. Miller, *Anal. Chem.*, **1995**, *67*, 4571.
19 Y. Yang, S.B. Hawthorne and D.J. Miller, *Environ. Sci. Technol.*, **1997**, *31*, 430.
20 S.B. Hawthorne, C.B. Grabanski, E. Martin and D.J. Miller, *J. Chromatogr. A*, **2000**, *892*, 421.
21 S. Kipp, H. Peyrer and W. Kleibohmer, *Talanta*, **1999**, *46*, 385.

22 K. Hartonen, G. Meissner, T. Kesala and M.L. Riekkola, *J. Microcolumn Sep.*, **2000**, *12*, 412.

23 T. Andersson, K. Hartonen, T. Hyotylainen and M.L. Riekkola, *Anal. Chim. Acta*, **2002**, *466*, 93.

24 R. Romero, R. Sienra and P. Richter, *Atmos. Environ.*, **2002**, *36*, 2375.

25 K. Kuosmanen, T. Hyoylainen, K. Hartonen, J.A. Jonsson and M.L. Riekkola, *Anal. Bioanal. Chem.*, **2003**, *375*, 389.

26 M.J. Simpson, B. Chefetz and P.G. Hatcher, *J. Environ. Quality*, **2003**, *32*, 1750.

27 T. Andersson, T. Pihtsalmi, K. Hartonen, T. Hyoylainen and M.L. Riekkola, *Anal. Bioanal. Chem.*, **2003**, *376*, 1081.

28 K. Kuosmanen, T. Hyoylainen, K. Hartonen and M.L. Riekkola, *Analyst*, **2003**, *128*, 434.

29 J.A. Field and R.L. Reed, *Environ. Sci. Technol.*, **1999**, *33*, 2782.

30 L. Wennrich, P. Popp and M. Moder, *Anal. Chem.*, **2000**, *72*, 546.

31 J. Kronholm, P. Revilla-Ruiz, S.P. Porras, K. Hartonen, R. Carabias-Martinez and M.L. Riekkola, *J. Chromatogr. A*, **2004**, *1022*, 9

32 F. Bruno, R. Curini, A. Di Corcia, I. Fochi, M. Nazzari and R. Samperi, *Environ. Sci. Technol.*, **2002**, *36*, 4156.

33 J.A. Field, K. Monohan and R. Reed, *Anal. Chem.*, **1999**, *70*, 1956.

34 M.S. Krieger, W.L. Cook and L.M. Kennard, *J. Agric. Food Chem.*, **2000**, *48*, 2178.

35 X.W. Lou, D.J. Miller and S.B. Hawthorne, *Anal. Chem.*, **2000**, *72*, 481.

36 L.N. Konda, G.. Fuleky, and G.. Morovjan, *J. Agric. Food Chem.*, **2002**, *50*, 2338.

37 C.S. Eskilsson, K. Hartonen, L. Mathiasson and M.L. Riekkola, *J. Sep. Sci.*, **2004**, *27*, 59.

38 E.M. Golet, A.Strehler, A.C. Alder and W. Giger, *Anal. Chem.*, **2002**, *74*, 5455.

39 B. van Bavel, K. Hartonen, C. Rappe and M.L. Riekkola, *Analyst*, **2000**, *124*, 1351.

40 S. Pross, W. Gau and B.W. Wenclawiak, *Fresenius' J. Anal. Chem.*, **2000**, *367*, 89.

41 K. Kuosmanen, T. Hyoylainen, K. Hartonen and M.L. Riekkola, *J. Chromatogr. A*, **2002**, *943*, 113.

42 E. Priego-Lopez and M.D.L. de Castro, *Talanta*, **2002**, *58*, 377.

43 C. Crescenzi, A. Di Corcia, M. Nazzari and R. Samperi, *Anal. Chem.*, **2000**, *72*, 3050.

44 K. Luthje, T. Hyoylainen and M.L. Riekkola, *J. Chromatogr. A*, **2004**, *1025*, 41.

45 R.C. Timpe M.D. Mann J.H. Pavlish and P.K.K. Louie, *Fuel Proc. Technol.*, **2001**, *73*, 127.

46 M.M. Jimenez-Carmona, V. Fernandez-Perez, M.J. Gualda-Bueno, J.M. Cabanas-Espejo and M.D.L. de Castro, *Anal. Chim. Acta.*, **1999**, *395*, 113.

47 V. Fernandez-Perez, M.M. Jimenez-Carmona and M.D.L. de Castro, *J. Anal. Atom. Spectrom.*, **1999**, *14*, 1761.

48 S.B. Hawthorne, A.J.M. Lagadec, D. Kalderis, A.V. Lilke and D.J. Miller, *Environ. Sci Technol.*, **2000**, *34*, 3224.

49 S. Hashimoto, K. Watanabe, K. Nose and M. Morita, *Chemosphere*, **2004**, *54*, 89.

50 A.A. Dadkhah and A. Akgerman, *J. Hazard. Mat.*, **2002**, *93*, 307.

51 J. Kronholm, B. Desbands, K. Hartonen and M.L. Riekkola, *Green Chem.*, **2002**, *4*, 213.

52 J. Kronholm, J. Kalpala, K. Hartonen and M.L. Riekkola, *J. Supercrit. Fluids*, **2002**, *23*, 123.

53 J. Kronholm, T. Kuosmanen, K. Hartonen and M.L. Riekkola, *Waste Management*, **2003**, *23*, 253.

54 A.J.M. Lagadec, D.J. Miller, A.V. Lilke and S.B. Hawthorne, *Environ. Sci. Technol.*, **2000**, *34*, 1542.

55 P.J. Hammond, UK Patent, GB2343890, **2002**.

56 M. Mannila and C.M. Wai, *Green Chem.*, **2003**, *5*, 387.

57 Q.Y. Land and C.M. Wai, *Green Chem.*, **2003**, *5*, 415.

58 A. Basile, M.M. Jimenez-Carmona and A.A. Clifford, *J. Agric. Food Chem.*, **1998**, *46*, 5205.

59 E. Ibanez, A. Kubatova, F.J. Senorans, S. Cavero, G. Reglero and S.B. Hawthorne, *J. Agric. Food Chem.*, **2003**, *51*, 375.

60 M.M. Jimenez-Carmona, J.L. Ubera and M.D.L. de Castro, *J. Chromatogr. A*, **1999**, *855*, 625.

61 A.A. Clifford, A. Basile and S.H.R. Al-Saidi, *Fresenius' J. Anal. Chem.*, **1999**, *364*, 635.

62 R.S. Ayala and M.D.L. de Castro, *Food Chem.*, **2001**, *75*, 109.

63 Q.Y. Lang, and C.M. Wai, *Talanta*, **2001**, *53*, 771.

64 A.A. Clifford, in *Handbook of Green Chemistry and Technology*, Clark, J. and Macquarrie, D. eds., Blackwell, **2002**, Ch 24.

65 M.M. Jimenez-Carmona and M.D.L. de Castro, *Chromatographia*, **1999**, *50*, 578.

66 A. Ammann, D.C. Hinz, R.S. Addleman, C.M. Wai and B.W. Wenclawiak, *Fresenius' J. Anal. Chem.*, **1999**, *364*, 650.

67 A. Kubatova, A.J.M. Lagadec, D.J. Miller and S.B. Hawthorne, *Flavor Fragr. J.*, **2001**, *16*, 64.

68 V. Fernandez-Perez, M.M. Jimenez-Carmona and M.D.L. de Castro, *Analyst*, **2000**, *125*, 481.

69 M.Z. Ozel, F. Gogus and A.C. Lewis, *Food Chem.*, **2003**, *82*, 381.

70 T.M. Pawlowski and C.F. Poole, *J. Agric. Food Chem.*, **1999**, *46*, 3124.

71 M.S.S. Curren and J.W. King, *J. Chromatogr. A*, **2002**, *954*, 41.

72 J.A. Brisbin and J.A. Caruso, *Analyst*, **2002**, *127*, 921.

73 S. Morales-Munoz, J.L. Luque-Garcia and M.D.L. de Castro, *Spectrochim. Acta B Atomic Spectr.*, **2003**, *58*, 159.

74 L. Gamiz-Gracia and M.D.L. de Castro, *Talanta*, **2000**, *51*, 1179.

75 J. Gonzalez-Rodrigucz, P. Peres-Juan and M.D.L. de Castro, *Anal. Bioanal. Chem.*, **2003**, *377*, 1190.

76 J. Gonzalez-Rodriguez, P. Peres-Juan and M.D.L. de Castro, *J. Chromatogr. A*, **2004**, *1038*, 3.

77 Z.H. Xu, G.S. Qian, Z.W. Li and S.M. Zhao, *Chinese J. Anal. Chem.*, **2003**, *31*, 1307.

78 V. Fernandez-Perez and M.D.L. de Castro, *Anal. Bioanal. Chem.*, **2003**, *375*, 437.

79 C. Li-Hsun, C. Ya-Chuan and C. Chieh-Ming, *Food Chem.*, **2004**, *84*, 279.

80 M. Sasaki, Z. Fang, Y. Fukushima, T. Adschiri and K. Arai, *Ind. Eng. Chem. Res.*, **2000**, *39*, 2883.

81 P. Krammer and H. Vogel, *J. Supercrit. Fluids*, **2000**, *16*, 189.

82 R.L. Holliday, J.W. King and G..R. List, *Ind. Eng. Chem. Res.*, **1998**, *36*, 932.

83 J.W. King, R.L. Holliday and G.R. List, *Green Chem.*, **2000**, *1*, 261.

84 Y. Yang, D.J. Miller, and S.B. Hawthorne, *J. Chem. Eng. Data*, **1997**, *42*, 908.

85 C.L. Guillemin, J.L. Millet and J. Dubois, *J. High Res. Chromatogr. Chromatogr. Commun.*, **1981**, *4*, 280.

86 R.M. Smith, and J. Burgess, *J. Chromatogr. A*, **1997**, *785*, 49.

87 D.J. Miller and S.B. Hawthorne, *Anal. Chem.*, **1997**, *69*, 623.

88 T.S. Kephart and P.K. Dasgupta, *Talanta*, **2002**, *56*, 977.

89 Y. Yang, L.J. Lamm, P. He and T. Kondo, *J. Chromatogr. Sci.*, **2002**, *40*, 107.

90 T. Kondo and Y. Yang, *Anal. Chim. Acta*, **2003**, *494*, 157.

91 Y. Yang, A.D. Jones, J.A. Mathis and M.A. Francis, *J. Chromatogr. A*, **2002**, *942*, 231.

92 R. Nakajima, T. Yarita and M. Shibukawa, *Buinseki Kagaku*, **2003**, *52*, 305.

93 T. Yarita, R. Nakajima and M. Shibukawa, *Anal. Sci.*, **2003**, *19*, 269.

94 P. He and Y. Yang, *J. Chromatogr. A*, **2003**, *989*, 55.

95 S.J. Marin, B.A. Jones, W.D. Felix and J Clark, *J. Chromatogr. A*, **2004**, *1030*, 255.

96 T. Andersen, Q.N.T. Nguyen, R. Trones and T. Greibokk, *Analyst*, **2004**, *129*, 191.

97 R. Tajuddin and R.M. Smith, *Analyst*, **2002**, *127*, 883.

98 L.J. Lamm and Y. Yang, *Anal. Chem.*, **2003**, *75*, 2237.

99 T. Kondo, Y. Yang and L.J. Lamm, *Anal. Chim. Acta*, **2002**, *460*, 185.

100 A. Jones and Y. Yang, *Anal. Chim. Acta*, **2003**, *485*, 51.

101 S. Saha, R.M. Smith, E. Lenz and I.D. Wilson, *J. Chromatogr. A*, **2003**, *991*, 143.

Part 4
Concluding Remarks

Green Separation Processes. Edited by C. A. M. Afonso and J. G. Crespo
Copyright © 2005 WILEY-VCH Verlag GmbH & Co. KGaA, Weinheim
ISBN 3-527-30985-3

Concluding Remarks

Carlos A. M. Afonso and João G. Crespo

As discussed thoroughly in this book, in many situations the focus of innovation does not lie on the search for novel structures for active ingredients, but on the optimization of production processes for bulk chemicals, intermediates and fine chemicals. The challenge today is also to concentrate engineering research in non-traditional reaction processes, using new synthetic methodologies such as microreactor technology, solventless reaction and solid-phase chemistry, and integrated reaction and product recovery; the same holds true for the design of new separation systems: reactive distillation and crystallization, fluid extraction using new green(er) solvents and membrane processing, among others.

From the ideas discussed here it is clear that many key areas concern the use of novel materials and solvents that are under development. Advances in nanotechnology will enable the solution-oriented design of nanomaterials. Enhanced understanding of physics and chemistry fundamentals on the nanoscale, will allow the design of new materials – stationary phases for chromatography, membranes with improved selectivity and life-time, nanoparticles, nanoassemblies and other custom-designed nanostructures.

Concerning the use of green(er) solvents it is worth mentioning:

– The use of water as a solvent: water is one of the cheapest and most environmentally friendly solvents; however, most conventional chemical transformations take place in organic solvents. By contrast, nature carries out chemical reactions with ease in water at pH values close to neutral. Therefore new paradigms in research have to be developed in order to implement the use of water in organic syntheses, dealing not only with the stability of reagents and catalysts, but also with the solubility of substrates and products. In addition, superheated water allows other opportunities in reaction, extraction and separation technology owing to the remarkable change in its solubilization properties. Apart from the technological demand for the use of superheated water on a large scale, more specific applications are expected to emerge in areas such as food processing and environmental remediation.

– Supercritical solvent technology: supercritical solvents, especially carbon dioxide, offer novel synthetic options owing to their unusual solvating properties

Green Separation Processes. Edited by C. A. M. Afonso and J. G. Crespo
Copyright © 2005 WILEY-VCH Verlag GmbH & Co. KGaA, Weinheim
ISBN 3-527-30985-3

coupled with ease of removal of the solvent, options for phase-transfer catalysis and absence of traces of organic solvents in the isolated products.

– Ionic liquids: ionic liquids have been described as green solvents, as a result of their low (or negligible) vapor pressure. However, the separation and recovery of the final product(s) from the solvent remains a problem to be totally solved. The combination of ionic liquids with supercritical carbon dioxide is also a potential emerging separation and reaction technology. The wide variety of ionic liquids available and the possibility of having a tailored synthesis of an ionic liquid for given applications open a large number of opportunities in this topic, principally through the use of chiral ionic liquids.

– Fluorinated solvents: owing to their unique affinity properties fluorinated compounds may be used as solvents or solid supports, offering novel, simple and greener approaches for performing organic transformations and separations, including the efficient reuse of catalysts and chiral resolution.

To conclude, we believe that Chemistry will have a central role in the transition towards a sustainable, competitive and knowledge-based economy and society. In combination with its own supplier and downstream sectors it may contribute decisively to improved economic and social welfare.

Index

Green Separation Processes. Edited by C. A. M. Afonso and J. G. Crespo
Copyright © 2005 WILEY-VCH Verlag GmbH & Co. KGaA, Weinheim
ISBN 3-527-30985-3

– reused 225
palladium complex 64
palmitoyl-POPAM dendrimer 313
parallel combinatorial library 99
parallel library 98
parallel synthesis 89
paramagnetic 293
paramagnetic particle 299
particle aggregation 295
particle permeability 190
partition coefficient 152
Pd catalyst 214, 224
Pd-catalyzed dimerization 244
PDMS membrane 265
peat 23
Peng-Robinson equation 132
pentanedione 16
per-evaporation 240
perfluorinated 243
perfluorinated hydocarbon 219
perfluorinated liquid 229
perfluorinated solvent 242
perfluoroalkane 219
perfluorohexane 219 f, 225
perfluoromethylcyclohexane 220
perfluorooctyl-1,3-dimethylbutyl ether 223
perfusion chromatography 187, 189
permanent magnet 293
permeability 121, 194, 273, 288, 301
permeability of the membrane 273
permeable particle 187
permeate-side concentration
   polarization 279
pervaporation 146 f, 271, 273, 275, 279,
   282 f, 285 f
pervaporation membrane 276, 286
   – diffusion coefficient 276
   – diffusion selectivity 276
   – permeability selectivity 276
   – sorption coefficient 276
   – sorption selectivity 276
pervaporation module 280
pervaporation of vanillin 280
pervaporation unit 281
pesticide 25, 107
petrochemical 14
petrochemistry 265
petro-plastic 15
pharmaceutical 67, 188, 267, 304
pharmaceutical industry 9, 48, 83
phase change 114
phase equilibria 128
phase equilibrium 211f
phase partitioning 116

phase separation 49
phase-transfer catalysis 344
phase-vanishing (PV) method 225
phenylacetylene 225
phospholipid coat 294
phosphonium salt 63
phosphorus trichloride 227
photolithography 36, 209
pilot plant 146
piperazine 100
polarizability 124
polarization of concentration 282
polluter pay principle 21
pollution 26, 107
   – air 107
   – soil 107
   – water 107
polluting emission 107
pollution prevention 60, 108
pollution source 106
poly(amidoamine) 319
poly(amidoamine) dendrimer 307
poly(arylether)-containing dendrimer 307
poly(propyleneamine) dendrimer 309, 319
poly(propyleneamine) product 307
poly(styrene-divinylbenzene) column 337
polyamide-polyphenylene sulfone 265
polyaromatic hydrocarbon (PAH) screening
   in soil 329
polycationic dendrimer 318
polydimethylsiloxane (PDMS) 285
polyether block amide (PEBA) 285
polylactic acid 15 f
polymer 15, 133, 261
polymer bead 100
polymer industry 208
polymer matrix 295
polymer synthese 83
polymeric membrane 257
polymeric microsphere 187
polymeric support 90
polymer-supported ultrafiltration 316
polyoctylmethylsiloxane (POMS) 285
polypropylene 100
polysaccharide 294
polystyrene copolymer 91
polystyrene-based acid-cleavable resin 95
polystyrene-bound peptide 97
polyurea dendrimer 314, 316
polyvinsyl alcohol 297
polyvinyl alcohol membrane 147, 284 f
polyvinylidene fluoride membrane 317
porous catalyst 37
porous catalytic wall 37